HEIDEGGER ON SCIENCE

HEIDEGGER ON SCIENCE

EDITED BY

TRISH GLAZEBROOK

STATE UNIVERSITY OF NEW YORK PRESS

Published by
STATE UNIVERSITY OF NEW YORK PRESS, ALBANY

© 2012 State University of New York

All rights reserved

Printed in the United States of America

No part of this book may be used or reproduced in any manner whatsoever without written permission. No part of this book may be stored in a retrieval system or transmitted in any form or by any means including electronic, electrostatic, magnetic tape, mechanical, photocopying, recording, or otherwise without the prior permission in writing of the publisher.

For information, contact
State University of New York Press, Albany, NY
www.sunypress.edu

Production, Laurie D. Searl
Marketing, Anne M. Valentine

Library of Congress Cataloging-in-Publication Data

Heidegger on science / [edited by] Trish Glazebrook.
    p. cm.
    Includes bibliographical references and index.
    ISBN 978-1-4384-4267-9 (hardcover : alk. paper)
    1. Science—Philosophy.   2. Heidegger, Martin, 1889–1976.   I. Glazebrook, Trish.

Q175.H377 2012
501—dc23
                                                                                     2011031098

10 9 8 7 6 5 4 3 2 1

For George and Marion

CONTENTS

Acknowledgments ix

Abbreviations and Translations xi

Introduction 1

I. READING HEIDEGGER ON SCIENCE

Why Read Heidegger on Science? 13
   Trish Glazebrook

Heidegger's Critique of Science 27
   William J. Richardson, S. J.

II. QUANTUM THEORY

Beyond Ontic-Ontological Relations: Gelassenheit, Gegnet, and
Niels Bohr's Program of Experimental Quantum Mechanics 47
   James R. Watson

Heidegger's Theses Concerning the Question of the
Foundations of the Sciences 67
   Ewald Richter
   Translated by Trish Glazebrook and Christina Behme

III. SCIENCE AND THE HUMAN EXPERIENCE

From Animal to Dasein: Heidegger and Evolutionary Biology 93
   Lawrence J. Hatab

Carnap and Heidegger: Parting Ways in the Philosophy of Science 113
   Patrick A. Heelan

Lost Belongings: Heidegger, Naturalism, and Natural Science   131
   David R. Cerbone

IV. TECHNOSCIENCE

Heidegger's Philosophy of Science and the Critique of Calculation:
Reflective Questioning, *Gelassenheit*, and Life   159
   Babette E. Babich

Gelassenheit: Beyond Techno-Scientific Thinking   193
   Ute Guzzoni

Opening Ways of Transformation   205
   Gail Stenstad

V. REVISITING *BEING AND TIME*

Heidegger and the Empirical Turn in Continental Philosophy
of Science   225
   Robert P. Crease

A Supratheoretical PreScientific Hermeneutics of
Scientific Discovery   239
   Theodore Kisiel

Heidegger's Philosophy of Science: The Two Essences of Science   261
   John D. Caputo

Developments and Implications   281
   Trish Glazebrook

List of Contributors   297

Index of Heidegger Terms   301

Index   305

ACKNOWLEDGMENTS

I would like to thank the contributors for their good work and patience during the assembling of the volume. I would particularly like to thank Babette Babich for her advice throughout. Thanks goes also to Michael Bauer and *New Scholasticism* for their part in granting permission to reprint Father Richardson's paper, and to Brill Publishing on behalf of Martinus Nujhoff, Dordrecht who originally published John Caputo's paper. I also owe gratitude to Jane Bunker at the State University of New York Press for her continuing support of this project, and Andrew Kenyon who shepherded the book to completion. Laurie Searl was also crucial and patient during this process, and Matt Story has been indispensable. Reviewers made extremely helpful comments that have greatly strengthened the volume. The Social Sciences and Humanities Research Council of Canada and Dalhousie University funded my research in part. Many members of the North American Heidegger Circle provided useful discussion of material contained in my contributions to the volume. Nonetheless, any shortcomings remain my responsibility. Finally, I am grateful to my son, Laird, who showed admirable forbearance when working on this volume cut short our time to play.

COPYRIGHTS

"Heidegger's Philosophy of Science: the Two Essences of Science" by John Caputo is reprinted from *Rationality, Relativism and the Human Sciences*, eds. J. Margolis, M. Krausz and R. M. Burian (Dordrecht: Martinus Nijhoff, 1986), 43–60, with permission.

"Heidegger's Critique of Science" by William J. Richardson, S. J. is reprinted from *New Scholasticism* 42, no. 4 (1968), 511–36, with permission.

ABBREVIATIONS AND TRANSLATIONS

References provide citations of a German text followed by the corresponding page number in an English translation, as indicated below. All references to the *Gesamtausgabe* are given by "GA" and the volume number. For example, GA 65, 357/250 means Volume 65 of the *Gesamtausgabe*, page 357, which corresponds to page 250 in the specified English translation.

Where no English reference is given, authors have provided their own translation. Where authors have provided their own translation, although a translation is available, reference to the published translation is given in order that interested readers might examine the context of Heidegger's remarks, although the author's translation may not match the English text *verbatim*.

BW   *Basic Writings*, ed. David Farrell Krell (San Francisco: HarperCollins, 1993)

EM   *Einführung in die Metaphysik*, 5. Auflage (Tübingen: Max Niemeyer Verlag, 1987)
        An Introduction to Metaphysics, tr. Ralph Manheim (New Haven: Yale University Press, 1959)

FD   *Die Frage nach dem Ding*, 3. Auflage (Tübingen: Max Niemeyer Verlag, 1987)
        Section B.I.5.a)–f$_3$) (S. 50–83) is translated as "Modern Science, Metaphysics, and Mathematics" in BW, 271–305, which reprints with minor changes and deletions the translation at *What Is a Thing?* trs. W.B. Barton, Jr. and Vera Deutsch (Chicago: Henry Regnery Co., 1967). Translation citations are to BW.

G   *Gelassenheit*, 10. Auflage (Pfullingen: Verlag Günther Neske, 1992)
        Discourse on Thinking, trs. John M. Anderson and E. Hans Freund (New York: Harper & Row, 1966)

GA 1   *Frühe Schriften* (Frankfurt: Vittorio Klostermann, 1978)

GA 2	*Sein und Zeit* (Frankfurt: Vittorio Klostermann, 1977) *Being and Time*, trs. John Macquarrie and Edward Robinson (New York: Harper & Row, 1962)
GA 3	*Kant und das Problem der Metaphysik* (Frankfurt: Vittorio Klostermann, 1991) *Kant and the Problem of Metaphysics*, tr. Richard Taft (Bloomington: Indiana University Press, 1990)
GA 5	*Holzwege* (Frankfurt: Vittorio Klostermann, 2003) »Der Ursprung des Kunstwerkes,« 7–68; "The Origin of the Work of Art," in PLT, 17–87. »Die Zeit des Weltbildes,« 75–113; "The Age of the World Picture," in QCT, 115–54.
GA 6.1	*Nietzsche I* (Frankfurt: Vittorio Klostermann, 1996) *Nietzsche, Volume 2: The Eternal Recurrence of the Same*, ed. David Farrell Krell (New York: Harper & Row, 1984)
GA 6.2	*Nietzsche II* (Frankfurt: Vittorio Klostermann, 1997) *Nietzsche, Volume 3: The Will to Power as Knowledge and as Metaphysics*, ed. David Farrell Krell (New York: Harper & Row, 1987) *Nietzsche, Volume 4: Nihilism*, ed. David Farrell Krell (New York: Harper & Row, 1982)
GA 7	*Vorträge und Aufsätze* (Frankfurt: Vittorio Klostermann, 2000) »Die Frage nach der Technik,« 7–36; "The Question Concerning Technology," in QCT, 3–35. »Wissenschaft und Besinnung,« 39–65; "Science and Reflection," in QCT, 155–82. »Bauen Wohnen Denken,« 147–64; "Building, Dwelling, Thinking," in PLT, 145–61. »Das Ding,« 167–187; "The Thing," in PLT, 165–86. ». . . dichterisch wohnet der Mensch . . . ,« 191–208; ". . . Poetically Man Dwells . . . ," in PLT, 213–29. »Logos (Heraklit, Fragment 50),« 213–34; ("Logos (Heraclitus, Fragment B 50),"), trs. David Farrell Krell and Frank A. Capuzzi in *Early Greek Thinking* (San Francisco: HarperCollins, 1984), 59–78.
GA 9	*Wegmarken* (Frankfurt: Vittorio Klostermann, 1967) »Was ist Metaphysik?« 103–22; "What Is Metaphysics?" in BW, 93–110. »Vom Wesen des Grundes,« 123–75; "On the Essence of Ground," in P, 97–135.

»Vom Wesen der Wahrheit,« 177–202; "On the Essence of Truth," in BW, 115–38.
»Vom Wesen und Begriff der *physis*. Aristoteles' Physik B, 1,« 239–301; "On the Essence and Concept of *physis* in Aristotle's Physics B, I" in P, 183–230.
»Brief über den Humanismus,« 313–64; "Letter on Humanism," in BW, 217–65.
»Zur Seinsfrage,« 385–426; "On the Question of Being" in P, 291–322.

GA 10   *Der Satz vom Grund* (Frankfurt: Vittorio Klostermann, 1997) *The Principle of Reason*, tr. Reginald Lily (Bl;oomington: Indiana University Press, 1991)

GA 12   *Unterwegs zur Sprache* (Frankfurt: Vittorio Klostermann, 1985) *On the Way to Language*, tr. Peter D. Hertz (New York: Harper & Row, 1971), which does not include »Die Sprache,«; "Language," tr. Albert Hofstadter in PLT, 189–210.

GA 13   *Aus der Erfahrung des Denkens* (Frankfurt: Vittorio Klostermann, 2002)

GA 16   *Reden und andere Zeugnisse eines Lebensweges 1910–1976* (Frankfurt: Vittorio Klostermann, 2000)

GA 17   *Einführung in die phänomenologische Forschung* (Frankfurt: Vittorio Klostermann, 1994)

GA 20   *Prolegomena zur Geschichte des Zeitbegriffs* (Frankfurt: Vittorio Klostermann, 1979) *History of the Concept of Time: Prolegomena*, tr. Theodore Kisiel (Bloomington: Indiana University Press, 1985)

GA 21   *Logik. Die Frage nach der Wahrheit* (Frankfurt: Vittorio Klostermann, 1976)

GA 24   *Die Grundprobleme der Phänomenologie* (Frankfurt: Vittorio Klostermann, 1975) *The Basic Problems of Phenomenology*, tr. Albert Hofstadter (Bloomington: Indiana University Press, 1982)

GA 25   *Phänomenologische Interpretation von Kants Kritik der reinen Vernunft* (Frankfurt: Vittorio Klostermann, 1977) *Phenomenological Interpretation of Kant's* Critique of Pure Reason, trs. Parvis Emad and Kenneth Maly (Bloomington: Indiana University Press, 1997)

GA 26 Metaphysische Anfangsgründe der Logik im Ausgang von Leibniz (Frankfurt: Vittorio Klostermann, 1978)
The Metaphysical Foundations of Logic, tr. Michael Heim (Bloomington: Indiana University Press, 1984)

GA 27 Einleitung in die Philosophie (Frankfurt: Vittorio Klostermann, 2001)

GA 29/30 Die Grundbegriffe der Metaphysik. Welt—Endlichkeit—Einsamkeit (Frankfurt: Vittorio Klostermann, 2004)
The Fundamental Concepts of Metaphysics: World, Finitude, Solitude, trs. William McNeill and Nicholas Walker (Bloomington: Indiana University Press, 2001)

GA 38 Über Logik als Frage nach der Sprache (Frankfurt: Vittorio Klostermann, 1998)

GA 41 Die Frage nach dem Ding. Zu Kants Lehre von den transzendentalen Grundsätzen (Frankfurt: Vittorio Klostermann, 1984)
Section B.I.5.a)–f$_3$) is translated as "Modern Science, Metaphysics, and Mathematics" in BW, 271–305, which reprints with minor changes and deletions the translation at What Is a Thing? trs. W.B. Barton, Jr. and Vera Deutsch (Chicago: Henry Regnery Co., 1967). Translation citations are to BW.

GA 42 Schelling. Vom Wesen der Menschlichen Freiheit (1908) (Frankfurt: Vittorio Klostermann, 1988)

GA 45 Grundfragen der Philosophie. Ausgewählte »Probleme« der »Logik« (Frankfurt: Vittorio Klostermann, 1984)
Basic Questions of Philosophy: Selected "Problems" of "Logic," trs. Richard Rojcewicz and André Schuwer (Bloomington: Indiana University Press, 1994)

GA 49 Schelling: Zur erneuten Auslegung seiner Untersuchungen über das Wesen der menschlichen Freiheit (Frankfurt: Vittorio Klostermann, 1991)

GA 53 Hölderlins Hymne »Der Ister« (Frankfurt: Vittorio Klostermann, 1993)
Hölderlin's Hymn "The Ister," trs. William McNeill and Julia Davis (Bloomington: Indiana University Press, 1996)

GA 56/57 Zur Bestimmung der Philosophie (Frankfurt: Vittorio Koostermann, 1999)

GA 58	*Grundprobleme der Phänomenologie* (1919/20) (Frankfurt: Vittorio Klostermann, 1993)
GA 65	*Beiträge zur Philosophie. (Vom Ereignis)* (Frankfurt: Vittorio Klostermann, 2003) *Contributions to Philosophy (From Enowning)*, trs. Parvis Emad and Kenneth Maly (Bloomington: Indiana University Press, 1999)
GA 79	*Bremer und Freiburger Vorträge* (Frankfurt: Vittorio Klostermann, 1994) »Die Kehre,« 68–77; "The Turning," in QCT, 36–49.
H	*Holzwege*, 7. Auflage (Frankfurt: Vittorio Klostermann, 1994) »Der Ursprung des Kunstwerkes,« 1–74; "The Origin of the Work of Art," in PLT, 17–87. »Die Zeit des Weltbildes,« 75–113; "The Age of the World Picture," in QCT, 115–54.
ID	*Identität und Differenz*, 5. Auflage (Pfullingen: Verlag Günther Neske, 1976) *Identity and Difference*, tr. Joan Stambaugh (New York: Harper & Row, 1969)
SD	*Zur Sache des Denkens*, 3. Auflage (Tübingen: Max Niemeyer Verlag, 1988) *On Time and Being*, tr. Joan Stambaugh (New York: Harper & Row, 1972)
PLT	*Poetry, Language, Thought*, tr. Albert Hofstadter (New York: Harper & Row, 1971)
P	*Pathmarks*, ed. William McNeill (Cambridge: Cambridge University Press, 1998)
QCT	*The Question Concerning Technology and Other Essays*, tr. William Lovitt (New York: Harper & Row, 1977)
SF	*Zur Seinsfrage* (Frankfurt: Vittorio Klostermann, 1956)
SZ	*Sein und Zeit*, 16. Auflage (Tübingen: Max Niemeyer Verlag, 1986) *Being and Time*, trs. John Macquarrie and Edward Robinson (New York: Harper & Row, 1962)
US	*Unterwegs zur Sprache*, 11. Auflage (Pfullingen: Verlag Günther Neske, 1997)

On the Way to Language, tr. Peter D. Hertz (New York: Harper & Row, 1971), which does not include »Die Sprache,« 11–33; "Language," tr. Albert Hofstadter in PLT, 189–210.

VA *Vorträge und Aufsätze*, 8. Auflage (Pfullingen: Verlag Günther Neske, 1997)
»Die Frage nach der Technik,« 9–40; "The Question Concerning Technology," in QCT, 3–35.
»Wissenschaft und Besinnung,« 41–66; "Science and Reflection," in QCT, 155–82.
»Bauen Wohnen Denken,« 139–56; "Building, Dwelling, Thinking," in PLT, 145–61. »Das Ding,« 157–79; "The Thing," in PLT, 165–86.
». . . dichterisch wohnet der Mensch . . . ,« 181–98; ". . . Poetically Man Dwells . . . ," in PLT, 213–29.
»Logos (Heraklit, Fragment 50),« 199–221; ("Logos (Heraclitus, Fragment B 50),"), trs. David Farrell Krell and Frank A. Capuzzi in *Early Greek Thinking* (San Francisco: HarperCollins, 1984), 59–78.

WD *Was Heisst Denken?*, 5. Auflage (Tübingen: Max Niemeyer Verlag, 1997)
What Is Called Thinking?, tr. J. Glenn Gray (New York: Harper & Row, 1968)

INTRODUCTION

Discussion of science runs throughout Heidegger's writings from 1916 to 1976. Starting with Father Richardson's 1968 contribution, included below, treatments of specific issues in Heidegger's analysis of science have appeared throughout the decades since. The first book-length treatment was not, however, until Joseph Kockelmans published *Heidegger and Science* in 1985.[1] That volume is important for its contribution toward understanding the influence of the phenomenological tradition of Hegel and Husserl in particular on Heidegger, and hence its analysis is not directed primarily at science in the sense of philosophy of science understood more traditionally. Yet Heidegger's assessment of the sciences has consistently and persistently received attention from a significant proportion of his readers. The question of science in Heidegger is thus important for his interpreters, but underdeveloped. This volume addresses that gap.

More precisely, it seeks to show how broad the gap is by charting some landing sites. It is accordingly not aimed at providing an all-encompassing explication and assessment of Heidegger's views on science, but at demonstrating some of the rich intellectual resources available in his writings for thinking through and evaluating the function of science in the human project of knowing. Readers seeking an over-arching logic within which each chapter can be conveniently situated will be disappointed. For I began this collection with the specific intent of avoiding thematic meta-narrative, and, guided only by principles of diversity and inclusivity of voice, intended it to be preparatory to further debate about both science and Heidegger's critique of it.

Yet the chapters address themes around which they can be collected and organized. The book is in a sense then organic, or at least phenomenological, insofar as the themes have emerged have from the contributions themselves, rather than having been determined beforehand by my own *a priori* assessment of what warrants discussion. Emergent themes represent perceptions of what aspects of Heidegger's engagement with science are more generally significant than any single opinion might indicate (i.e., a kind of lay of the land with respect to contemporary debates in Heideggerian philosophy of science). There are contributions from North Americans and from Germans. Women's contributions can be found here in greater proportion than in Heidegger scholarship more generally. There is a mix of senior scholars whose work founded and defined Heidegger's reception, mid-career contributors who work within but also in tension with that tradition, and younger scholars who bring different generational concerns and interests.

To introduce the volume, I provide a brief synopsis of each chapter on its own, followed by an account of the section divisions and the rationale for organizing the chapter in this way. There are specific areas of scientific inquiry that are deliberately not addressed in this volume. For example, the contributions do not discuss artificial intelligence, computer science, and cybernetics, although Heidegger identified cybernetics in "The End of Philosophy" as a new fundamental science and an extreme possibility for science that has implications for the nature of Dasein. Hubert Dreyfus has, however, already given these topics substantial attention and shown their potential for Heideggerian analysis. Likewise, Heidegger's contribution to psychiatry is important. Indeed, he helped revolutionize the science. Through the work of Medard Boss, his phenomenology led to the development of existential therapy as an alternative to the psychoanalytic approaches of Freud and Jung, and provides one avenue to direct psychiatry away from behavioristic models. Yet again, this aspect of Heidegger's work has already received significant attention, especially since the Zollikon seminars, which include conversations and letters exchanged between Heidegger and Boss, were published in German in 1987 and in English translation in 2001. My aim is to demonstrate possibilities for further inquiry into Heidegger's philosophy of science that are not already so well established.

CONTRIBUTIONS

The volume begins with my justification for it: Why read Heidegger on science? Reasons detailed include that he has much to say in dialogue with traditional philosophy of science, that his critique of science is foundational for his critique of technology, that his analysis offers rich resources to environmental philosophers, and that his work grounds arguments for the social responsibility of the sciences. This chapter also details how "modern" and "science" can be understood in Heidegger, including *Naturwissenschaften* versus *Geisteswissenschaften*, and responds to criticisms that his work on science is outdated or poorly informed, or that he was simply anti-science.

Father Richardson's 1968 paper broke ground as the first paper published on Heidegger and science. Somewhat dated by its language and treatment of verification versus Popper's arguments on falsification, it nonetheless raises many of the themes treated more recently in other contributions to this volume. Richardson evaluates Heidegger's critique of scientific method, with particular attention to the rigor of research, the function of the mathematical, the role of experiment, the nature of truth, the connection between science and metaphysics, and the assessment of modernity. His paper also challenges this volume in its claim that Heidegger offers no philosophy of science.[2]

James Watson argues that Heidegger's characterization of modern science misses the mark with respect to quantum theory. In *Die Frage nach dem Ding*, Heidegger's understanding of modern science shifts to accord with Bohr's characterization of experiments in quantum theory (i.e., there is no reality independent of measurement and observation—world becomes picture). But medieval knowledge was also projective. The transition from medieval knowledge to modern science is not the result of experimental methodology, but displacement of divine revelation by secular commitment to mathematical reasoning. Quantum theory is different again insofar as it makes Newtonian realism untenable. Heidegger takes quantum theory to be simply an extension of classical physics because he has missed Bohr's insight that the interaction between quantum objects and instruments of measurement puts objectivity itself into question. Quantum theory disrupts Heidegger's contention that objects are reducible to the totalizing and controlling matrix of standing-reserve, a view only possible within the positivist perspective of realism where Heidegger remained. This perspective is also at work in the logic and ideology of capitalist economies, but not in Bohr's program of experimental science that generates probabilistic results because it has rendered impossible the certainty of representation.

Ewald Richter discusses quantum physics to argue for an alternative discourse on science. He explores the way in which beings are thematized in factical experience versus scientific inquiry, to distinguish truth from correctness as a way of uncovering what for Heidegger is the danger of technology. Thus, an alternative to scientific thinking not only safeguards beings, but also the human being. Richter argues that the futural orientation of quantum theory, through its use of statistical mechanics, gives time a directionality that it owes to a more original time, which it is not the task of physics to explain. Thus, from within science itself becomes visible something that exceeds scientific account and requires an alternative epistemology.

Lawrence Hatab focuses on Heidegger's 1929–1930 lecture course *The Fundamental Concept of Metaphysics* to take up the question of biology. He explores common ground between Heidegger's early phenomenology and evolutionary theory, and assesses what the former can contribute to resolving the limitations of scientific approaches in the latter. Heidegger argues that animals live in an impoverished world compared with humans, and Hatab uncovers a tension in Heidegger's account concerning whether the difference is one of degree. Moving to the question of how evolutionary biology can account for the emergence of culture out of nature, he rejects reductive naturalism, in particular, genetic reductionism and the "selfish gene," as unable to account for the actual behavior of organisms. He argues that the emergence of culture makes understanding humans on a continuum with other animals problematic. Rather, *Dasein*'s radical openness and differentiated otherness surpass its biological existence, a surpassing made possible by language insofar as it opens for *Dasein* its world.

Patrick Heelan examines the role of metaphor in scientific discourse. He argues that phenomenological critiques of the analytic, empiricist view of science and Heidegger's hermeneutics of experience have made it possible to understand the different roles assigned in contemporary science to theory and praxis. The latter is assigned to ontological understanding for the purpose of human culture, while theory is assigned to technological design for the purposes of environmental control. Science is thus Janus-like, one side looking to computational, technological control as a resource for multiple praxes, the other looking toward human as ultimately constitutive of ontological scientific knowledge. This bivalence underscores the prevalence of metaphor in scientific discourse, especially medical science and clinical practice, although modern culture and the analytic, empiricist view nonetheless mask the presence of metaphor. He shows that in phenomenological analysis, metaphor is as fundamental for scientific discourse as literality is for the analytic, empiricist view. But theory is mathematical, while both practice and the praxical are empirical, so it makes no sense to predicate mathematical models literally of the phenomenological lifeworld. At best, scientific theory and lifeworld praxis can come together in some unambiguous way, guided by professional experts who are aware that they are seeking no more, and no less, than a praxical consensus about a set of relevant, soluble lifeworld issues.

David Cerbone develops the challenge Heidegger poses to scientific naturalism. He uses Quine's analysis of the ontology of science to demonstrate Heidegger's claim that science annihilates the thing. If Quine's response to this annihilation is analogous to eliminative materialism (i.e., so much the worse for the thing), Heidegger's is to suggest that there is something that cannot be captured in scientific accounts, and furthermore, that this excess, which Cerbone identifies as "the poetic," that contrasts against the scientific because it heeds the thing in its particularity, is a significant part of the human lifeworld that is much more important in shaping human experience than scientific discourse. For in its poetic function, language brings human being into dwelling with beings. Cerbone's treatment here of the poetic can be read with Hatab's analysis of language as distinguishing Dasein in its uniqueness and Heelan's comments that the presence of metaphor in scientific discourse is masked toward a common theme: The discourse of the sciences, useful and in many ways successful though it may be, is impoverished and reductive when taken to be the final word on human experience and understanding.

Babette Babich moves the volume more deeply into the relation between science and technology. She examines the difference between the calculative thinking of the sciences that takes problem solving as its aim. Calculative thinking dominates modernity such that even philosophy of science imitates science through its use of representational thinking that precludes the critique that thoughtful reflection makes possible. Such reflection is called for because of the planetary assault on nature enacted technologically and only possible because of the ordering of nature through mathematization. The reflective thinking that Heidegger calls *Gelassenheit* does not solve problems as such because it does not generate results that provide conclusive answers. This is the poverty of reflection, but also therein the possibility of reflection on the meaning of technology.

How, then, is *Gelassenheit* then to be understood as an alternative to representational thinking? Ute Guzzoni's contribution consolidates and fills out Heidegger's analysis. Innovatively, she explicates *Gelassenheit* as a turn to the visible, a pictorial thinking that is "sensitive-reflective" rather than representational. Although one might consider these two ways of thinking oppositional, she argues that this approach is inappropriate—to conceive sensitive-reflective thinking strictly in terms of its contrast against techno-scientific thinking is to reduce it to the terms of the latter. Rather, *Gelassenheit* must be understood on its own terms. Rather than the grasping characteristic of representational, conceptual thinking, pictorial thinking is a waiting that is nonetheless not passive insofar as it entails attentiveness—an orientation toward and preparation for what might come to pass in thinking. Guzzoni thus makes sense of difficult Heideggerian terms like releasement, openness, granting, and mystery. The remainder of her chapter draws from Proust, van Gough, Adorno, and Marcuse to explicate pictorial thinking in terms of *Sinnlichkeit* (i.e., sensuousness and physicality), but also resonant with *Besinnung*, reflection.

Gail Stenstad pushes further the consequences of Heidegger's arguments for transformative thinking. She distinguishes it from ethics or any kind of normative theorizing, and examines two movements in thinking that open the possibility of such transformation: movement from the question of being that thinks through the historical unfolding of being; and movement from living with beings through analysis of technology, technicity, and machination. She argues in her third section for the openings to transformation made possible by bringing these two movements together.

Robert Crease's contribution also treats *Sein und Zeit* in its argument for an empirical turn in continental philosophy of science away from treating science as an impoverished form of revealing and toward closer contact with the actual practices of the sciences. Crease develops an ontic analogue of Heidegger's notion of formal indication that can characterize scientific research, especially with respect to its temporality. The papers, by Hatab, Watson, Richter, and Heelan can be read as taking just the kind of empirical turn Crease advocates insofar as they address practice in actual sciences rather than monolithic "science."

Theodore Kisiel's chapter examines how the hermeneutics of science is central for Heidegger's account. This chapter appears for the first time in this volume and is especially significant because of the detailed analysis of §69 of *Sein und Zeit*, and Kisiel's situating of Heidegger's analysis of the sciences in the context of the Nazi appropriation of the university. He draws on the 1937 lecture to the Faculty of Medicine at the University in Freiburg, "Die Bedrohung der Wissenschaft," the threat to science. Heidegger's assessment indicates that the sciences and the University are threatened by National Socialist appropriation, which he sees as world-historical suicide. This lecture was first published in German in 1993 and has been little researched. Kisiel's analysis is therefore important and significant for understanding Heidegger's critique of science in its historical context, but also for understanding the significance of the fact that that science is the modern realization of the human will to knowledge. As Kisiel puts it, the threat to science is its "inability and unwillingness *to renew and transform itself from within*." It has become nothing more than another technological worldview. Against his claims in the Rectoral Address, Heidegger has in this lecture become disillusioned about the role of the university in Dasein's definitive constitution as the inquirer, articulated explicitly in *Sein und Zeit*.

John Caputo's contribution is the second of the chapters herein collected that has previously appeared in print. It details the two essences of science at work in Heidegger. The first is the hermeneutic essence found in *Sein und Zeit* that he argues is inseparable from an allegedly pure logic of science and found in the historical life of the scientist. The second is the deconstructive essence of science that indicates an understanding of human being and world, of being and truth. The latter serves to show that Heidegger is not simply anti-science, but rather sought to critique and delimit science in its deconstructive sense. This chapter is included because it engages critical analysis of science on the basis of Heidegger's work, in contrast to critical readings of Heidegger's understanding of the sciences. It is placed here because its focal treatment of *Sein und Zeit* begins the chronological ordering of the papers according to which of Heidegger's texts are at issue.

In the final chapter, I develop two issues that arise directly out of Heidegger's critique of science: ecophenomenology, and the social obligations of the sciences. Heidegger's critique of science has immediate consequences for environmental philosophy that clearly demonstrate the inadequacy of the term *environmental ethics*. His reading of Aristotle provides a conceptual basis for scientific and technological practices that might be ecologically sound, and his notion of dwelling suggests that human ways of knowing can respect, care for and safeguard nature. Concerning the social obligations of the sciences, Heidegger argues that science warrants nontechnical, nonscientific reflection in order to evaluate its thinking and paradigmatic epistemological function. The chapter explicates Heidegger's view and its direct consequences, and applies the developed view to pressing issues of global capital that are central both to environmental sustainability and social justice in studies in international development.

STRUCTURE AND RATIONALE

The first issue is the very question of what is the value of reading Heidegger on science. Richardson's arguments that Heidegger does not do philosophy of science stand in both tension and accord with reasons I lay out for pursuing this rich line of inquiry into Heidegger's thinking. In both these chapters, quantum physics figures in the discussion, and the second theme of the book addresses this issue. Again the chapters demonstrate both agreement and disagreement: Watson argues that Heidegger's account of modern science is deficient with respect to quantum theory, while Richter argues that Heidegger's analysis uncovers a deficiency in physics itself. whereas section is included because of widespread opinion that quantum theory challenges Newtonian physics, against Heidegger's assessment that it does not change science essentially.

One of the most important insights about science noted by Heidegger is that it is a hermeneutic practice. The third section thus situates science in human experience by focusing on particular aspects of language that raise issues about what it means to be human in the scientific age. Hatab argues that language makes possible the development of culture beyond the biological reality of the human organism, Heelan explores metaphor in scientific discourse to argue that scientific theory and experience in the lifeworld come together in identifying praxical issues for which resolution is possible, and Cerbone's uses the poetic to explore the limitations of the scientific perspective. His analysis of the impoverishment of the latter is a bridge to the next group of chapters that address humanism using Heidegger's analysis of technology. Babich asks explicitly what it means to be human in the current context of global technics, while Guzzoni and Stenstad look for alternatives to the representational thinking of the sciences.

The three chapters in the final group each look back to *Being and Time*. They could therefore have constituted an early section in the volume. Indeed, originally I intended simply to present the chapters chronologically on the basis of which of Heidegger's texts were central to the analysis, in order to remain true to the strategy of avoiding meta-narrative. Challenges emerged, however, in that several chapters address developments in Heidegger's thinking across texts, and because authors showed shared interest in common themes. Moreover, it became clear that these three chapters in particular track the development of Heidegger's thinking concerning science throughout his life. Crease's argument for an empirical turn in continental philosophy of science reflects Heidegger's early interest in ontic sciences and regional ontologies. Kisiel develops the hermeneutics of science from *Being and Time* into the discussions of the function of the sciences in the university in Heidegger's confrontation with National Socialism in the 1930s, whereas Caputo examines the deconstructive essence of science in Heidegger's later critique of technology against the hermeneutic essence laid out in *Being and Time*. Thus, Kisiel's chapter connects the analyses of *Being and Time* with the later work on technology. Together, these chapters suggest that the political experiences of the 1930s motivated Heidegger's transition from the sciences as essentially hermeneutic to the deconstructive essence of science. Furthermore, Caputo argues that developments beyond *Being and Time* should not displace its insights, but rather that both the earlier and the later analyses of science are necessary for assessing its consequences, limits, and implications.

Moreover, Caputo makes clear how Heidegger's assessment of technology, sometimes taken to be his most important contribution to philosophy and surely the most widely known aspect of his work, has its roots in his critique of science. I excluded of cybernetics and psychotherapy on the basis that they are already well documented by Heidegger's readers, yet surely this is even more true of Heidegger's assessment of technology, on which I included three chapters; Caputo's chapter shows the difference between reading that critique as philosophy of technology, and interpreting it as Babich, Guzzoni, and Stenstad do, as part of his larger philosophy of science.

The closing chapter brings the volume back to Richardson's claim that philosophy of science "could be done in a Heideggerian framework" and brought "to a richer fruition than Heidegger himself has been able to do" by exploring possibilities for the development of his work on science in the contexts of environmental studies and international development studies. This closing contribution thus finalizes a movement throughout the volume from questions *about* science to the question *of* science, that is, from theoretical considerations of method and practice to questions about the role and function of science in the human experience.

In conclusion, the chapters in this volume demonstrate that the question of science pervades Heidegger's work, and is deeply connected to the issues of truth, language, technology, and metaphysics that are central to the development of his thinking. They show that several questions central to philosophy of science in the analytic tradition, for example, the questions of realism, method, metaphor, and the ways in which quantum theory challenges traditional paradigms, are to be found in Heidegger's thinking long before they appear in the analytic literature. Perhaps most importantly, they culminate in evidence that Heidegger has contributions to make to pressing questions of global significance that have immediate consequence in the lifeworld of both so-called "developed" and "developing" nations.

They also controversially suggest, taken as a whole, that for all his insistence that he is doing metaphysics, much of what is pivotal in Heidegger's thinking concerns, and could even be said to be driven by, epistemological questions about the nature of knowledge. Indeed, he has held since early days that *Dasein* is fundamentally and quintessentially the inquirer—in later years, when his focus is on thinking, whether *Denken* or *Besinnung*, or *Entbergung* or *Gelassenheit* in contrast to *Gestell*, it remains the case that *Ereignis* is an event in which the human relation to being is determined and experienced, whether confined to the representational or opened to possibilities of poetic alternatives, as one of thinking, much as in the earlier years the question of being was for Heidegger the question of the meaning of being. Heidegger's question of science shows he is not enough of a realist to be able to separate metaphysics from epistemology.[3]

Thus Richardson is right—this volume indeed moves far beyond Heidegger's own work. Yet each of these contributions is deeply immersed in Heidegger's texts. It may be that more recent perspectives on Heidegger's critique of science see his work differently, i.e., as more centrally and pervasively focused on the question of science than Richardson saw. Or that the contributors see the philosophy of science differently, that is, as not encompassing only technical questions of scientific practice, but as evaluative enquiry about science and the sciences themselves. First and foremost, this volume is evidence that Heidegger's work is a positive research program in the philosophy of science. Its primary intent is to show that further work on Heidegger and science is called for, and promises substantial reward.

NOTES

1. Joseph J. Kockelmans, *Heidegger and Science*, (Washington, DC: Center for Advanced Research in Phenomenology & University Press of America, 1985).
2. Cf. Patrick Heelan, "Heidegger's Longest Day: Twenty-five Years Later" *From Phenomenology to Thought, Errancy, and Desire: Essays in Honour of William J. Richardson* (Dordrecht: Kluwer, 1995).
3. Cf. Trish Glazebrook, "Heidegger and Scientific Realism" *Continental Philosophy Review* 34, no. 4 (2001), 361–401.

I

READING HEIDEGGER ON SCIENCE

WHY READ HEIDEGGER ON SCIENCE?

Trish Glazebrook

Heidegger wrote extensively concerning science for more than sixty years. Four aspects of his analysis in particular demonstrate the breadth and scope of his sustained critique of science, and indicate specific trajectories for its further development. First, he has much to say to traditional philosophers of science concerning the experimental method, the role and function of mathematics and measurement, the nature of paradigms and incommensurabilty, and realism versus antirealism. Second, his assessment of technology is incipient in and arises from his reading of the history of physics, so theorists who overlook this aspect of his work may find they are working with a deficient theoretical framework when attempting to come to terms with his critique of technology. Third, he offers rich conceptual resources to environmental philosophers, especially those who work at the intersection of environment and international development. Fourth, his arguments for reflection on science support a renewed sense of social obligation on the part of the sciences that should be of especial interest to science, technology, and society theorists.

I have examined these first two issues elsewhere.[1] Rather than repeating that work here, I situate this volume against traditional philosophy of science only by showing briefly how his concern with science begins with a tension in his thinking between realism and idealism. On the second issue, I show here only how Heidegger's thinking concerning Ge-stell arises directly from his prior thinking about basic concepts and the mathematical in science. The issues of ecophenomenology and the social obligations of the sciences are continuations of fertile and promising lines of thinking Heidegger opened. Thus, reading Heidegger on science brings one to these issues, and I have addressed them in the final chapter of this volume by developing Heidegger's thinking in contemporary contexts.

Heidegger's critique of science thus speaks to diverse audiences, and prompts a rethinking of the relation between human being and nature that has epistemological, ontological, and political consequences not only in philosophy but also for policy and practice. Before detailing these aspects of his analysis, however, preliminary qualifications of what he means by "*Wissenschaft*" and "modern" are called for.[2] Furthermore, concern that his analysis might be outdated, and dismissal of his critique on the basis that he is simply "anti-science," warrant response, lest the value and significance of his interrogation of science be prematurely forfeited.

"SCIENCE," "MODERN," AND CRITIQUING HEIDEGGER'S UNDERSTANDING

The word "science" can be difficult to pin down in both Heidegger's work and other discourses. The sciences simply do not unify easily. A totalizing conception of even natural science is inherently problematic, given diversity of method. For example, although mathematical physics is primarily a theoretical inquiry that collects empirical data through experiment in order to test and support hypotheses, geology and biology are both field sciences that use observation not only to establish evidence but also to generate research directives. Disciplinary tags like "political science" and the "social sciences" further complicate what "science" means. These disciplines are not scientific in the sense of using experimental methods, yet they can broadly be taken as scientific insofar as their research methods entail standards of rigor, and their evidentiary strategies rely on quantification. Nonetheless, to ignore the role and value of qualitative methods in the political and social sciences is to construe them reductively and fail to conceptualize their practices appropriately. Naming these disciplines "sciences" may serve little other purpose than establishing their validity on a par with the natural sciences that set definitive and paradigmatic epistemic standards in modernity.

The German distinction between *Naturwissenschaften* and *Geisteswissenschaften* is likewise not without difficulties. The term "*Geisteswissenschaften*" was coined in 1849 in reference to Mill's "moral sciences," which require methods of understanding significantly different from those of the natural sciences.[3] The "sciences of spirit," to translate the term literally, are directed at cultural projects like art, religion, and politics, and include disciplines like history, archaeology, languages and education, as well as philosophy,[4] and theology and jurisprudence have also come to fall under this disciplinary rubric. Consistent with the Cartesian separation between *res cogitantes* and *res extensae*, it may seem that *Geisteswissenschaften* deal with the nonphysical or mental and psychical, while *Naturwissenschaften* treat the physical. Yet since human being has both mental and physical aspects, human self-understanding needs both approaches. Indeed, psychology can be classified as both, so the separation between "natural" and "moral" sciences is not always exclusive. Alternately, mathematics is strictly neither. Heidegger himself most often uses "*Wissenschaft*" throughout his writing in reference to physics, but also to biology in the late 1920s. In other places, he refers to theology, philology, archaeology, art history, and history as *Wissenschaften*.[5] He is moreover well known for his argument in *Basic Problems of Phenomenology* that philosophy itself is inherently scientific, such that the expression "scientific philosophy" is a pleonasm. (GA 24, 15–19/11–15) Thus, it appears that Heidegger intends by "*Wissenschaft*" radically diverse realms of human enquiry and knowledge at different points in the development of his thinking.

Nonetheless, Heidegger is focally concerned with physics, and physics is typically what he intends by "science," especially "modern science." This preoccupation may have been intensified by the central role *physis* plays in his reading of the Greeks, and by the particular influence of Aristotle, whose *Physics* B.1 he examines in close detail in 1939. Alternately, these interpretive enquiries might in fact themselves have been prompted by his already explicit interest in science. As early as 1917, he uses Galileo to show that knowledge in modernity begins methodologically with projection of concepts rather than empirical observation. In *Being and Time*, when he makes the return in §69 to the question of phenomenological method promised in §7, the mathematical projection of nature is the focus of analysis. The 1917 essay and the treatment of Galileo and Newton in *Die Frage nach dem Ding* bookend the discussion in *Being and Time* with such similar language and analysis that his insights in 1927 are unlikely to have been directed at anything other than physics—modern physics is the enactment and origin of the mathematical projection of nature. That "science" means for him not exclusively but first and foremost physics indicates not a commitment to reductionism, in which all natural sciences are taken to boil down to physics, but his insight that the conceptual framework Galileo and Newton bring to bear on nature is determinative of modern ontology and epistemology.

His engagement with science may accordingly seem outdated, given recent moves to displace the paradigmatic function of physics in favor of alternative conceptual models.[6] The role of the ontology and epistemology of physics in determining the modern lifeworld should not, however, be underestimated. Much development policy is, for example, informed by conceptions of objectivity implemented by early modern physicists and still pervasive. Development theorists have long argued for "appropriate technologies," over and against noncontext-sensitive initiatives introduced on the assumption that the universality of knowledge allows its applications to function effectively independent of cultural, and in other ways particular, situation. The latter approach has exacerbated problems with respect both to sustainability and social justice. Likewise, feminist theorists do not support scientific methods that produce different results depending on serendipitous factors like personal bias, but nonetheless argue that science is not a value-free enterprise.[7] The physicist's ideal of objectivity has exceeded its context in scientific knowledge production, and been imported into policy and practice in nonconstructive ways. Heidegger's view of science is consistent with these criticisms, and he argues moreover, as detailed below, that the notion of objectivity impedes analysis of the ways in which science itself is a situated project. The separation between science and ethical obligation that arises in consequence of the ideology of objectivity has historically supported racist, sexist, imperialist, and unsustainable attitudes and practices.[8] Heidegger's lifelong critique remains significant and timely because his insight that the ontology and epistemology of physics inform the modern experience leads him to question the value of both the mathematical projection of nature and the epistemological ideal of objectivity in his ongoing critique of representational thinking.

The meaning of "modern" in the phrase "modern science" is also slippery. Co-teaching with Shimon Malin, a physicist at Colgate University, I quickly realized that we were using the term quite differently. He meant twentieth-century physics. Philosophical analyses of "modern science" generally intend rather Galilean-Newtonian physics, as "modern philosophy" likewise begins with Descartes. Philosophically speaking, modernity starts in the mid-seventeenth century. Because this is also true for Heidegger, a second reason emerges for thinking that perhaps his analysis is outdated. Several developments in twentieth-century physics challenge assumptions basic to Galilean–Newtonian physics, and accordingly many scientists and science analysts take the so-called "new" physics to be fundamentally different.

For example, the Newtonian universe is fundamentally deterministic, but chaos theory suggests that some events or processes are nondeterministic. One such process is radioactive decay: The decay of a single particle cannot be predicted, despite half-life calculations. Similarly, a pendulum hung from a bar that is pushed back and forth along one axis by a motor will suddenly leave that axis of swing and move erratically; the moment at which it will do so cannot be predicted in advance. Chaos theorists claim that such unpredictability is not epistemological, that is, the consequence of insufficient data concerning the system's initial state, but inheres ontologically in the system. Likewise, experiments testing Bell's inequalities in quantum physics challenge Newton's deterministic model by demonstrating that correlations in particle spin exceed the predictions of statistical probability. There is much debate about how to interpret these results. One suggestion is that local causality is breached, that is, contrary to special relativity, information has traveled faster than the speed of light; others argue that some hidden variable is at work. Furthermore, quantum theory has proven difficult to reconcile with gravitational theory. The fundamental forces operating in the universe, that is, gravity and the forces holding atomic particles together, are not yet understood in relation to each other. String theory potentially resolves this problem, despite disputed details, competing variations, and controversy concerning its status as a theory.[9] Supersymmetry also offers hope for reconciling at least three of the four fundamental forces, but falls prey to the so-called "hierarchy problem" in which its predictions exceed empirical indicators. Although human understanding of the cosmos is by no means complete, chaos, quantum, and string theory, as well as supersymmetry, are significant developments in the human understanding of the physical universe. They all converge on one point: Newtonian physics is not the last word on the nature of the universe.

Heidegger says nothing about chaos and string theory. Concerning quantum theory, Father Richardson argues that Heidegger's conception is inadequate because he never acknowledges its radical break with the Galilean–Newtonian paradigm.[10] Yet, as Kockelmans notes, "Heidegger had a remarkable knowledge of both physics and biology and . . . was able to conduct a penetrating discussion on important topics with leading scientists."[11] Heidegger does in fact see significant differences between Newtonian and quantum physics, e.g. the latter's reliance on statistical mechanics (VA, 56–7/172–3), but clearly believes that they are *essentially* the same: in both, "nature has in advance to set itself in place for the entrapping securing that science, as theory, accomplishes." (VA, 57/172–3) Physics projects an interpretive framework in which nature appears as "a coherence of forces calculable in advance." (VA, 25/21) This is just as much the case for quantum theory as for Newtonian physics, and indeed a central issue in string theory and supersymmentry is precisely to establish the coherence of fundamental forces. Furthermore, he argues that what is distinctive of modern physics is that it is mathematical, (FD, 50/271, *et passim*) and indeed, like quantum theory, neither chaos nor string theory nor supersymmetry can "renounce this one thing: that nature reports itself in some way or other that is identifiable through calculation and that it remains orderable as a system of information" (VA, 25/23). Chaos theory only became practicable when computer systems achieved adequacy for its massive calculations, and the mathematics of string theory entails extra dimensions for which empirical evidence continues to be evasive. Contemporary physics is very much a case of mathematics preceding physical interpretation. The idea that the universe "is written in the language of mathematics" is as old as the Pythagoreans, and made definitive for modern physics by Galileo.[12] None of the new physics of the twentieth century challenges this mathematical projection. If Heidegger is right 1) that "modern physics is the herald of Enframing" (VA, 25/22) insofar as "nature . . . is identifiable through calculation and . . . remains orderable as a system of information" (VA, 26/23); 2) that Enframing is the essence of technology as the "gathering [that] concentrates man upon ordering the real as standing-reserve [Bestand]" (VA, 23/19); and 3) that technology enacts "the organized global conquest of the earth" (GA 6.2, 358/248), then his alleged failure to account for the new physics is no basis for rejecting his views as outdated and inadequate. Rather, his critique stands as an urgent challenge to the contemporary scientific establishment to think through how science is complicit in its ideology and method with global environmental destruction.

One final reason for questioning the validity of his analysis needs response. Heidegger is not opposed to science *per se* insofar as he does not reject the human project of understanding nature. The most well-known basis for dismissing him as simply "anti-science" is the claim he makes repeatedly in *Was Heisst Denken?* that "science does not think" (WD, 4/8, *et passim*). But he also says often in this text that "most thought-provoking of all is that we are still not thinking" (WD, 2/4, *et passim*). His objection is not so much to science as to scientism, that is, the preclusion of other ways of thinking by the representational thinking of the sciences, and the marginalization, displacement, and devaluation of other methodologies and bodies of knowledge by the scientific standard of objectivity that has become epistemologically dominant in modernity. He argues the latter point originally in the mid-1930s. In §76 of the *Beiträge*, he observes that scientific ways of thinking have permeated other disciplines, and he distinguishes historical science (i.e., the journalistic collecting of facts) from the discipline of history, which takes an interpretive stance toward facts and endows them with meaning. Of course one always has an interpretive basis for what counts as a fact to be collected, but the point is that a scientistic view of history, in claiming objectivity, denies its perspectival stance. If history and philosophy are infected by scientism, then the possibility of critically understanding the place of science in modern thought is precluded (GA 65, 151-5/104-5). In other words, if the sciences allegedly uncover truths that are universal and thus ahistorical, and the discipline of history itself becomes scientific, then such history cannot uncover the historical significance of the sciences—that they are not contingent but rather a human destiny (i.e., a situated realization of the urge to know that determines what it means to be human in modernity). The sciences therefore are not intrinsically or inherently destructive for Heidegger. Rather, it is uncritical acceptance of their role and function in determining modernity that is threatening. They are an historical project, and as such, their ontology and epistemology warrant delimitation. In "Science and Reflection" he thus calls for critical interrogation and evaluative assessment of the sciences, much as he argues for poetic assessment of technology in the technology essay.

Having given some account that what Heidegger means by "science" is primarily natural science and paradigmatically physics despite the complexities of the term, and having responded to criticisms that his view is outdated, or that he can be dismissed as simply "anti-science," I can now proceed directly to the positive account of why to read Heidegger on science. I first argue that his phenomenological approach is useful to the philosopher of science working in the Anglo-American tradition insofar as it uncovers how the realist and antirealist are working at cross-purposes. Second, I show the significance of his reading of the history of physics for technology theorists whose efforts are informed by his questioning of technology. The third reason (that he makes rich conceptual resources available to environmental philosophers, especially those working at the intersection of environment and international development), and the fourth (that his arguments for reflection on science call for a renewed sense of the social obligations of the sciences) are treated in the final chapter in this volume.

REALISM AND IDEALISM

Heidegger's engagement with science arises in large part out of a tension in his thinking. He begins his career with a thorough commitment to realism. In 1912, he came out strongly with what is now called instrumental realism in "Das Realitätsproblem in der modernen Philosophie." He argues that the "healthy realism [gesunden Realismus]" of empirical, natural science has produced such "dazzling results [glänzenden Erfolge]" that science stands as an "irrefutable, epoch-making fact [unabweisbare, epochemachende Tatbestand]" (GA 1, 3–4). He poses a problem for philosophers: Although philosophy since Berkeley has moved toward the claim that "even the mere positing of an external world independent of consciousness is inadmissible and impossible" (GA 1, 3), the sciences are convinced that their analysis goes beyond sense data to objects that exist independently of research. Scientific methodology entails an ontological commitment ("background realism" in traditional, analytic philosophy of science) by which only philosophers are troubled. Thus in Being and Time, Heidegger calls philosophical demands for proof of realism scandalous.[13] The task of the sciences is to explain experience, and in his analysis, they do what is referred to in traditional philosophy of science as "saving the phenomena": They take their objects at face value empirically. Dasein is not an isolated subject that must secure access to an equally independent object. Rather, being-in-the-world is Dasein's "basic state" (SZ, 52/78), such that Dasein is already submerged amongst objects in its practices, including science, and need not worry how to bridge the gap between subject and object (cf. SZ, 60/87). Heidegger does not attempt to establish a correspondence between ideas and what they represent, but extends a praxical assumption of the quotidian (i.e., background realism) into the sciences. For they do not arise *ex nihilo*, but from the lifeworld. The early Heidegger is, then, a naive realist.

At the same time, however, *Being and Time*'s existential analytic is a renewed inquiry into transcendental subjectivity in response to the Kantian insight that experience is structured by categories of understanding. Dasein interprets its world on the basis of already determined structures of understanding (SZ, 151/191). Heidegger's realism is thus in conflict with his acknowledgment of the *a priori* nature of understanding: Access to reality cannot be had independent of structures of mind. How can the sciences describe and explain objects experienced as independently constituted, if understanding is projective? This question is answered by the ontological difference (i.e., the difference between being and beings: "The being of entities 'is' not itself an entity").[14] In *Being and Time*, "entities *are*, quite independently of the experience by which they are disclosed," yet "Being 'is' only in the understanding" (SZ, 183/228). Similarly, in *The Metaphysical Foundations of Logic*, "the cosmos can be without humans inhabiting the earth, and the cosmos was long before humans ever existed" (GA 26, 216/169), yet "there is being only insofar as Dasein exists."[15] Beings exist without Dasein, but being does not. Being confers not ontological status, but intelligibility. Accordingly, Heidegger does not take the objects of science to be mere theoretical constructs, yet by the late 1920s his realism is no longer naïve: Although the entities described by science do not depend ontologically on human knowing, there is no access to them outside an interpretive framework. Thus he holds the realist thesis that scientific objects exist independently of human consciousness, but also the antirealist thesis that objects outside human consciousness are unintelligible. Things in the scientist's world may therefore turn out to be fictitious (e.g., phlogiston or caloric), but the hermeneutic nature of scientific understanding does not imply global error of the kind threatened by Descartes' evil genius or the film *The Matrix*.

Accordingly, Heidegger looks like an internal realist, that is, an antirealist who accepts the reality of objects within conceptual schemes on the basis that representations constitute the real. This is, however, idealism. Heidegger is not committed to the thesis shared by the idealist and the internal realist that *a priori* structures of understanding *constitute* the real. therefore, he can agree with the analytic philosopher of science that scientists must return to the phenomenon for final arbitration of a theory's success, for theories can be more or less hermeneutically violent and should be open to revision. Both he and the traditional philosopher of science are working under Kantian insight into transcendental subjectivity. But Heidegger is also writing out of the German tradition of phenomenology, and against its idealism. Thus, his analysis shows that the realist and the antirealist are at cross-purposes insofar as the latter's thesis is epistemological, while the former's is ontological. The antirealist need not be an idealist: Acknowledgment of the hermeneutic nature of scientific inquiry can be coupled with a commitment to the transcendent reality of the objects of science. The fact that the objects of science do not reduce ontologically to the conceptual scheme in which they figure does not mean it makes sense or is useful to talk about them as independent of any conceptual scheme.

THE QUESTION CONCERNING TECHNOLOGY

The question of science is further significant for Heidegger because therein begins his critique of technology. In 1917, he tells a reductive yet insightful story about the history of science in which he contrasts Aristotle's method for studying *ta physika* against Galileo's approach to the problem of freefall:

> The old contemplation of nature would have proceeded with the problem of fall such that it would have tried through observation of individual cases of falling phenomena to bring out what was now common in all cases, in order . . . to draw conclusions about the essence of falling. Galileo does not start with the observation of individual falling phenomena, but on the contrary with a general assumption (an hypothesis) which goes: bodies fall—robbed of their support—so that their velocity increases proportional to time ($v = g \cdot t$), that is, bodies fall in uniformly accelerated motion. (GA 1, 419)

Whereas Aristotle makes generalizations on the basis of observation, Galileo hypothesizes a universal law and then seeks its experimental validation. Modern science is thus axiomatic—it begins with axioms, which (Heidegger notes twenty years later) Newton also calls "laws," of motion (FD, 71–2/291–2). Heidegger does not fully assess the consequences of this methodological difference between ancient and modern science until his 1939 lectures on Aristotle's *Physics*.

In these lectures, Heidegger points to Aristotle's definition of *ta physei onta*: "they have within themselves a principle of movement (or change) and rest."[16] Artifacts have no such internal principle of motion, except insofar as they are made from some natural material that retains its principle of movement; for example, as Antiphon points out, if one planted a wooden bed, and anything grew, it would be wood, not a bed. Definitive of artifacts for Aristotle is their formal conception in the mind of the artist prior to production.[17] That is to say, artifacts, unlike natural entities, have their origin and developmental principle not in themselves, but in the artist. Hence Galileo and Newton dispense methodologically with the Aristotelian distinction between two separate kinds of knowledge, production (*techne*) and the study of nature (*physis*), when they begin their physics with hypotheses, i.e., ideas in the mind of the physicist. Herein lies incipient Heidegger's later claim that "Modern science is grounded in the essence of technology,"[18] expressed in 1976 in the form of a question: "Is modern natural science the foundation of modern technology—as is supposed—or is it . . . already the basic form of technological thinking?"[19] In light of the 1917 text and the 1939 treatment of Aristotle's *Physics*, his answer to this question cannot but be the latter: modern science is already the basic form of technological thinking.

In the technology essay, Heidegger names the "basic form of technological thinking" "Ge-stell." "Ge-stell" is a development of what begins in *Being and Time* as "basic concepts" but is already complicated in that text by analysis of the transition from *Zuhandenheit* and *Vorhandenheit*. He subsequently names this concept "the mathematical" in *Die Frage nach dem Ding*. Understanding this central moment in the technology essay therefore requires understanding the development of his on-going assessment of science. What Heidegger intends by "basic concepts" in *Being and Time* is explicit in *The Basic Problems of Phenomenology*: Basic concepts define regional ontologies by representing the object of a specialized science. They delimit, for example, "the 'world' of the mathematician" by signifying "the realm of the possible objects of mathematics" (SZ, 64–5/93). This is the sense in which sciences are positive: They "have as their theme some being or beings . . . posited by them in advance" (GA 24, 17/13). Biology, for example, begins with an understanding of what life (*bios*) is, whereas zoology takes as its starting point an *a priori* conception of the animal (*zoon*). Common ground exists here between Heidegger's analysis and Kuhn's account of paradigm shifts in 1962, insofar as Heidegger argues in 1927 that a crises occurs in a science when its basic concepts undergo revision.[20] Yet in *Being and Time* already, what a science projects to make theoretical enquiry possible goes beyond mere delimitation of its subject area. In the move from everyday dealings to the theoretical attitude, the understanding of being changes over from readiness-to-hand (*Zuhandenheit*) to presence-at-hand (*Vorhandenheit*). Analysis of the basic concepts of the theoretical attitude uncovers much more than a science's subject area here: it shows "the clues of its methods, the structure of its way of conceiving things, the possibility of truth and certainty which belongs to it, the ways in which things get grounded or proved, the mode in which it is binding for us, and the way it is communicated" (SZ, 362–3/414). Basic concepts thus do much more than simply posit the object-area of a specialized science. They provide an interpretation of being that defines the metaphysics, epistemology and methodology of the theoretical attitude. Thus they determine a human orientation toward beings that can extend far beyond any particular, specialized science.

In *Die Frage nach dem Ding*, Heidegger revisits this issue of what is posited a priori in science through analysis of "the mathematical." In identifying "the mathematical" as the definitive aspect of modern science, he does not just mean that science uses calculation. Rather, he argues that *"ta mathemata"* meant for the Greeks "what we already know [things] to be in advance, the body as bodily, the plant-like of the plant, the animal-like of the animal, the thingness of the thing, and so on" (FD, 56/251). The mathematical is what is brought to enquiry by the understanding. It is the "the fundamental presupposition of the knowledge of things" (FD, 58/254). Like basic concepts in *Being and Time*, the mathematical is not just metaphysical, but also epistemological insofar as it establishes the nature of knowledge. The "fundamental presupposition" of modern science is that the sciences have their foundation not in their object but in reason. When Descartes grounds knowledge in the self-certainty of the *ego cogito*, he establishes representational thinking as the ground of objectivity. Thus Descartes validates philosophically the Galilean–Newtonian method of beginning with rational hypotheses on the basis of which evidence-providing experiments can be devised. Accordingly, *Ge-stell*, representational thinking (which in the case of the essence of technology takes all beings in advance as standing-reserve), is at work at the heart of modern science, which determines the real a priori as what can be represented as object. Hence modern science is already inherently technological insofar as it functions on the basis of representational thinking, but the reduction of the real to standing-reserve can only happen because the real is already reduced to the representational object. Further analysis of objectivity makes this clearer.

Objectivity is the certainty that knowledge is impartial and disinterested, that is, that one has not committed error as Descartes characterizes it in the fourth of his *Meditations*: a libidinal economy in which will exceeds understanding. To avoid such error is to know how things are, rather than how one wants them to be. Thus, science appears to describe "the way the world is," rather than providing perspectival and situated analysis. Accordingly, technology, if it is understood as nothing more than applied science, that is, the ordering by instrumental reason of nature uncovered impartially by theoretical reason, "threatens to sweep man into ordering as the supposed single way of revealing [*Entbergung*]" (VA, 36/32), and thereby precludes other ways of understanding. Modern physics may precede technology by a couple of centuries, but it prepares the way for technology because it already has the essence of technology, Ge-stell, at its heart (VA, 25–6/21–2) insofar as it founds knowledge on the certainty of subjective representation (i.e., objectivity). Scientific objectivity thus prepares the way for the essence of technology to hold sway. That is, science and technology each begin with an a priori projection of a concept of nature, much as ancient *technê* began with an idea of the thing to be produced. Modern science conceives *ta physika* as objects, that is, spatiotemporally extended bodies subject to "a coherence of forces calculable in advance" (VA, 25/21). Technology brings to nature an a priori conception of *Bestand* in which nature is revealed as resource (i.e., as a source of energy that can be stockpiled.) Modern physics is "the herald of Enframing" (VA, 25/22) because objectivity already contains a commitment to nature's quantifiability that plays out in technology as its economic reckonability. Accordingly, analysis of "basic concepts" and "the mathematical" are formative for Heidegger's later position on "Ge-stell."

Science and technology are thus for Heidegger both truths, that is, ways of revealing in which beings are uncovered by human understanding. To read Heidegger on technology without coming to terms with his analysis of science is therefore to work with a deficient theoretical framework. For his critique of technology arises in consequence of his analysis of the history of science: it is in his ongoing treatment of science as the mathematical projection of nature that his conception of Ge-stell has its origin. Nor can his account of the historical emergence of modern technology, only hinted at in the discussion of the relation between science and technology in the technology essay, be understood apart from his long-standing critique of the modern scientific mathematization of nature. Scholars who do not follow this path in the development of his thinking risk falling into the postmodern trap of nihilistic technics, that is, the play of empty forms. Derrida, for example, uses the metaphor of the bottomless chessboard to express the infinite possibility of interpretations.[21] For Heidegger, the chessboard is not bottomless. In the case of neither science nor technology is human being thrown into an abyss of nothingness. Heidegger began his *Antrittsrede* in 1929 with the nothing beyond beings that is rejected by science, and he developed that beginning into his concluding question, "Why are there beings at all, and why not rather nothing?" (GA 9, 122/110). In both science and technology, human being is thrown against the plethora of nature. His subsequent analyses of science and technology suggest that there is something beyond both that is reducible to neither: nature. Science and technology are ways of revealing that project an interpretation (an "as-structure") onto nature: Natural entities are interpreted first and foremost as object for science, and as resource for technology. These projections are deeply and historically related insofar as modern technology is possible because its essence, *Ge-stell*, is *already at work* in modern science. It is the mathematical projection of objectivity onto nature in science that heralds the subsequent technological projection of reckonable resource. Science and technology are deeply complicit and inseparable ways of understanding nature for Heidegger, and he believes other ways are possible that do not set upon nature in such over-whelming assault.

NOTES

1. Trish Glazebrook, *Heidegger's Philosophy of Science* (New York: Fordham University Press, 2000); "Heidegger and Scientific Realism," *Continental Philosophy Review* 34, no. 4 (December 2001), 361–401; and "From *physis* to nature, *technê* to technology: Heidegger on Aristotle, Galileo and Newton" *The Southern Journal of Philosophy* 38, no. 1 (2000), 95–118.

2. For a discussion of the German "Wissenschaft" in distinction to the English word "science," see Babette Babich, "Nietzsche's Critique of Scientific Reason and Scientific Culture: On 'Science as a Problem' and 'Nature as Chaos,'" in Gregory M. Moore and Thomas Brobjer, eds. *Nietzsche and Science* (Aldershot: Ashgate, 2004), 133–53.

3. Gadamer notes that the word was made popular by translator's of Mill's *Logic* (H-G. Gadamer, *Truth and Method*, tr. J. Weisenheimer and D. G. Marshall, New York: Continuum, 2000, 3). Cf. the discussion in Wikipedia available online at http://de.wikipedia.org/wiki/Geisteswissenschaft, and *Deutsches Wörterbuch* von Jacob Grimm und Wilhelm Grimm (Leipzig: S. Hirzel, 1971) available online at http://www.woerterbuchnetz.de/woerterbuecher/dwb/wbgui?lemid=GG05820.

4. Cf. Mill A *System of Logic* (London: Longmans, Green and Co., 1900), 554: "moral sciences" are the sciences of the thoughts, feelings, and actions of human beings, including the study of the laws of mind, human nature, society, political economy, and history.

5. Cf. esp. GA 25, 25; GA 65, 142/99 and §76.

6. One such conceptual model comes from biology; cf. Rupert Sheldrake, "The Laws of Nature as Habits," *The Reenchantment of Science*, ed. David Ray Griffin (New York: State University of New York Press, 1988). Another, perhaps less promising, from geology; cf. Frodeman's "Preface" in *Earth Matters: the Earth Sciences, Philosophy, and the Claims of Community*, ed. Robert Frodeman (Upper Saddle River, NJ: Prentice-Hall, 2000), vii–xiii, as well as Victor Baker, "Conversing with the Earth: The Geological Approach to Understanding," and Christine Turner, "Messages in Stone: Field Geology in the American West," in this anthology.

7. On the gender politics of objectivity, see Sandra Harding, *Whose Science? Whose Knowledge?* (Ithaca, NY: Cornell University Press, 1991) and *Is Science Multicultural? Postcolonialisms, Feminisms, and Epistemologies* (Bloomington: Indiana University Press, 1998), and Nancy Tuana, ed., *Feminism and Science* (Bloomington: Indiana University Press, 1989).

8. On imperialism and racism, see Trish Glazebrook, "Global Technology and the Promise of Control," *Globalization, Technology and Philosophy*, eds. David Tabachnik and Toivo Koivukoski (Albany: State University of New York Press, 2004), 143–58. On sustainability, see Trish Glazebrook, "Art or Nature? Aristotle, Restoration Ecology, and Flowforms," *Ethics and the Environment* 8, no. 1 (2003), 22–36, and "Gynocentric Eco-Logics," *Ethics and the Environment* 10, no. 2 (2005), Special Issue, ed. Christopher Preston, 75–99.

9. David Gross, who received a Nobel prize for his work on strong nuclear forces and has been a strong advocate of string theory, in his closing remarks at the twenty-third Solvay Conference in Physics in Brussels in December 2005, suggested that something absolutely fundamental and profound is still missing from string theory. Leonard Susskind, who invented it, still defends it against multiple criticisms (see his interview in *New Scientist* 2530, December 17, 2005, 48–50; cf. the letter from Michael Duff 2533, January 7, 2006, 16).

10. This paper is included below, so I do not detail his arguments here. I have responded directly to his criticism in Glazebrook, *Heidegger's Philosophy of Science*, 247–51.

11. Joseph Kockelmans, *Heidegger and Science* (Lanham, MD: University Press of America, 1985), 17.

12. *Discoveries and Opinions of Galileo*, tr. Stillman Drake (London: Anchor Books, 1957), 238.

13. SZ, 205/249. His view warrants comparison with Fine's "Natural Ontological Attitude," or what is called "background realism" in traditional philosophy of science. Cf. Glazebrook (2001), 376–382.

14. SZ, 6/26. Cf. John Haugeland, "Truth and Finitude: Heidegger's Transcendental Existentialism," *Heidegger, Authenticity, and Modernity*, eds. Mark A. Wrathall and Jeff Malpas (Cambridge, MA: MIT Press, 2000), 43–77: 47.

15. GA 26, 195/153. I have modified the translation to render it more consistent with Heidegger's phrase "es gibt."

16. *Physics, Books I–IV*, trs. P. H. Wicksteed and F. M. Cornford (Cambridge, MA: Harvard University Press, 1929), 192b8–11; GA 9, 247–8/228.

17. *Parts of Animals*, tr. D'Arcy Wentworth Thompson, *The Basic Works of Aristotle*, ed. Richard McKeon (New York: Random House, 1941), 640a32; cf. *Nicomachean Ethics* tr. H. Rackham (Cambridge, MA: Harvard University Press, 1934), 1140a13.

18. WD, 155/135. I have modified the translation since translating *"Wesen"* as "nature" obscures my point. "Essence" is commonly used to translate Heidegger's *"Wesen."*

19. Martin Heidegger, "Neuzeitliche Naturwissenschaft und Moderne Technik," tr. John Sallis, *Research in Phenomenology*, 7 (1977), 1–4: 3.

20. Thomas Kuhn, *The Structure of Scientific Revolutions* (Chicago: University of Chicago Press, 1962); cf. SZ 9/29.

21. Jacques Derrida. "Différance," *Margins of Philosophy*, tr. Alan Bass (Chicago: University of Chicago Press, 1982), 3–27.

HEIDEGGER'S CRITIQUE OF SCIENCE[1]

William J. Richardson, S. J.

On the longest day he ever lived, Heidegger could never be called a philosopher of science. But he is a philosopher—an important one—and no genuine philosopher can afford to ignore the problems of science. Perhaps we will understand his attitude toward science best if we realize that his most explicit analysis of the nature of science occurs in an address he delivered in Freiburg in 1938 to a group of scientists, doctors, and art historians. His was the last lecture in a series devoted to the theme of the "Foundations of the World-View that Characterizes Modern Times." Let this suggest at once that Heidegger's critique of the sciences is inseparable from, because unintelligible without, his interpretation of the fundamental nature of modern times as such. For the scientific attitude is simply an aspect of one's view of the World, or, more precisely, the scientific attitude is a function of the way the scientist deals with the beings he encounters within the world of his experience. To try to understand what this implies for Heidegger is the task we propose to deal with here. In order to make as clear as possible the thought of a man whose reputation is not based on his clarity of exposition, let us keep the matter as uncluttered as possible by asking simply:

1. What does Heidegger understand the method of modern science to be?
2. What critique does he make of this method?
3. How are we to evaluate this critique?

METHOD OF MODERN SCIENCE

The notion of "science," of course, is an old one—older than the word *scientia* it transliterates, as old (at least) as the *epistemê* of the Greeks. If it has for modern man a meaning all its own, this is because science developed a specifically new method of inquiry with the advent of modern times (VA, 54–5/169–70). What characterizes the method of modern science and distinguishes it from ancient science for which Aristotle supplied the basic formula is the notion of "research."

Heidegger sees three ingredients as essential to the notion of modern scientific research. The first may be described in general terms as rigor of procedure. When the scientist addresses himself to the investigation of a particular object, this very address already defines the area (*Bezirk*) within which the investigation takes place. More precisely, the scientist sketches out in preliminary, tentative fashion what is to be looked for in the object. I say "provisional" here in a very literal sense, for the initial blueprint is a *pro-viso*, a looking forward to what the object is expected to reveal itself to be. Heidegger uses the word *Entwurf*, a throwing-forth, or, as we would say, a "*pro-ject*," the way we speak of the pro-ject of a long-range development plan for the university (H, 71/118). Now it is according to this project that the scientist examines the object more closely in an effort to verify the project by experiential contact with it. And it is in the close correlation of project and confirmation that the rigor of method consists.

It is in this context that we can understand the role of mathematics in the evolution of this method. If we go back to the primitive notion of *ta mathêmata*, we realize that the word did not designate for the Greeks, initially at least, the science of measure. Rather, *ta mathêmata* (from *manthanô*, "to learn") signified those things that were already learned. Heidegger takes this to mean those things that are known in advance of actual experiential contact with objects. I understand him to mean what we would call after Kant *a priori* knowledge of objects, although he does not actually use that formula. In any case, it is because knowledge of quantitative measure is the type *par excellence* of *a priori* knowledge of objects that the term *ta mathêmata* came to refer, even for the Greeks, to the knowledge of quantitative measurement (H, 71–2/118–9; cf. FD, 49–83/271–305).

Now physics, the first-born and still the paradigm of modern scientific disciplines, has a special affinity with the science of quantitative measurement. It might be well to remark here that in the context of present analysis the paradigmatic physics is Newtonian physics. What is to be said of quantum physics must be said *mutatis mutandis*. The word "physics" derives from *physis*, "nature." It implies an *a priori* (i.e., "mathematical" in the primitive sense) knowledge of what "nature" means, hence a pro-visional projection of what "nature" is to reveal itself to be as the physicist goes about his task of investigation. What blueprint of "nature" does the physicist project? That of the visible movement about him. Visible movement, however, says change of place and succession in time, situation in place-time. As a result, only those phenomena swim into the physicist's ken that can be filtered through the screen of spatio-temporal terms, or rather space-time is the way for him that energy appears. Such is the provisional project with which he begins his work.

How does he proceed? By confirming this project in actual experiential contact with his object, by establishing a bond as possible between his anticipation of experience and its realization. But because the projected view of nature is such that it discerns only visible movement through different points of the spatio-temporal field, and because "points" of space and "moments" of time, or constellations of space-time, are conceived as fundamentally quantitative entities (or at least as calculable relationships), the privileged instrument for confirming the anticipatory project will be techniques of quantitative (especially numerical) measurement and calculative relationship—in other words, mathematics in the strict sense as we understand it today (H, 72–3/119). Now this correlation between the kind of blueprint projected by the physicist and the mathematical calculation that characterizes his procedure in confirming it is what Heidegger understands by the "exactitude" (*Exaktheit*) of scientific method. We can see what he means, then, when he says: ". . . The mathematical research of physics is not exact because it calculates accurately, but it must calculate accurately because the tie that binds it to the object of research has the character of exactitude . . ." (H, 73/119–20). If another scientific discipline (e.g., history) does not admit of the exactitude of the physical sciences, this does not mean that its method is any less rigorous, but only that the correlation between project and confirmation does not have the character of "exactitude." The reason is simple: the area of its investigation is profoundly incalculable.

The second ingredient essential to the method of scientific research is experimentation. Nature, we said, is projected by a Newton as the complex of visible movement in the world about him. He must respect this movement with its incessant change, and for all the demands of accurate calculation, he must sedulously refuse to destroy this movement even in his thought. Individual facts are meaningful insofar as they form part of, and bear testimony to, the movement as such. Yet the meaning for the physicist of a given sequence of facts in the movement is not as fluid as the sequence itself. Meaning implies a certain stability in the movement, a certain constancy of recurrence. Now stability and con-stancy both derive from the Latin *stare*, "to stand," and suggest that the meaning of facts becomes clear to the extent that the movement, without compromise of its fluidity, takes a stand (*Stand*) over from and opposed to (*gegen*) the scientist, and thus becomes a *Gegenstand*, an "object" of investigation. The meaning of facts becomes clear to the extent that the movement becomes more and more "objectified."

Now the determination of the meaning of facts (through the objectification of the movement in which they occur) takes place as the stable pattern of facts is discerned in the constancy of their recurrence. When the pattern is conceived as endowed with a certain "necessity" (even a statistical one), it is called a "law" (Cf. VA 54–5/170). In scientific research, then, the meaning of facts becomes clear through the formulation of laws (H, 73–4/120).

It is in the actual elaboration of these laws that experimentation plays an essential role. An experiment begins with the laying down pro-visionally of a law. With the term "laying down" (*Zugrundelegung*) Heidegger is translating literally the Greek word, *hypothesis*: Placing (*thesis*) a provisional law underneath (*hypo-*) the facts to be explained. In planning the experiment the scientist conceives of certain conditions under which he can follow closely a given sequence of events and by antecedent design submit this sequence to his calculating control. The purpose is to see whether or not the facts thus controlled confirm the law that has tentatively been laid down. If they do, the experiment succeeds and the law is established. If not, the procedure begins all over again with the laying down of another tentative law.

In all this, two points are worth noting:

1. The tentative law (hypothesis) with which the experiment begins is no more than an elaboration of the scientist's original projection of the blueprint for nature as such. That is why the hypothesis is never a merely arbitrary construct (*hypotheses non fingo*, Newton said), but a special functioning of a broader *a priori* that conditions his entire approach to the object (H, 75/121–2).

2. The verification process is made possible only by a closely controlled, calculative (therefore mathematical) measurement that is itself but a refinement of what we spoke of as rigorous confirmation of an original project.

It is precisely in this respect that modern scientific method differs so radically from the method of Aristotelian science. For Aristotle, too, had insisted on *empeiria* ("experience") as essential to his method. But Aristotle's *empeiria* amounted to no more than close observation and description of phenomena. It did not demand the correlation of project and mathematical verification. And even Roger Bacon, so often heralded as the forerunner of the modern scientist, comes no closer to the modern notion of experimentation than Aristotle did. He insists on the importance of *experimentum*, to be sure, but this (for Heidegger) was by way of reaction against the theological method of his medieval contemporaries, (i.e., the argument from authority.) ". . . If Roger Bacon calls for *experimentum*—and this he certainly does—he does not mean the experiment of science conceived as research. Rather he demands instead of the *argumentum ex verbo* the *argumentum ex re*, instead of the analysis of dogmatic statements, the observation of things themselves, that is, the Aristotelian *empeiria*" (H, 75/122).

To turn for a moment from physics to the historical sciences, Heidegger finds in the historical method an analogue to the physicist's use of experiment in the historian's examination of sources (i.e., the quest for, evaluation of, exploitation and interpretation of them). To be sure, the historian does not seek to establish laws in the same way the physicist does, but he wants to do more than merely narrate a series of events. As much as the physicist, it would seem, the historian seeks to discern meaning in the sequence of events (i.e., he seeks to explain individual facts by stable patterns of constancy that emerge from his scrutiny of the past). But stability and constancy emerge from the welter of events only to the extent that these patterns become for him objects that he can control. Where for the physicist the instrument of control is the experiment, for the historian the instrument of control is the examination of sources (H, 76/122–3).

We have seen the first two ingredients of what Heidegger conceives to be the method of modern science:

1. rigor of procedure and

2. controlled experimentation (or examination of sources).

The second is the implementation of the first. He suggests another element: *Betriebcharakter* (i.e., "institutionalization"). The sense seems to be as follows. The method already described must necessarily limit the scope of its particularized projects to be verified by experiment, precisely in order to be able to control the verification of them. This means that the researcher must carefully select a special aspect of his object that he wishes to investigate—in other words, he must specialize. But this specialization, indigenous to research as such, quickly proliferates into more and more specialized refinements that call in turn for more and more specialized investigation. The researcher's universe is indeed an expanding one. The result is that the individual researcher becomes more and more inadequate to the demands that this inevitable expansion imposes on him. He must seek the help of other researchers, rely on other instruments of calculation than his own mathematical powers (electronic computers and the like). In short, he must ally with other researchers and technicians to organize institutes of research. But this gradual institutionalization is indigenous to research as such. ". . . In itself, research is not institutional because its work is done in institutes, but institutes are necessary because science, by reason of its nature as research, is institutional of its very nature . . ." (H, 77/124).

CRITIQUE

Such in brief is Heidegger's delineation of the nature of scientific method. Whether or not it is adequate must be left to the scientists to say, but it hardly says anything new. In what sense does it constitute a critique? Recall that he analyzes the scientific method in an effort to meditate upon the "Foundations of the World-view that characterizes Modern Times." In effect, he says that the scientific attitude is a function of man's view of the world as such and of the interpretation of the way beings encountered within the world are what they are. In other words, it is grounded in a philosophical attitude or stance by which man interprets the whole of his experience. Since, after Aristotle, a meditation on beings as beings (*on hêi on*), that is, whatever is insofar as it is, is called "metaphysics," the philosophical attitude/stance is in effect a metaphysical attitude/stance, so that the scientific approach to whatever is grounded in a metaphysical interpretation of the very same thing. It is with the metaphysical stance implied in the modern scientific attitude that Heidegger finds fault, rather than with the attitude itself. We must see this in more detail.

Supposition

The scientific method implies that the scientist conceives of himself as a subject and of the beings with which he deals in research as objects that are posed before him to be investigated. This seems innocuous enough. But recall that this relationship between subject (the scientist) and object (the research material) is made possible only by antecedent projection of a blueprint of the being under investigation, and it is only to the extent that the being conforms to this projected blueprint that it becomes an object at all. In other words, the antecedent project filters out every element in a being that makes it what it is, except that aspect by which it becomes an object for the scientific gaze. It is only when the phenomenon that is being interrogated has been reduced to its mere object-character (let's call it "objectness") that the researcher can proceed top the task of measurement, that is, verification of the project (whether by experimentation or examination of sources). Because the object-ness of the phenomenon under investigation is the only aspect of it that is relevant for the scientist in terms of the rigor of his method, the risk is that the phenomenon *is* for him only to the extent that it is an object to be measured in research, and the scientist risks the assumption that Max Planck made explicit: "The actual is whatever can be measured" (VA, 54/169).

In any case, the constriction of the scientific enterprise to the scope of the subject–object relationship remains as significant today for the examination of microphysical particles as for classical mechanics. What has changed with the advent of subatomic physics is the manner in which the object-ness of objects is experienced, but not the subject–object polarity as such (VA, 57/172–3). Heisenberg's principle of indeterminacy, for example, many very well render obsolete certain theories of the objectifying process such as the Kantian one, but what it makes clearer than ever is the correlative character of the subject–object conjunction, so that without presupposing the subject–object polarity as such, the "principle of indeterminacy" is unthinkable (GA 9, 402/304). The method of modern science presupposes, then, a metaphysical stance that sees the beings encountered in the world to be no more than objects for a subject.

Metaphysical Foundation

This metaphysical stance according to which every being (i.e., everything that is) is conceived as either a subject or an object (i.e., an object for a subject) Heidegger calls "subject-ism" because it arose at that moment in the history of the West when man first experienced himself as subject. This took place with the arrival on stage of René Descartes. It was with Descartes that the epoch of history that we call "modern times" began. It was in Descartes' experience of beings and the "world" that the foundations of the worldview of modern time slay. The modern scientific attitude as such is no more than an epiphenomenon of Cartesian subject-ism.

Why is the Cartesian experience so characteristic of modern times? What, for that matter, was the experience? What characterizes modern man most profoundly as Heidegger sees it is the vindication of his own liberty. This took place with the declaration of independence from the ties of faith that bound his medieval forefathers. The universe becomes for him by and large an anthropocentric rather than theocentric one, or, as Heidegger expresses it, ". . . he frees himself unto himself . . ." (H, 81/127). This has momentous implications for Descartes, especially for his conception of truth.

For medieval man, we are told, received his certitudes (whether the teachings of dogma or the assurance of eternal life) from his faith. Medieval scholars might want to qualify this, of course, but that's another matter—this is the way Heidegger sees it, and that is our only concern for the moment. Once modern man chose independence in the name of liberty, how was he to replace his vanished certitudes? It could only be in and through himself. ". . . This is possible only to the extent that self-liberating man guarantees for himself the certitude of the knowable . . ." (H, 99/148). This was the context, then, in which Descartes set for himself the task of discovering a new foundation for truth that would be unshakable (*fundamentum inconcussum veritatis*)—foundation because it would be the ground upon which all truths would ultimately rest, unshakable because man himself would be the final arbiter of it.

We know well enough how Descartes proceeded. The "unshakable foundation of truth" was for him the *cogito ergo sum*: "I—in thinking—am." The statement itself was, after all, true, for it corresponded to the situation of fact. Moreover, the statement was not only true but certain, for in making it the thinker not only knew the situation of fact but knew that he knew it and was stating it accurately, for even to doubt it was itself a type of thinking that implied the existence of the one who thought. This radical truth, then, was unshakable. And not only was it unshakable but it was the foundation of all other truths, for any other truth was formulated by a type of thinking which rested on the basic truth which consisted in the thinker's certainty of his own existence.

All of this is quite classical. But it is the consequence of this starting point of Descartes that for us is significant. The ego, whose existence for Descartes is certified in the act of thinking, is for the first time in the history of thought conceived as a subject (*subjectum*), something that is thrown under, or better (according to the Greek *hypokeimenon* that it translates) "lies under" everything else—in this case that underlies all truth. In other words, the ego is aware of its own existence, is a *subjectum* for the very same reason that it is a *fundamentum* of all truth. But if we go one step further, we see that everything that is not the thinking subject becomes something about which the ego-subject thinks, something that is pro-posed to it, in fact pro-posed by the subject for thought. In other words, it is an "object" of thought. The object becomes thinkable by the subject only to the extent that it is controlled by the subject, for without this control, certitude (knowing that we know) is impossible. The result is that in this new perspective everything that is, including the ego itself, becomes either a subject of thought or an object of thought.

One can see, then, why the method of modern science is only an epiphenomenon of Cartesian subject-ism. For the rigor of research procedure is grounded in a more fundamental quest of the researcher for certitude, for knowing that he knows. The correlation of projection (of blueprint) and verification (through experiment) is no more than the manifestation in his laboratory world of the underlying paradigm of the subject–object relationship whereby an object is proposed to the subject to the extent that the subject proposes it to himself, and, indeed, in such a way that he can control it enough to gain certitude about it.

In any case, the world itself in this new perspective becomes a worldview (i.e., world-as-viewed) for it becomes nothing more than the sum total of the objects of thought, so to speak the collective object for any given subject. In a word, the first consequence of Descartes' discovery of the unshakable foundation of truth in the self-awareness of a thinking ego is that all reality becomes divided into subjects and objects.

Critique

Fair enough, but so what? Why should this be a matter for criticism? In the simplest terms: because for Heidegger "reality" is much more than that, an agglomeration of subjects and objects. If everything that is becomes pressed into the categories of subject and object, then something in them is lost—the wondrous depth, the beauty, the deep-down (nonobjectifiable) freshness of things is overlooked. The marvelous mystery of their presence to man, of his presence to them, even of his presence to himself and to the world, is disregarded. It is this presence of beings (including man) that Heidegger understands by their Being (*Sein*), that which accounts for the fact that they *are* (present), the Is of what is. Briefly, with the emergence of the subject–object polarity with Descartes, the Being of beings is forgotten.

This needs a word of explanation. From the moment of his philosophical awakening at the age of eighteen when he read for the first time Franz Brentano's doctoral dissertation on *The Manifold Sense of Being in Aristotle* (1862), where "being" translates the German *Seiendes* and the Greek *on*, the question about Being was posed. In 1962, he writes:

> . . . On the title page of [this] work, Brentano quotes Aristotle's phrase: *to on legetai pollachôs*. I translate: "A being becomes manifest, (i.e., with regard to its Being) in many ways." Latent in this phrase is the *question* that determined the way of my thought: what is the pervasive, simple, unified determination of Being that permeates all of its multiple meanings? . . . How can they be brought into comprehensible accord? This accord can not be grasped without first raising and settling the question: whence does Being as such (not merely beings as beings) receive its determination.[2]

This was his first philosophical question, and his last; to this day, it remains his only question. But to understand the direction that his philosophical way has taken, we must appreciate the impact that Husserl had on his impressionable mind when he first devoted himself to the study of philosphy at the university. For to Husserl, a phenomenologist, that which is (a being) was that which appears (a *phainomenon*), that which becomes manifest or manifests itself to the phenomenologist. As far as Heidegger was concerned, too, for a being "to be" meant to be manifest (present as manifest), that is, revealed. Being came to mean that process that lets beings be manifest, revealed, present to the phenomenologist and him to them.

Now for a being to be revealed (or reveal itself) means that somehow it is lit up before man. Being is the lighting process as such. Furthermore, when a being is lit up it emerges out of darkness, as if the veil of darkness (*-velum*) that previously concealed it is torn aside and it becomes *re-velatum*, unconcealed. Being is the process of re-velation, nonconcealment as such. Furthermore, the Greek word for the concealment is *lêthê*, and the alpha-prefix serves to suggest the privation of this *lêthê* (concealment). What is revealed, unconcealed, is *a-lêthes*, or "true," the process of revelation/nonconcealment is *a-lêtheia*, truth. Being, then, is the process of *a-lêtheia*, truth, as such. To interrogate the meaning of Being, then, means to meditate on truth as the process of nonconcealment by which beings are revealed, become manifest, are present to man.

In the long course of his peregrination, Heidegger approaches the problem in a thousand different ways, but the fundamental drift of his thought is always the same. To the extent that the metaphysician considers beings as beings, he profits from the lighting process of Being. But the lighting process itself is not that which it lights up. Being is not a being. But what appears to the metaphysician is what is lit up (i.e., beings), and Being itself, as the lighting process, recedes from him, that is, hides itself within the beings that are lit up and as them, although always remaining different from them. It is easy to understand, then, how the metaphysician, although benefiting from the lighting process, can remain oblivious of it (i.e., he forgets, so to speak), the very process that makes the beings he contemplates accessible to him. Because the entire metaphysical enterprise rests on this accessibility of beings as foundation, then to interrogate the lighting process as such is to probe the very foundations of metaphysics.

This is what Heidegger set out to do. Identifying metaphysics with ontology in the early years, he describes his effort as a "fundamental ontology," but later on he speaks of it only as the overcoming of metaphysics, and with it, of course, science, too. But he will succeed in this task only to the extent that he helps the metaphysician (or the scientist who functions always within a given metaphysical framework) to realize that truth, the deepest, most original truth that founds all other truths, is not merely a conformity between man's mind and the situation of fact, and still less some form of certitude (i.e., conformity known as such), but rather the process of *a-lêtheia*, nonconcealment, by which beings are revealed, become present.

One can appreciate, then, the sense of his critique of Descartes and, by implication, that scientific attitude which Cartesian subject-ism first made possible. To the extent that both are grounded in the concept of truth as certitude, they are oblivious of truth as a process of nonconcealment (i.e., of Being itself), which alone lets the object be manifest to subject, alone gives the subject access to objects, alone establishes the domain, describes the horizon, wherein the encounter between subject and object can take place. For the scientist as for the metaphysician, to neglect Being as *a-lêtheia*, for him to overlook the mystery of Presence that surrounds and permeates him is to reduce beings to the level of so many empty shells of objectness, moved about and controlled like pawns in a game, victim of his own impoverishment, stripped of his true dignity, which for Heidegger is to be a being open to Being, present to the ineffable mystery of Presence. It is against the impoverishment of science (and of the scientist himself) that Heidegger protests, not against its unquestionable wealth.

For the impoverishment as he sees it has taken a terrible toll on contemporary society in the West, and Nietzsche's nihilism testifies to the fact. Nietzsche proclaimed the nihilism of all values in nineteenth-century Europe. But "values" had become current coin, Heidegger claims, as an attempt to ascribe importance to beings once their genuine source of importance (i.e., the mystery of their presence, their Being) was forgotten in the subjectism of Descartes. Thus value became the goal of all intercourse with beings. Soon the intercourse was considered as culture, the values as cultural values, goal of all human creativity, which, in turn, is placed at the service of man in achieving certitude of himself as a subject. From here it was easy to reduce values themselves to the level of mere objects submitted to man's control in an effort to establish a special place for himself in the world conceived as collective object, become now a hierarchy of values. At this point, values become as shallow and empty as the mere objectness they mask. Nietzsche experienced deeply this emptiness. For Nietzsche, values had ceased to have any meaning, and because he took God to be the symbol of the entire hierarchy of values of which He was necessarily the head, God himself was dead. Because values (God included) meant nothing at all, Nietzsche could speak of this nothingness of values as a nihil-(nothing)-ism. In such a context, to proclaim the death of God was simply to declare a situation of fact, and the scientific attitude, to the extent that it fed upon the deicidal subjectivism, must share the blame for His demise.

The tragedy of it was, however, that Nietzsche himself could not break the iron circle. In trying to overcome the nihilism of values, which was grounded in a metaphysics that had conceived truth as certitude and forgotten *alêtheia* (Being) as nonconcealment, Nietzsche himself fell into merely another type of subjectism in which man would strive to be a superman by establishing dominion over the earth. Superman, then, became just a subject of another kind—albeit a superior subject—and the dominion would be nothing more than a superior type of control over beings—beings, however, still conceived as objects. In this conquest, science would of course play an increasing role but to the extent that for Nietzsche Being-as-*alêtheia* (the only source of importance for beings) still counted for nothing, the result was but a new nihilism all the more profound. This type of nihilism, initiated by Descartes and culminated in Nietzsche, where forgetfulness of Being masquerades as technical progress—or technological achievement—Heidegger calls "technicity" (*Technik*).

It is such a nihilism, forgetfulness of *alêtheia* (truth as nonconcealment) under the guise of technicity, that Heidegger himself is striving to overcome through the meditation on the meaning of Being. The effort involves coming to grips with the problem of science in a positive way on more than one occasion. In his first major treatise, *Being and Time*, in which he makes a phenomenological analysis of the human *Dasein* as being-in-the-World, Heidegger takes time to show how the scientific attitude arises out of the structure of *Dasein* as transcendence (passage) beyond beings to Being, which is there conceived as the horizon of the World (see e.g., SZ, 408–15). Following him here, however, would take us too far into the details of *Dasein*-analysis to be feasible within the present compass. Let us mention, however, two subsequent occasions when he addresses himself formally to the scientific mind.

The first occurs in the famous inaugural address in Freiburg when he assumed Husserl's chair of philosophy in 1929. It bore the title, *What is Metaphysics?*. The audience that day was university-wide, hence almost to a man composed of scientists of one sort or another, all come to hear the first lecture of their illustrious colleague. And what did he talk to them about? About science? Not at all. About "nothing"! For he wishes to share with them his own concern for the problem of Being and he is perfectly aware that to them this means "nothing." Naturally! For the scientist examines beings—and nothing else; his research is guided by beings—and nothing more. And yet he speaks of what concerns him (beings) in reference to "nothing." What about this "nothing," then, this no-thing, that is not-a-being (*Nichts*)? Heidegger then proceeds to get his hearers to at least experience this "something" that is not-a-being, Non-being, and thus appreciate the importance of raising a question about it.

Non-being is disclosed through the phenomenon of anxiety. This is not the only experience in which Non-being is revealed to man—Heidegger mentions the phenomenon of boredom and joy in the presence of the beloved—but it is the experience analyzed in *Being and Time* and so he returns to it here (GA 9, 111/100). What does he mean by anxiety? It is that strange uneasiness that steals over a man from time to time when he seems to lose his sure grasp of the beings around him, yet leaves him somehow aware of "something" that is not a being. This is how it differs from fear, because fear is always in view of something, as a child is afraid of a dog. But in this uncanny emptiness of anxiety, things about us seem to dissolve—or, at least, their meaningfulness dissolves. We become ill at ease, disoriented, alienated from the world-about. But precisely what are we anxious about? That's the trouble—we can't say. About nothing! Yet not *absolutely* nothing. About "something" quite real, but that is no "thing" like any other "thing," nor is it situated here nor there nor anywhere. It is no thing and nowhere. We call it *the* No-thing, that is, Non-being (*Nichts*). But even as these beings seem to slip away from a man in experiencing the No-thing, indeed because they slip away, they light up for him with a startling strangeness that leave shim struck with wonder at the simple fact that they "are." Thus, in the ". . . effulgent night of Non-being [disclosed by] anxiety, there occurs for the first time the original open-ness of beings as such: that they are beings and not Non-being . . ." (GA 9, 114/103).

Much has been made of Heidegger's so-called nihilism in meditating on Non-being (*Nichts*), but the sense of it is clear. It is simply an attempt to bring the scientist to an experience of Being as he understands it, but if one approaches it in the context of beings as the scientist must, the most that can be said about it, initially at least, is that it is not-a-being, Non-being. Fourteen years later (1943) in an epilogue added to the fourth edition of the address, he says explicitly: ". . . Non-being, as other than beings, is the veil of Being . . ." (GA 9, 312). And again in 1956:

> . . . The question "What is Metaphysics?" was analyzed in a philosophical inaugural lecture delivered before all the assembled faculties. The question is posed therefore in the circle of all sciences and addresses itself to them. But how? Not with the presumptuous purpose of improving on their work or, indeed, disparaging it. . . . The sciences assume that the [thinking] about beings exhausts the whole domain of what can be researched and interrogated, that outside of beings there is "nothing else." . . . From the viewpoint of scientific [thinking] that knows only beings, what is wholly and completely not a being (namely Being) offers itself as Non-being. . . . It asks: what about that which is wholly other than each and every being, that which is not a being? . . . (GA 9, 418/316)

Heidegger again addresses himself to the problem of science in summer 1953 when he participated in a more general discussion in Munich entitled "The Arts in the Technical Age." He entitles the discourse "Science and Meditation," which in effect becomes a meditation for science on the meaning of Being. In it he tries to show the evolution of the notion of science out of Greek thought by defining science in general terms as the "theory of the actual." He begins with an analysis of the word "actual" (*das Wirkliche*), then the word "theory" (*theoreô*: Betrachtung) but comes soon enough to the "theory of the actual" for the modern mind as the proposing of an object that is to be controlled by calculative measurement—a reprise in fresh terms of the correlation between project and confirmation we have already seen to be the basic structure of scientific research.

What he adds here is an insistence that there is an unapparent content (*unscheinbare Sachverhalt*) latent in the matter for investigation that cannot be calculated or verified because it cannot be made into an object. And this invisible content, he claims, is unavoidable by the scientist. In physics, for example, what is unavoidable is nature itself that keeps emerging into presence for him with an abiding dynamic novelty even as he endeavors to immobilize it into a stable pattern of constancy (i.e., into an object he can control whether by direct means or, as is more and more the case microphysical phenomena, indirect). And the physicist *depends* on this emerging presence before him in order to go about his task (VA, 57–9/173–4).

If this avails for physics, it avails as well for the other sciences, too, each in its own way. Psychology and psychiatry—and I presume we should understand here too all of the social sciences—objectify man in terms of the processes that make him either sick or well. Freud's investigation would be a classic example of this. For Freud, man is and can be no more than an object of clinical research. But these investigations depend on the continued emergence—the coming-into-presence—of human ek-sistence as such, man's presence to the mystery of Presence, that Heidegger calls *Dasein*. History as a science (*Historie*) depends on history as a process—a process continually coming-to-presence—(*Geschichte*), which it doesn't fashion but must presuppose. Likewise philology in all its forms (semantics, linguistics, etc.—that whole array of disciplines which deal scientifically with what has been said through human speech) depends on the Ur-phenomenon of Language itself as the emerging-into-presence of some primordial Utterance (see VA, 59–60/174–5).

All the sciences, then, depend on the presencing of some unapparent content that they cannot objectify yet still less can gainsay. The paradox is that just as this unapparent content is unavoidable (*unumgängliche*) so, too, it is inaccessible (*unzugängliche*) to science. To take physics again, the scientist is bound to the object-ness that is proposed to him and cannot get beyond it with the instruments of his science.

> ... Physics as science cannot make any statements about [the nature of] physics. All statements of physics speak the language of physics. Physics itself cannot possibly be the object of a physical experiment. So, too, for philology . . . [It] is never the object of philological treatment. The same is true for all the sciences . . . We can never determine by mathematical calculation what mathematics itself is. (VA, 61/176)

The unapparent content of scientific research, then, is unavoidable yet inaccessible. Indeed, ". . . that element in the sciences which in any given case in unavoidable (nature, man, history, language, etc.) is *as* unavoidable [likewise] for the sciences and through them inaccessible" (VA, 62/177). This unapparent content remains inaccessible for two reasons. First, because the limited sphere of a scientist's research (visible nature for the physicist, man for the psychologist, etc.) must be presupposed in his initial project before the instruments of his discipline can function. To be sure, the scientist is constantly re-examining the fundamental concepts of his discipline but the re-examining is always within the discipline itself and constricted by the conditions of the antecedent project that give rise to the discipline in the first place. In this sense, then, the unapparent content of scientific investigation is that which comes-to-presence before the scientist upon which the project is projected—nature as such, man as such, language as such, and so forth. Such content is inaccessible to the instruments of a scientific discipline, because these are effectual only in the subject–object polarity interior to the discipline itself. *This* content lies beyond the limits of the discipline, transcends all of its projects.

There is a second sense by which the unapparent content of science is unavoidable but inaccessible. It consists in the fact that the real source [*Wesen*] of this unapparent content is not nature as nature or man as man but rather the process of coming-to-presence itself; Presence as such, which presences for the physicist as nature, for the philologist as language, and so on. *This* foundation of science lies hidden from the scientist—not as an apple is hidden in a basket but as a river is hidden in its source (VA, 63/179). It is, in a word, the process of *alêtheia* by which beings are revealed in their presence, it is Being-as-nonconcealment. The meditation on this foundation of science Heidegger takes as his own lifetime task. His critique of science, for all its prophetic tone, is no more than an invitation to the scientist to share the meditation with him.

CONCLUSION

How are we to evaluate all this, if, indeed, it is possible to assign a "value" to it at all? Two remarks seem to be in order. In the first place, it is a mistake, one would think, to take the negative tone with which Heidegger sometimes declaims against the forgetfulness of Being that characterizes our scientifically oriented culture as a repudiation of science itself. In no way does he eschew the just claims of science and the testimony of its marvelous achievements, in order to preach the gospel of a swift return to the woods. He even admits that forgetfulness of Being to which science bears witness is not merely due to the ineptitude of scientists, but due as well to the fact that Being as *a-lêtheia* conceals itself as source in the very beings to which it gives rise. Technicity, then, is an epoch in the history of Being, of Being-as-history, the manner in which Being discloses itself in our age to man in general and to the scientist in particular. What he criticizes is the aggrandizement of science, the failure to admit the limitations of the scientific method, the refusal to admit the deeper level of presence beneath object-ness in beings that cannot be submitted to controlled experiment. In a word, he criticizes the impoverishment of the scientist himself as man, to the extent that his infatuation with science leaves him less of a man, whose greatest prerogative does not consist in technical progress of technological achievement but in openness to the mystery of Presence (i.e., to Being [*a-lêtheia*] itself). In this respect, it would seem that Heidegger has something good to say and scientists, who are first of all men, may profit from his message.

What Heidegger offers is not a philosophy of science but a humanism for a scientific age. His thought would have importance, then, if it suggested a way to reconcile the two divided cultures, scientific and humane. Be that as it may, many men of solid achievement—men like Hess (psychology)' Binswanger and Boss (psychiatry)' Bauch (art history); Schadewalt, Ruprecht, Staiger, Alleman (philology and literature); Astrada (history); Weiszäcker and Heisenberg (physics)—estimable gentlemen all—acknowledge a genuine debt to him that Americans might do well to respect.[3]

But beyond this salutary admonition of Heidegger, does his concept of Being as *a-lêtheia* offer the scientist any real help? Here, one would have to be more reserved. Heidegger himself distinguishes several kinds of truth: there is the truth of propositions (i.e., based on conformity between proposition and situation of fact)—this is logical truth; there is the truth of beings, that is the revelation (nonconcealment) of a particular being as what it is (*alêthes*)—ontic truth; there is the truth of Being, the process of *alêtheia*, the lighting process itself by which these beings are lit up with ontic truth—this he calls ontological truth (GA 9, 131/104). In insisting on the meaning of truth as *a-lêtheia*, he has, to be sure, rendered service, but his own efforts have been devoted to exploring the meaning of ontological truth, while what the scientist must be concerned with is ontic truth. In other words, Heidegger, for all his importance as a philosopher is simply not a philosopher of *science*.

The contemporary scientist, then, will be pained by the poverty of Heidegger's analysis of ontic truth as such. Concerned as he is to show how the scientific attitude is based upon the subject-ism of Descartes, he tends to take as the paradigm of scientific method Newtonian physics, dealing with matter in motion through Euclidean space. The result is that his notion of the subject–object relation in science seems to be basically that of spatial exteriority and separateness of parts outside of parts. But contemporary physical science is concerned less with the motion of parts in space than with the structure of physical objects and the stability in these structures, with light and radiation, symmetries, laws of conversion, invariance principles, principles of ordering, and so on. What is decisive in the subject–object relation here is not opposition and separation (i.e., spatial separation) but interdependence between subject and object, therefore unity with distinction. The classic formulation of this is, of course, the Heisenberg Principle of Indeterminacy.

A conception of ontic truth that would be of genuine service to this type of science would have to go much further than Heidegger has gone in explaining the minutiae of the research technique. One would have to explain much better how mathematics is related to the objects of scientific investigation: How is its formulism to be interpreted? What do the variables represent? What do the numbers represent? To what extent is mathematics merely a language for expressing the essence of relations? In another direction, what exactly is observed in an experiment: a scientific object or merely a numerical symbol of an object? Furthermore, what role does a scientific system as a whole play in the experimental enterprise? Finally, how are abstract concepts in the scientist *related* to Being as *a-lêtheia*? Is Being known only when abstract conceptual knowledge is suppressed? Or is it known through a higher awareness that preserves abstract conceptual structures intact and simply enriches them? If so, what is this "higher awareness?"[4]

To none of these questions has Heidegger addressed himself in his published work. He would probably say that this is not his business—that it is for others to do. As far as we are concerned, however, it would seem sufficient to say, after remarking the poverty of Heidegger's thought as a philosophy of science, that the exploration of these questions which would make possible a philosophy of science could be done in a Heideggerian framework. The exploration would not comprise his insight. Rather, such an exploration, with full deference to the validity of Heidegger's experience as suggesting a humanism for a scientific age, could implement the insight, explicate it, and bring it to a richer fruition than Heidegger himself has been able to do.

NOTES

1. Reprinted from *New Scholasticism* 42, no. 4 (1968), 511–36.
2. Martin Heidegger, "Preface" to W. J. Richardson, S. J. *Heidegger: Through Phenomenology to Thought* (The Hague: Martinus Nijhoff, 1963), xi.
3. See Martin Heideggers *Einfluss auf die Wissenschaften*, ed. Carlos Astrada et al. (Bern: A, Francke, 1949), and Martin Heidegger zum siebsigsten Geburtstag (Pfulligen, 1959).
4. In formulating these critical remarks, the writer owes much to his private discussion with Rev. Patrick Heelan, S. J., author of *Quantum Mechanics and Objectivity, A Study of the Physical Philosophy of Werner Heisenberg* (The Hague: Martinus Nijhoff, 1965).

II

QUANTUM THEORY

BEYOND ONTIC-ONTOLOGICAL RELATIONS

Gelassenheit, Gegnet, and Niels Bohr's Program of Experimental Quantum Mechanics

James R. Watson

Beginning with Heidegger's 1925 Marburg lectures, an ensemble of markers trace a series of multiple and perilous paths to the Gelassenheit and Gegnet markers of the period from 1944 to 1955. The latter are especially important as reflections on and consequences of the "worlds" encountered by the post-Kehre Heidegger on his way beyond ontic-ontological relations. In ways that are not yet clear, Heidegger's *Gelassenheit* and *Gegnet* markers reflect the profound yet ambivalent effects of quantum mechanics on Heidegger's thinking. In what follows, I attempt to show why Heidegger's characterization of modern science accurately reflects the "ideology" of classical physics and thus misses the mark of the emerging characteristics of quantum mechanics. What makes this a difficult perception, however, concerns the fact that his Gelassenheit and Gegnet markers are heavily indebted to the reflections of Werner Heisenberg and Niels Bohr.

From 1925 to 1929, Heidegger's formulations of a hermeneutical phenomenology moved in the dimension of an encountering understood as the basis for derivative modes such as knowing and explanation. Hermeneutical phenomenology is the phenomenal exhibition and interpretation of the structure of encountering beings in the ontic-existentiell self-evidence of a world becoming "transparent in a Dasein by way of the positive vision of the phenomenon of in-being." The meaning of the proposition, "An entity always is only for a consciousness" means, accordingly, that "a world is encountered" and the entity "thus directs us to interpret the structure of encounter, the activity of encountering" (GA 20, 296–98/216–17).

At the same time, however, it is clear that being-in-*the*-world is not entailed by the proposition that *a* world of beings must be encountered prior to any attempt at knowing and explanation. Heidegger's equivocation between being-in-a-world and being-in-the-world, and thus the entire problematic of the worldhood of the world, persists beyond, so to speak, *Sein und Zeit*. *Mit-Sein*, being-with both among human beings and equipment, the specific togetherness of existence, can be universalized only if the inescapable ontico-ontological reciprocity of hermeneutic phenomenology is jettisoned. The ontical priority of the question of Being in *Sein und Zeit* follows upon *both* the thrownness of Dasein, its already being-in-a-world, and its not yet realized possibilities of being, which is to say, on its specific ontico-ontological condition (GA 2, 18; SZ, 13/34).

In *Sein und Zeit*, evidence of reluctance to follow the path of ontico-ontological reciprocity can be seen not only in Heidegger's characterization of work but also in his differentiation of Sorge, concern in the sense of un-easiness, into Besorgen and Fürsorge, respectively characterizing our concern with and for things and persons. This differentiation already indicates Heidegger's acceptance of the classical picture of modern science as a theoretical-mathematical projection on the natural world, a program to which things must submit for their elaboration and interconnection. Heidegger's differentiation proceeds from this picture, followed by his painting of an alternative picture possibility of existence as something more fundamental and from which science and theoretical activity in general are derived. The ontological thrust of the ontical descriptions is thus aimed at grounding what-is (being-in-the-world) rather than setting forth "alternative" possibilities of being-in-a-world. A strong sense of continuity (and community) pervades all of *Sein und Zeit*. Alienation from this ground is how Heidegger understands modern natural science in 1927 and then again in 1938. This tension, if not contradiction, of ontical priority and the pull of grounding will persist in Heidegger.

Related to his reading of Kant, there are, however, indications of a shift in Heidegger's understanding of modern natural science in 1935-1936. The text *Die Frage nach dem Ding* begins with a distinction between scientific and philosophical lectures: The philosophical approach differs from a scientific presentation in that philosophy "executes a continuous shifting of standpoint and level" (GA 41, 1/2). Philosophy examines approaches, actual and possible, to things but it does not explain things. Heidegger's concern is still that of grounding in the sense of limiting the applicability of science to the concerns of "a historical people" (GA 41, 10/10). So far we find nothing unexpected, including the political intimations. Yet, in the course of this lecture text, there is a shift indicating a somewhat different perspective on science. Transcendence or leaping beyond things [*Überspringen*] is possible only with the encountering of human and things, an encountering in which the question of the thing is simultaneously the question of who we are. Only in the encounter of humans and things is there questioning and only with the encountering questioning is there anything at all. The highest principle of all synthetic judgments is thus "the between" [*das Zwischen*], the open between us and thing, the forming region of interplay where and when we and things have definite characteristics (GA 41, 244–46/242-44). In his analysis of Heidegger's text, Gendlin emphasizes that for Heidegger the so-called "pro"-nouns "this" and "that" are really "the most original and earliest mode of saying anything and thereby selecting and determining a thing. Only after our interplay with things do they come to have a resulting nature of their own. The noun becomes possible only on the basis of our pointing. Our demonstrative definitions precede more developed definitions, i.e., 'things' arise only in the context of their relation to us and our pointing them out" (GA 41, 260/258). This formulation accords with Bohr's characterization of quantum experimentation and his claim that independently of measurement observation there is no "deep reality."

In 1935–1936, Heidegger seems to be aware of the radical change in the structure of physical theory brought about by the special theory of relativity. No longer can physics be understood as a science of nature independent of observers. In 1938, however, Heidegger notes that physics can be experimental only because it is essentially mathematical and transforms knowledge of nature into research.[1]

The alienation of natural science thus lies in the very essence of research—a projecting of both a fixed ground plan of natural events and a rigorous manner "in which the knowing procedure must bind itself and adhere to the sphere opened up" (GA 5, 71/118). Logically, then, we arrive at the second essential characteristic of research: the methodology of experimentation. On these points, Heidegger makes no essential distinction between classical Newtonian mechanics and quantum mechanics. He characterizes both as the laying down of a law, with reference to their ground plan of the object sphere, as a basis for controlling the observations of motions and thus for controlling them in advance by calculation (GA 5, 74/QCT, 121). In essence, modern natural science takes over and assumes the role of Being in the determination of beings. Mathematical physics is a metaphysical variant of *Seinsvergessenheit*.

The time of the world as picture begins with the Cartesian space-time of objective representations, of beings represented to us with certainty within a universal space-time grid. Within this projection, the world emerges as the sum total of representations of beings as objects. Our modern world is the world drawn by alienated, human-centered science. The Being of beings is replaced by the picture-view of modern natural science and its mimetic doublings in the realms of the arts and life sciences. Grounded in human subjectivity, modern natural science, driven by will, advances as a domineering project of mastery. Heidegger tells us this project is possible because the doctrine of revealed truths no longer binds us to a hierarchical scheme of beings, to "a specific rank of the order of what has been created—a rank appointed from the beginning—and thus caused, to correspond to the cause of creation (*analogia entis*)" (GA 5, 83/130). It is remarkable that Heidegger does not interpret the *argumentum ex verbo* regulation of medieval knowledge as following upon a projection or a worldview. Why does Heidegger not regard church doctrine as a carefully prepared projection of a region for the appearance of beings? Because, says Heidegger, a Greek, medieval, and Catholic worldview are equally absurd (although Plato did determine the beingness of what-is as eidos and thus laid down the presupposition for the world becoming a picture).

But Heidegger is not very convincing on this point, especially because a stronger case for the linkage of knowledge and technology-as-domination can be made for the medieval church view and its enforcement than for modern natural science after Newton.[2] I say "after Newton" because modern natural science in its classical Cartesian–Newtonian form is largely medieval in its ideological character. Descartes believed that God created the world in accordance with mathematical ideas. The truths of classical physics were "viewed" by Descartes as revealed truths. A one-to-one correspondence between every element of physical theory and physical reality was possible, according to Descartes, only because the mathematical language of the former mirrors the perfect mind of the creator. The price Descartes was willing to pay for this one-to-one correspondence was the ontological separation of the subjective world of lived experience from the world of physical reality. The observed results of scientific experiments could not be trusted because they involve imperfect human perception. The way to the Cartesian–Newtonian paradigm was the hard road of mastering the passions, the body, and nature by the godlike mathematical soul. The failures of Scholastic Physics were thus essentially theological: Analogical reasoning was inferior to mathematical reasoning. In Hermann Weyl's words:

> When the authoritative world-view of the Church came to grief in the Middle Ages and waves of skepticism threatened to wash away everything stable, man's belief in the truth clung to geometry as to a rock. At that time it could be set down as the highest ideal of all science that it can be pursued *more geometrico*.[3]

Worldviews are never neatly separated by clear demarcations, nor does their interweaving exhibit a purely synchronous development. The authoritative worldview of the Church "came to grief" but did not disappear with the rise of mathematical reasoning as the highest ideal of science. The truths uncovered in accordance with mathematical ideas were still held by

Descartes to be "revealed truths." Nadeau and Kafatos call this seventeenth-century metaphysical view the "hidden ontology of classical epistemology."[4] Copernicus, Kepler, Galileo, Newton, and Einstein all held this view. It is a mistake to construe this view as following on a scientific commitment to experimentation because none of the major figures of this complex of overlapping traditions held their beliefs on the basis of controlled observations. For Copernicus, the sun was the center of the universe simply because it was a better symbol of God than the earth, a symbol that God himself would prefer. Kepler's extension of Copernicus—the laws of planetary motion—was experimental only in a derivative sense. The elliptical orbit of planets was simply a consequence of a model of the universe that Kepler regarded as more mathematically harmonious than Copernicus' model. Despite the absence of experimental confirmation, Galileo remained committed to his mathematical law or constant describing the acceleration of bodies in free fall. Like Einstein after him, Galileo held to the article of faith stating that all movement must be subject to the law of number corresponding to the eternal and immutable truths in the mind of God.[5] The law of number and physical reality form a sacred union in classical physics, relegating scientific experimentation to a subservient secular role.

In 1954, Heidegger referred to the recent works of Werner Heisenberg on the problem of causality in quantum theory. According to Heidegger, Heisenberg treated this problem as a "purely mathematical problem of the measurement of time" (VA, 43/161). In its baldness, this is a curious statement against the backdrop of what Niels Bohr and Heisenberg were proposing since the late 1920s as a radically new conception of nature. The Copenhagen interpretation of quantum theory radically departed from classical physics by proposing that the basis of scientific theoretical structure is not microscopic space-time realities but the concrete sense realities of everyday life. It is true that Heisenberg, along with Max Born and Pascual Jordan, developed the theory of matrices in the belief that science can only deal with quantities measurable in experiments. But this new mathematical theory was prompted by the realization that the classical assumption of "real" atoms and molecules in the sense of exactly definable and determinable was untenable given the experimental findings that physical quantities can be known only through acts of observation. Heisenberg's indeterminacy principle and matrices broke with the classical logic premised on Aristotle's law of excluded middle. By the 1930s physicists were divided into the two camps: realism (Planck, Schrödinger, DeBroglie, and Einstein) and the Copenhagen Interpretation to its new logical framework of complementarity (Bohr, Heisenberg, Dirac, Pauli, Jordan, and Born).[6]

Given this complex situation, it is even more curious that Heidegger goes on to state, "since the beginning of the modern period in the seventeenth century, the word 'real' has meant the same thing as 'certain'"

(VA, 43/162). Heidegger seems to understand all of modern science in the mode of classical physics, and he also seems to imply that quantum theory is nothing but a development and extension of classical physics. In his August 1954 lecture, "Science and Reflection," Heidegger contends that if

> objectness [were] to be surrendered, the essence of science would be denied. This is the meaning, for example, of the assertion that modern atomic physics by no means invalidates the classical physics of Galileo and Newton but only narrows its realm of validity. But this narrowing is simultaneously a confirmation of the objectness normative for the theory of nature, in accordance with which nature presents itself for representation as a spatio-temporal coherence of motion calculable in some way or other in advance. (VA, 50/169)

Here Heidegger reads quantum theory along lines laid down by the realist camp. Thus, Max Planck is quoted: "The actual is whatever can be measured" for the purpose of supporting Heidegger's reading of modern science as "the theory of the real." What Heidegger avoids or misses is the crucial point Bohr made in 1929:

> the very recognition of the limited divisibility of physical processes, symbolized by the quantum of action, has justified the old doubt as to the range of our ordinary forms of perception when applied to atomic phenomena. Since, in the observation of these phenomena, we cannot neglect the interaction between the object and the instrument of observation, the question of the possibilities of observation again comes to the foreground. Thus, we meet here, *in a new light*, the problem of the objectivity of phenomena which has always attracted so much attention in philosophical discussion.

Two years earlier, Bohr made another related point, one that completely undermines Heidegger's extension of the classical ontological and epistemological model to atomic physics:

> It must not be forgotten, however, that in the classical theories any succeeding observation permits a prediction of future events with ever-increasing accuracy, because it improves our knowledge of the initial state of the system. According to the quantum theory, just the impossibility of neglecting the interaction with the agency of measurement means that *every observation introduces a new uncontrollable element*. Indeed, it follows from the above considerations that the measurement of the positional co-ordinates of a particle is

accompanied not only by a finite change in the dynamical variables, but also the fixation of its position means *a complete rupture in the causal description of its dynamical behavior*, while the determination of its momentum always implies a gap in the knowledge of its spatial propagation.[7]

Unlike his 1938 lecture, "The Age of the World Picture," "Science and Reflection" distinguishes classical from nuclear physics, noting that the latter can neither be traced back nor reduced to the former. Nevertheless, Heidegger insists that what does not change for the new physics is

> the fact that nature has in advance to set itself in place for the entrapping securing that science, as theory, accomplishes. However, the way in which in the most recent phase of atomic physics even the object vanishes also, and the way in which, above all, the subject-object relation as pure relation thus takes precedence over the object and the subject, to become secured as standing-reserve [*Bestand*], cannot be more precisely discussed in this place. (VA, 53/172–73)

What follows instead of a precise discussion of the way by which the subject–object relation becomes a *pure* relation taking precedence over both subject and object is Heidegger's answer to a question he will formulate shortly before his death in 1976: "Is modern natural science the foundation of modern technology—as is supposed—or is it, for its part, already the basic form of technological thinking, the determining fore-conception and incessant incursion of technological representation into the realized and organized machinations of modern technology?"[8] Heidegger's 1954 answer comes in brackets, perhaps marking its provisional character. It tells us that rather than disappearing, the subject–object relation now attains its most extreme dominance as predetermined from out of *Gestell*, one in which both subject and object "are sucked up" as *Bestand*.

However, the subject-object polarity of classical epistemology and its goal of bringing the object to stand in constancy are rooted in the assumption of a godlike perspective from which physical reality can be known as it is in itself. If, on the other hand, there is always an interaction between beings under observation and measuring instruments, including the human eye, Planck's quantum of action precludes any possibility of a one-to-one correspondence between elements of theory and what is observed. What, then, is the meaning of subject and object sucked up as inventory? Does the predetermination of this instrumentalizing event preclude a radical transformation of the subject–object relation within modern science itself? If so, the essence of technology understood as *Gestell* determines the development of science such that modern natural science can be nothing other than the fundamental structure of technological thinking. But if this is indeed Heidegger's position in 1954–1955, why would he suggest, shortly before his death, that we deliberate on the relation of modern science to modern technology? Perhaps the question has something to do with the equivocation between being-in-a-world and being-in-the-world, an equivocation taking the form of a separation of truth from Dasein's self-understanding in Heidegger's post-1939 thought.[9]

Our conceptual experience of the world is shaped in part by centuries of overlapping and conflicting worldviews. If, however, as proposed by classical physics, the sum of the parts in physical reality compose the whole of what-is, Heidegger's thesis of a preconceptual understanding of Being would amount to nothing more than a thesis of a prescientific understanding of reality that could, in principle, become conceptual with the completion of science as a complete theory of reality. This also would imply that the connection of instrumental referential totalities and derivative theoretical entities in *Sein und Zeit* was itself pre- and proto-technological.[10] Modern science as a complete theory of reality would thus complete the project of hermeneutical or existential phenomenology, making modern science the last phase of active nihilism from a post-Kehre perspective. This would also mean that the western metaphysical roots of fundamental ontology are determinate and fateful in the process of scientific-technological globalization. Another beginning is possible only as a retrieve because there is only one world progressively exhausted by the domination of technological calculation. In this sense, Heidegger's Kehre is necessitated by the very logic underlying the unfolding of fundamental ontology both in-itself and in the Western world after the publication of *Sein und Zeit*. But what Heidegger took as a coalescence of events and *Geschichte* during the turbulent 1930s and 1940s, was actually a trap set by what was already prefigured in Heidegger's thought as the essence of science-technology. Regarding the specific conflation of this essence with the evil forces of America and the Soviet Union, we need not discuss here.[11]

However, there is a subsequent shift with the later *Gelassenheit* marker, a shift that departs from the worldhood of the world to the possibility of *a* new rootedness. It is in the context of this shift and the two 1944 markers—*Gegnet* and *Gelassenheit*—that I discuss certain crossings of Heidegger's thought, without dwelling on the political ellipses and their philosophical ramifications, and Bohr's reflections on "the art of experimentation."

In 1955, however, Heidegger is speaking of a "completely new relation of man to the world and his place in it" (G, 17/50). This new relation emerges, on the one hand, as the forces of technology create conditions outstripping our capacity for decision and, on the other, as probable global destruction does not happen. What is uncanny pertains to what is hidden in the improbability of the nonoccurrence of the probable beyond our calculative abilities. Furthermore, what Heidegger sets forth in this more or less popular address, was already prepared in 1944–1945 with the shift away from the problematics of *the* world, representation, horizon, and willing to the notion of the *Gegnet*. Things appearing in the *Gegnet* are no longer objects in relation to egos. The improbability of the probable not happening thus moves us from the transcendental horizon of the world and the referential totality of instruments to another side of the *Gegnet*, to the side of "that-which-regions" (G, 48–50/72–73). Following the lead of the English translation and its absolute use of the adjective for the purposes of avoiding a noun or reification of the regioning, the side of the *Gegnet* turned toward our re-presenting is not exclusive of the side other than that of willing and representing. A world is thus a totality only within the horizon of *that* world, and then only because *that* world is a world of representation. In other words, the shift in Heidegger's thinking from the transcendental horizon of the world to the *Gegnet* implies, whether explicitly stated or not, the recognition of a plurality of worlds, one of which has certainly conquered, dominated, and most often destroyed the worlds of others. What is improbable, therefore, concerns this European world and the fact that it has neither yet destroyed all the others . . . nor itself.

Decisively, so to speak, Heidegger also makes clear that *Gelassenheit* does not necessarily require as the historical or logical prerequisite of first being bound by the European transcendental horizon of *the* world (G, 50/73). Nevertheless, he still maintains, "The program of mathematics and the experiment are grounded in the relation of man as ego to the thing as object" (G, 56/79). Yet, although classical physical science is in word and deed part and parcel of the European transcendental horizon of *the* world, the experimental findings of quantum mechanics indicate a break away from the unfolding of the historiographical character of the subject–object relation. Indeed, Heidegger tells us: "The historical rests in that-which-regions, and in what occurs as that-which-regions" (G, 57/79). It is possible, then, to dwell in more than one world as the occurrence of that-which-regions calls from one world to another. From the standpoint of the classical Cartesian-Newtonian world, the experimental results of quantum mechanics evince a puzzling world of wave–particle duality, null measurements, particle spin, and nonlocal effects. This world can neither be shared nor understood by inhabitants of the Cartesian–Newtonian world as long as they remain bound to their "revealed truths" ideology. Heidegger seems to understand this when he notes in "Science and Reflection" that nuclear physics can neither be traced back not reduced to classical physics. On the other hand, his claim concerning the exacerbation of the subject–object relation to that of *Bestand* in the domain of the new physics evinces a refusal to come to terms with the implications of quantum mechanics.

The theory of the quantum of action, quantum mechanics, began with Max Planck's study of the statistical aspects of electromagnetic radiation.[12] This story is well known. However, what Planck "discovered" and how he did it are instructive regarding Heidegger's distinction between challenging things to emerge and allowing them to emerge. Planck began with the assumption that particles cannot vibrate any way they please and then mathematically restricted their energies to certain multiples of their vibration frequencies or colors: $E=nhf$ where E is the particle's energy, n is any integer, f is the frequency of the particle's vibration, and h is a constant to be chosen.

In accord with Heidegger's idea of *herausfordern*, we can use the example of challenging the earth in order to extract iron ore. Intensifying our challenge by employing the tried and true torture of heating the ore for the purpose of casting it, we get it to glow red. Not content with such limited objectives, however, we decide to challenge it to the utmost for the purpose of making it into a pure standing reserve (*Bestand*) and as such no longer capable of appearing as it is in itself. Classical physics tells us we can do this by submitting the iron ore to further torture heating. In accordance with universal (Newtonian) laws, we should get the continuous blue glow signifying ultimate mastery. But this is not what happens. Planck's quantum of action follows nature's insistence that no energy transitions will occur except as specific whole chucks of energy. As something more than zero, Planck's constant spelled the end for the ideological dream of mastery and Heidegger's contention that objects are in fact reduced to the controlling matrix of standing reserve.

Not only does nature insist on discreteness, it also dictates indeterminacy when we simultaneously attempt to measure its different yet combined voices. When and where the differences are great enough, measurement does not match the unity of nature's conjugations. What is remarkable about Heisenberg's Indeterminacy Principle is that it applies only to the measurement of conjugate attributes (i.e., to attributes that are unlike one another). Thus the uncertainty relations (Indeterminacy Principle) preserve what for classical theory and logic can only be the contradictory coexistence of unmeasured wave and measured particle duality. The limits thus placed on *herausfordern* by the stability (individuality) of particles are also reflected in the fact that contrary to the positivist scheme, our descriptions of nature must employ polyvalent words and concepts. The duality of wave and particle is only a riddle from the Cartesian perspective. Within the mutual implication of subject and object, on the other hand, the interplay between nature and ourselves cannot be completely compensated for. In Heisenberg's words:

> Natural science does not simply describe and explain nature; it is part of the interplay between nature and ourselves; it describes nature as exposed to our method of questioning. This was a possibility of which Descartes could not have thought, but it makes the sharp separation between the world and the I impossible.[13]

It is precisely the necessary ambiguity of our contextually bound descriptions of nature in the mutual but finite plasticity of nature and ourselves that provides for what Patrick Heelan sees as the possibility for scientific entities to "come to share the reality of a World-for-us, and that they do so to the extent that the subject succeeds in embodying itself in appropriate readable technology."[14]

The formalism of quantum mechanics is very exact *within the limits* dictated by the Uncertainty/Indeterminacy Principle and the quantum of action. The connection of mathematics and physical reality in quantum mechanics does not provide an explanatory framework for understanding the cosmos; rather this connection derives from the experimental evidence of the stability or individuality of atomic fields and serves only for predictions concerning the outcomes of further experiments. The "discovery" of the quantum of action does not lend support to Heidegger's (1936–1938) sixth proposition about science:

> 6. Every science, even the so-called "descriptive" ones, *explains*: What is unknown in the region is led back, in various ways and ranges, to something known and understandable. Research provides the conditions for explanation. (GA 65, 146/101)

However, what was "known" in 1900 about black-body radiation contradicted what was "known" about ordinary objects in everyday experience. After Maxwell's "discovery" of the light–matter connection, the Newtonian picture of reality told us that black bodies should glow bright blue at all temperatures. The experienced (seen-known?) red glow of iron, for example, seemed to be a perceptual limitation/imperfection. Energy, according to classical physics, has no limitations. When Planck restricted energy to finite packets and came up with calculations matching experimental findings, his restriction met with suspicion (or rejection) since it contradicted established physical theory. The "idea" of packets of energy was something unknown and alien to the contextually situated understanding of nature of the time. What was "discovered"? Heidegger informs us: "Beings as a region lie in advance for science, they constitute a positum, and every science is in itself a 'positive science (including mathematics)" (GA 65, 145/101). Does the concept "being" comprehend both the "object" of classical physics and the quantized "packet of energy" along with the "wave–particle duality" of quantum mechanics? Is there any phenomenological evidence that the latter were laying in advance of their "discovery"? Is it actually the case that a quantitative projecting open of nature preceded the experiments of the last century and thus constituted them as "against all mere experiri" (GA 65, 164/113)? Consider the following apposite examples.

In 1964, John Bell of the Centre for European Nuclear Research decided to tackle the argument between Einstein and Bohr by *first* devising the mathematical relationships between two particles like those in the Einstein–Podolsky–Rosen (EPR) thought experiment,[15] and then by explaining how certain kinds of measurement could settle the issue of whether quantum theory was complete (Bohr) or incomplete (Einstein). Einstein, after finally accepting the uncertainty relations as a fact of nature, remained convinced that every element of physical theory must have a counterpart in physical reality. To establish this, the EPR thought experiment suggests it is possible to circumvent the quantum measurement problem by using experimental information about one particle to deduce complementary properties such as position and momentum of another particle that originated with the first from the same quantum state. Because the two paired photons move apart from one another at the speed of light, the measurements made on one particle after a sufficient time has passed cannot effect the other particle because no signal can travel faster than the speed of light. Despite the limitation of the quantum measurement problem, EPR argues that if we measure the position of one of these particles, such a measurement will not affect the other because they are space separated. This accords with the principle of local causes: energy transfers between space-like separated regions cannot occur at superluminal speeds. EPR then argues that because we can calculate the momentum of the particle that was not measured and because we already know the position of the particle that was measured, we can deduce both the momentum and position of the particle that was not measured. Thus we can circumvent the rules of measurement observation in quantum physics. In other words, Bohr's thesis that particles have no dynamic attributes before they are measured must be wrong. Quantum theory is thus an incomplete theory. The EPR thought experiment was designed with the assumption of the principle of local causes for the purpose of confirming the realist assumption of a physical reality independent of the observer and acts of measurement.

What came to be called Bell's Interconnectedness Theorem was based on his discovery of a loophole in von Neumann's proof that the world cannot be made of ordinary objects possessing dynamic attributes of their own. The problem was that David Bohm and Louis de Broglie and other neo-realists had built ordinary object models of quantum reality in apparent violation of von Neumann's proof. But Bell discovered that these realist models all contained objects whose attributes are context sensitive, changing their attributes in response to their environment. Subsequently, Bell devised a proof for the invalidity of all models of reality having the property of locality. This proof unravels the Newtonian scientific-ideological worldview. Newton had a premonition of this:

> That one body may act upon another at a distance through a vacuum without the mediation of anything else . . . is to me so great an absurdity, that I believe no man, who has in philosophical matters a competent faculty for thinking, can ever fall into.[16]

Indeed! *Unmediated* action-at-a-distance is a scandal for the ideological project of mastery[17]—what Heidegger describes as the process of challenging (*herausfordern*), ordering (*Bestellen*), and ultimate reduction to inventory (*Bestand*).

Instantaneous connections across infinite distances between all particles that have ever been in contact with one another—nonlocal effects—do not provide for the possibility of the ultimate perfection of classical epistemology and its correlative principle of certainty. Quite the contrary. We cannot command the nonlocal nature of reality to appear or submit to some kind of exquisite examination ordeal. We do not "disturb" or "create" phenomena by acts of measurement or observation. Such metaphors derive from the realist assumption that the world is describable *independent* of observation and measurement.[18] If, on the other hand, experiments indicate a nonlocal universe, then any description of physical reality must accord with these findings if they are to avoid the pitfalls and dangers of ideology (world as picture or *the* world).

In 1972, John Clauser and Stuart Freedman performed the experiment that verified the statistical predictions on which Bell based his theorem.[19] However, the Clauser–Freedman experiment did not establish that the statistical predictions of quantum mechanics resulted from superluminal communications. In 1975, Jack Sarfatti proposed a theory, based on Planck's quantum of action, of superluminal transfer of negentropy (order) in which there is no transport of energy.[20] Then, in 1982, Alain Aspect led a team that performed the crucial experiment confirming that superluminal communication was a fact of nature.[21]

Despite the persistence of Heidegger's realist and classical perspective on modern natural science, I suggest that his *Gelassenheit* and *Gegnet* markers implicitly invoke the influence of the Copenhagen interpretation of the quantum formalism.[22] This influence is explicit in the case of Heisenberg and less so in the case of Bohr. The key text in this regard is the 1944-1945 record of a conversation between a scientist, a scholar, and a teacher, and only published in 1959 as one of the two essays in *Gelassenheit*. The American-English translation appears in 1966 as *Discourse On Thinking*. At the point in the discussion where the teacher, scholar, and scientist turn to the nature of the relation between *Gegnet* and *Gelassenheit*, the point is made that this non-ontic and non-ontological relation cannot be understood as causal, transcendental-horizonal, or humanistic. Instead this relation should be understood as one of determining the thing as thing and human nature itself. This is why we must wait in a releasement to that-which-regions for things and our nature (G, 54-59/76-81).

For classical modern physics, the interaction between objects and the experimental apparatus arrangement can be ignored or compensated for because the object is not effected by this interaction. In quantum physics, however, the interaction "forms an inseparable part of the phenomenon." Moreover, repetitions of the "same" experiment will necessarily yield different recordings of the "object" being observed. Comprehensive accounts of experience thus require statistical rather than deterministic laws because a deterministic account of quantum phenomena is precluded by their limited divisibility. Any recording or measurement of a quantum phenomenon involves an exchange or interaction that is uncontrollable in principle.[23] Although Bohr does not mention this, it is clear that the interaction of object and experimental apparatus involves a mutual exchange—what is given (recorded) also gives (recorder)—a mutual determination.

The ambiguity in Heidegger's account of science may have something to do with what Vilém Flusser called the automation of science. Referring to Flusser's 1964 essay, "The Crisis of Science," Martin Pawley points out that, for Flusser, "the uncertainty of pure science had been masked throughout the modern period by a Romantic vulgarization of the enormous pragmatic success of applied science."[24] It is certainly clear from Heidegger's side what this enormous pragmatic success meant during the 1933-1939 period of preparation and mobilization. For applied science, especially the coordination of technology and science affected by capitalist and socialist economic rationalization, the age of the world picture is still very much with us. But this program is not Bohr's program of experimental science. The quantum formalization does indeed defy (traditional) pictorial representation even though "the very word 'experiment' refers to situation where we tell others what we have done and what we have learned."[25]

According to Bohr, who surely will not have the last word on this issue, ambiguous communication requires, at least for the present, retaining and using the subject–object partition. However, we cannot draw a rigid line of partition. Under the global powers that be, there can be little doubt as to Heidegger's contention that this partition has been rendered as a pure relation having precedence over both subject and object. And there can be little doubt that all of us, including scientists, are affected by this context of power. The situation is perilous. However, what Heidegger cautions as the necessity of waiting in a releasement (from willing) and thus within that-which-regions is, I argue, very close to Bohr's 1929 insistence although we cannot visualize the quantum formalism, we must in some sense formulate concepts with some intuitive content. This problem urgency means that we can no longer make a sharp separation between object and subject:

> From these circumstances follows not only the relative meaning of every concept, or rather of every word, the meaning depending upon our arbitrary choice of view point, but also that we must, in general, be prepared to accept the fact that a complete elucidation of one and the same object may require diverse points of view which defy a unique description. Indeed, strictly speaking, the conscious analysis of any concept stands in a relation of exclusion to its immediate application.[26]

This relation of exclusion is precisely the one economic rationalization attempts to occlude.

Total ideology, the final configuration of metaphysics (GA 6.2, 179/149), and science-reflection within the event-horizon of knowledge are completely incompatible. Heidegger's understanding of modern natural science remained within the perspective of realism and positivism, according to which, physical reality is fully disclosed in mathematical descriptions. Positivism itself is premised on classical epistemology: "there must be a one-to-one correspondence between every element in a physical theory and every aspect of the physical reality described by that theory."[27] Contextually, classical natural science developed within the framework of European asceticism and its fanatical attempt to overcome the pleasures of the earthly body. The globalization of this sadomasochistic obsession with the onward marching Christian soldiers of the Crusades, the slave traders of the ancient and medieval worlds, and the ordained slaughters of the native populations of the New World is now so well established that even the loyal, albeit transplanted, subjects of the cleansed New World jogging under the consumer imprint can only enjoy the booty of centuries of murder, rape, torture, and theft through the masochistic techniques of the old asceticism.[28] Consuming others and their achievements requires self-consumption. This is the subject–object relation as pure relation and the ultimate reduction of subject and object to inventory. With the near total sacrifice of the "wild man" to civilization, the vale of tears of the *contemptus mundi* finds its ultimate expression in the pithy but exculpatory "shit happens." The forgetting of Being is also the selective forgetting of specific beings.

The "hidden ontology of classical epistemology," namely that "the truths of classical physics as Descartes viewed them were quite literally 'revealed' truths,"[29] is clearly a fundamental structure of what Heidegger understands as technological thinking. What I have attempted to show, however, is that the new physics, which developed out of the old by breaking with it on the basis of experimental findings, can no longer picture or represent *without loss*[30] what exists between measurements and observations. The only (scientifically) determinable relation between events is a matter of probability. There can be no certainty of representation in the domain of the new physics. The experimental findings of the new physics indicate that "our" classical world of "subjects" and "objects" is a highly coded ideological construction, cultivated over centuries, and highly resistant to the event-horizon of scientific-experimental knowledge. And it is precisely in this tension between ideological totalization and the limits established by the new science that Heidegger's cautions about any reification of *Sein-Seyn-Ereignis-Gelassenheit-Gegnet-* . . . must be "observed."

> The thing is not "in" nearness, "in" proximity, as if nearness were a container. Nearness is at work in bringing near, as the thinging of the thing. (VA, 170/178)

NOTES

1. Heidegger's 1938 characterization of the mathematical retreats from his Kant reading in 1935–1936. Unlike the Platonic judgment that geometry is empirically valid only with *some* certainty, the Kantian doctrine holds the axioms of geometry as synthetic *a priori* judgments valid apodictally. For an excellent discussion of this difference, see Peter Mittelstaedt, *Philosophical Problems of Modern Physics*, trans. W. Riemer (Dordrecht: Reidel, 1976), 41–50. If, on the other hand, the highest principle of *all* synthetic *a priori* judgments is, according to Heidegger's 1935–1936 reading, the "open between us and thing," the mathematical of the new physics cannot stipulate *in advance* what is *already known*. This point is intimately related to what Bohr calls "the wonderful development of the art of experimentation" (Niels Bohr, *Atomic Theory and the Description of Nature*. London: Cambridge University Press, 1961: 93).

2. Nietzsche made a case for this in his *Genealogie der Moral*, especially in the third essay, sections 19–22.

3. Hermann Weyl, "On Time, Space, and Matter," *Phenomenology and the Natural Sciences*, eds. Joseph J. Kockelmans and Theodore J. Kisiel (Evanston: Northwestern University Press, 1970), 93–94.

4. Robert Nadeau and Menas Kafatos, *The Non-Local Universe: The New Physics and Matters of the Mind* (Oxford and New York: Oxford University Press, 1999), 8. This text is a reworking of their 1990 collaboration *The Conscious Universe: Parts and Wholes in Physical Reality*, which has been reissued in a revised second edition (New York: Springer-Verlag, 2000). I refer to these texts as NLU and CU.

5. NLU, 152–55.

6. [In this debate, the realists hold that the properties of quantum particles exist prior to measurement, whereas advocates of the Copenhagen interpretation hold that quantum properties depend for their existence on measurement. In the Copenhagen interpretation, complementarity is the general principle that certain values cannot be simultaneously known. A garden variety example is that you can't have your cake and eat it too; a physicist once gave me jokingly an everyday example: "when I have the time, I don't have the energy; when I have the energy, I don't have the time." Heisenberg's uncertainty principle is a specific formulation of complementarity stating that a quantum particle's position and momentum (i.e., where it is and how fast it is moving) cannot be simultaneously known. Originally, the problem was taken to be one of how to arrange the measuring equipment. A quantum particle is measured by smashing another particle into it, which will change both where it is and how fast it is moving. So no measurement can be made without affecting the properties of the particle being measured, and experimenters must decide which property to measure and arrange their equipment accordingly. Any other property's value is lost during measurement and no subsequent experimental set-up can retrieve it. Einstein raised an objection to the Copenhagen interpretation published as Albert Einstein, Boris Podolsky, and Nathan Rosen, "Can Quantum-Mechanical Description of Physical Reality Be Considered Complete?" *Physical Review* 47 (1935): 777–80, known as EPR. EPR is a thought experiment demonstrating that the Copenhagen interpretation of quantum mechanics can never generate a complete description of physical reality since some data must always be irretrievably lost according to the uncertainty principle. Bohr's response was published in the next volume of *Physical Review* (48 [1935], 696–702). He argues that "the dependence on the reference system . . . means a radical revision of our attitude as regards physical reality" (Bohr 1935, 702). Thus, Bohr indicated that the implications of the Copenhagen interpretation are not just *epistemological* (i.e. about the limitations imposed by having to make choices between different experimental arrangements), but rather *ontological*. Most controversially, this is taken to mean that quantum properties have no value until measured. Hence, the debate at issue here is between realists, who hold that the properties of quantum particles exist prior to measurement, and advocates of the Copenhagen interpretation who maintain that quantum properties depend for their existence on measurement. The implications of this are discussed below; cf. note 15 and the surrounding text. Both EPR and Bohr's response are reprinted in *Quantum Theory and Measurement*, eds. John Archibald Wheeler and Wojciech Hubert Zurek (Princeton, NJ: Princeton University Press, 1983), with helpful commentary. A clear although still technical analysis can be found in R. I. G. Hughes, *The Structure and Interpretation of Quantum Mechanics* (Cambridge, MA: Harvard University Press, 1989), whereas an excellent and extremely accessible account (rather than the reductive gloss provided in this

note) can be found in Shimon Malin, *Nature Loves to Hide: Quantum Physics and Reality, A Western Perspective* (New York: Oxford University Press, 2001). (*Ed.*)]

7. Bohr (1961), 93 and 68. The emphases in both quotations are mine.

8. This question arrived in the form of a letter addressed to the participants in the tenth annual meeting of the Heidegger Conference at Depaul University in Chicago that is printed in *Research in Phenomenology* 7, tr. John Sallis (1977), 1–4, and reprinted in *Radical Phenomenology: Essays in Honor of Martin Heidegger*, ed. John Sallis (Atlantic Highlands, NJ: Humanities Press, 1978), 1–2. For a sympathetic discussion of Heidegger's position in this regard, see William J. Richardson, "Heidegger's Critique of Science," *The New Scholasticism*, Vol. XLII, No. 4 (Autumn, 1968), 519–20, reprinted above.

9. See John Sallis, "The Origins of Heidegger's Thought," *Research in Phenomenology* 7 (1977), 53–55.

10. This point has been made in a slightly different manner by Karsten Harries, "Fundamental Ontology and the Search For Man's Place," in *Heidegger and Modern Philosophy*, ed. Michael Murray (New Haven: Yale University Press, 1978). This is also the reason why Joseph J. Kockelmans can claim the many worlds projected by the various sciences all derive "from the world immediately lived by the community of man" (Joseph Kockelmans, "Heidegger on the Essential Difference and Necessary Relationship Between Philosophy and Science," *Phenomenology and the Natural Sciences*, ed. Joseph J. Kockelmans and Theodore J. Kisiel (Evanston: Northwestern University Press, 1970), 165).

11. See my "Heidegger's Essentials: Appropriations and Expropriations," *Martin Heidegger and the Holocaust*, eds. Alan Rosenberg and Alan Milchman (Atlantic Highlands, NJ: Humanities Press, 1996), and Reiner Schürmann's brief but concise essay "Riveted to a Monstrous Site," *The Heidegger Case*, eds. Tom Rockmore and Joseph Margolis (Philadelphia: Temple University Press, 1992).

12. See Max Planck, "Zur Geschichte der Auffindung des physikalischen Wirkungsquantums," *Die Naturwissenschaften* 31 (1943).

13. Werner Heisenberg, *Physics and Philosophy* (New York: Harper Torchbooks, 1962), 81.

14. Patrick A. Heelan, *Space-Perception and the Philosophy of Science* (Berkeley: University of California Press, 1988), 211.

15. Einstein, Podolsky, and Rosen (1977). John S. Bell's formulation of his theorem appeared as "On the Einstein-Podolsky-Rosen Paradox," *Physics* 1 (1964), 195ff. [Cf. Bernard d' Espagnat, "The Quantum Theory and Reality" *Scientific American* 241, no. 5 (1979), 158–80 for an accessible account of the implications of EPR and Bell's inequalities. He shows clearly how local causality (i.e., the thesis that information cannot be propagated faster than light; rather there must an identifiable chain of events that link a cause with its effects) is breached by experimental results testing Bell's inequalities. Realists generally hold to what are known as "hidden variable theories": there is such a chain of events (the hidden variable) that has not yet been identified. (*Ed.*)]

16. As quoted by Nick Herbert, *Quantum Reality: Beyond the New Physics* (New York: Anchor Books, 1985), 213.

17. [Referred to in the debate as "spooky action at a distance." (*Ed.*)]

18. NLU, 92–93.

19. Stuart Freedman and John Clauser, "Experimental Test of Local Hidden Variable Theories," *Physical Review Letters*, 28 (1972), 938ff.

20. For a discussion of Sarfatti's theory, see Gary Zukav, *The Dancing Wu Li Masters: An Overview of the New Physics* (New York: William Morrow, 1979), 310–13.

21. Alain Aspect, Jean Dalibard, Gerard Roger, "Experimental Test of Bell's Inequalities Using Time-varying Analyzers," *Physical Review Letters*, 49 (1982), 1881ff.

22. This tension ambiguity in Heidegger's account of modern natural science is discussed in Trish Glazebrook, *Heidegger's Philosophy of Science* (New York: Fordham University Press, 2000), 247–53. On my account, however, the meaning of technological science shifts in Bohr's formulation away from Heidegger's understanding of Ge-stell. Thus the correlation of quantum mechanics and technology is different from what Heidegger took to be the identity of science and Ge-stell.

23. Niels Bohr, "Quantum Physics and Biology," *The Philosophical Writings of Niels Bohr*, Vol. IV, eds. Han Faye and Henry J. Folse (Woodbridge, CT: Ox Bow Press, 1998), 182.

24. Vilém Flusser, *The Shape of Things: A Philosophy of Design*, trans. Anthony Mathews (London: Reaktion Books, 1999), 12.

25. "Philosophical Science and Man's Position," Faye and Folse (1998), 173.

26. Bohr (1961), 96.

27. NLU, 10–11.

28. See David E. Stannard, *American Holocaust: The Conquest of the New World* (New York and Oxford: Oxford University Press, 1992), especially Chapter III: Sex, Race, and Holy War, and Ward Churchill, *A Little Matter of Genocide: Holocaust and Denial in the Americas 1492 to the Present* (San Francisco: City Light Books, 1997). For a refutation of these "distortions of history," see any of the edifying "metaphysical" works of William Bennett.

29. NLU, 8.

30. Concerning the theoretical ramifications of the anti-epistemological implications of quantum theory and the radical, irreducible loss in representation affecting all quantum systems, see Arkady Plotnitsky, *Complementarity* (Durham, NC and London: Duke University Press, 1994).

HEIDEGGER'S THESES CONCERNING THE QUESTION OF THE FOUNDATIONS OF THE SCIENCES

Ewald Richter
Translated by Trish Glazebrook and Christina Behme

HEIDEGGER'S RELATIONSHIP TO THE SCIENCES

In his 1966 essay, "The End of Philosophy and the Task of Thinking," first published in French, Heidegger asks the question what "task" is reserved for thinking in our time. He characterizes the present with the critical remark that our time is decisively formed by a "technological scientific rationalization" (SD 79/72). He does not misjudge the tremendous success, which for him is unquestionable, of modern science, especially contemporary natural science. He in fact showed a special interest in dialogue with natural scientists well into his last years.[1] His central theme with reference to the assumptions of modern sciences should therefore receive increased attention.[2] The most significant thing for evaluating the character of modern science is for him accounting for what legitimate (although "elevated" [*abgehoben*]) realm of thinking the modern sciences occupy, and how this realm stands in relation to the above remark concerning a thinking that is for the time being "reserved," and hence preparatory.

Under no circumstances should Heidegger be grouped with those who claim to have "higher inspirations" at their disposal, to which "proclamations" they are called. Quite to the contrary, he moves toward a thinking that is thoroughly "sober" [*nüchtern*] (SD, 79/72), the comprehension of which presupposes an intensive practice. His thinking is a meticulous "exhibiting" and "uncovering."

What the sciences are for him is given concise formulation in his lecture, "The Origin of the Work of Art": Science is "not an original happening of truth, but always the cultivation of a domain of truth already opened" (GA 5, 50/62). Without a detailed treatment of his work, it is impossible to evaluate to what extent this characterization as "derivative" is not a devaluation of science. Once sufficient attention has been paid, however, to how he definitively understands the sciences, especially the currently dominant natural sciences, his evaluation of the sciences will be evident.

THE PRIMORDIAL OPENNESS OF BEINGS

An important moment in Heidegger's thinking, aptly significant for contemporary scientists, lies in the question of how we gain access in the first place to the entities with which we engage, and that are the subject of theoretical statements. The assertions of scientific theory are contingent upon clarification of the problem of whether and how what is discussed in any such an assertion can be related to legitimately accessible entities. When a scientist proposes a mathematically formulated theory, it can be asked whether this theory rightfully claims to relate to the "things themselves." On what is our confidence based that the things remain unaltered by knowledge, so that the latter gives reliable information about the "things themselves"?

To answer these questions for Heidegger, we must be clear about what kind of being reveals itself in a primordial way of being that is actually directed toward "the thing itself." A subsequent question concerns whether it suffices to say that our knowledge builds primarily on "sense perception," which in this context works as a kind of verification. One would hardly deny that sense perception plays a substantial role in empirical knowledge. But the question remains whether this statement is synonymous with the claim that sensory perception is the primordial foundation of knowledge. The possibility of knowledge is often considered only in terms of subjective conditions. Could it not be the case that sense perception is dependent on conditions, but that these conditions are "already always" foundational in a particular way for sense perception, such that they are difficult to keep in sight, or even bring into view? Through consideration of such a foundation, could not questions concerning the extent to which we can experience an entity "as itself," as well as how we are able to recognize this entity as "the thing itself," be newly and promisingly reformulated?

These questions are raised for the moment in an anticipatory way. In order to emphasize that they treat an important point for Heidegger, I recall his phrase concerning the "passing over" of the "Being of what is proximally ready-to-hand" (GA 2, 266; SZ 201/245). This phrase that stands in direct connection with his critique of the misunderstanding of the fundamental condition of Dasein, "being-in-the-world" (GA 2, 268;

SZ 202–3/247). A philosophical narrowing of view to a "one-sided orientation" (GA 2, 266; SZ 201/245), as he saw it, regarding the role of "sense perception" is rarely properly appreciated. It is thus necessary to stress that for him sense experience is already "included" in primordial understanding. It is "*embedded*" in an "understanding of the significance of environmental things" [*Bedeutsamkeitsverstehen*]. For an explicit (although at the same time "going-along-with" [*mitgehende*] and, as such, a-theoretical) philosophical-hermeneutic interpretation shows that "the primordial and leading" factor in experience of the surrounding world is an "understanding of the significance of significant environmental things," through which the "relevant how of perception" (i.e., sense perception) is stipulated.[3]

Pre-theoretical, "factical" life "is and lives experiencing." A "ground of experience that continually contributes to factical life is there" [*ist da*]. Factical life "does not first constitute Dasein, but on the contrary is itself and lives experiencing in a world" (GA 58, 66). It is necessary to do justice to this primordial life and its understanding of significance. The disclosure, in which an entity with which we have to do can primordially "encounter" us, is indebted to such understanding and signifying. The "as-structure" of understanding is already decided in this "having to do with." In the Marburg lecture of 1925/1926, Heidegger remarks informatively on "primordial understanding" and its as-structure(GA 21, §12 [a]). Such an as-structure is already contained in "simple" seeing and taking ("having to do with") insofar as this is a "taking-as." The discussion of the primordial should not be taken wrongly to mean that sensations stand in the primordial position. When I "see" a thing, this does not mean that I feel sensations. Rather, I simply see the thing. If we think about the knowledge of things, then we have to say that in primordial experience and understanding, a "world" has arisen *before* any theoretical grasp of objects.

In §25 of the Freiburg lecture of 1919/1920, Heidegger asks about the "formation of determinate fundamental experience out of factical lived experience" (GA 58, 110). He refers to a *not* yet "theoretical-scientific objectifying." Especially clear in these passages is the subtlety of each statement concerning the question of the unthematized "as-structure." The second part of §24 is entitled "The significance of the reality-character of factical life" (GA 58, 104). Here, Heidegger explains that it is essential to eliminate all theorizing, and rather to see the sense "in which factical experience renews its experience and always has it in the character of significance." If, for example, while drinking tea, I "pick up my cup" and talk about "my cup standing before me," then I am completely in the factical, and "factical experience in the factical life-context" stands in "the unthematized character of significance." If we subsequently ask about a "formation of determinate fundamental experience out of factical lived experience" that "expresses itself in the style of the factical experience," then according to Heidegger we confront the "basic phenomenon of cognition" in its "different forms of expression, forms of daily, personal and public life" (GA 58, 112). It is a cognizant describing or remembering, and this taking cognizance [*Kenntnisnahme*] "goes along with life." But even here, as he further explains, there is a modification in sense of a thematization. There is no "as" of conceptual characterization at play, but there is still an "as of significance." Seen in this light, taking cognizance is a "strange phenomenon" that stands "virtually at the border" between factical experience and theorizing. Taking cognizance is an "articulation" without being one in so far as "it remains in the basic mode" (GA 58, 114) of cognition.

In what follows, Heidegger gives a more precise account, but the brevity of my analysis demands explication oriented to his conclusions (which cannot substitute for reading his texts). We are comparing the events as experienced on one hand and as narrated on the other. What entitles us to call the narration a modifying articulation? Heidegger talks here about a thematization of the experience "as a whole." This whole establishes a context for what is factically experienced (GA 58, 118). A context of expectation is "stabilized" and taken as "explicit tendency." Heidegger indicates the danger of such modifying articulation, insofar as it can lead to a rupture of context that gives rise to the temptation to use the "fragments" as building blocks in a completely new, and now conceptual, construction of theoretical knowledge.

In light of these remarks, we must proceed cautiously and strictly if factical life and an understanding of significance are to be explicated philosophically and hermeneutically (i.e., a-theoretically). Hermeneutic phenomenology, with its methodological maxim ("to go back to the things themselves") has, in contrast to objectification and reflection, "to understand 'the pure motives of the sense of pure experience' emerging from full and lively lived experience."[4]

It would thus be thoughtless to assume that seeing and talking in dealings with "environmental things" can best be analyzed by a systematic reconstruction that is guided by theoretical science, in order to achieve an "optimal" attitude. Without question, attempt at a "planned, methodical reconstruction" *ab ovo* can be made from a scientific-theoretical perspective (clearly indicated as a "subsequent" viewpoint). But here we deal with something entirely different; that is, an undertaking that has a particular (on closer reflection, special) goal. In such a theoretical undertaking, no claim to begin with novel reconstruction of an unequivocal, presuppositionless language can be made. The attempt at "methodical reconstruction," as undertaken for example by Paul Lorenzen, does not claim to "go back behind life." Lorenzen knows he has to start his attempt as though "out at sea," and search there for floating driftwood, with the help of which a "comfortable ship" (unequivocal language) can be systematically built one plank at a time.[5] In our context, the possibility of such reconstruction is not denied. It is important to stress, however, that an initially vague albeit appropriate understanding is not thereby perfected—even after the fact; rather, after the "methodical reconstruction," a change in linguistic understanding has occurred that establishes the entire reconstruction according to a special authority.

In contrast, when Heidegger refers to primordial lived experience that takes place in manifold connections of significance [*Bedeutsamkeit*], the crucial point is that the "openness" of "beings themselves," that is, of "meaning" [*Bedeutsamen*] in relations of significance, is what is primordial. In primordial "meaning" [*Bedeuten*], Dasein has information about its world; the information itself is the disclosure of the prevailing facts of the matter" (GA 21, 150). The "state of being" of beings with which we are primordially familiar is given a definite character as a way of being (the readiness-to-hand of the ready-to-hand, which he also calls "equipment"), and as the what-being of the ready-to-hand being.

It would be a huge misunderstanding to interpret the latter account as meaning that all knowledge begins with an imprecise understanding that lays the ground for dealings that are as of yet barely informed. Such a view uncovers trivial differences, while Heidegger is discussing different kinds of human behavior toward beings, and different understanding of ways of being. If things that are already evident are represented in a novel stance toward beings that attributes or denies them definite qualities, then such things have been removed from their primordial relations of significance in order to say something about them *as such*. This transition is discussed in *Being and Time*: "Something *ready-to-hand with which* we have to do or perform something, turns into something '*about which*' the assertion that points it out is made. . . . Both *by* and *for* this way of looking at it, the ready-to-hand becomes veiled as ready-to-hand" (GA 2, 209; SZ, 158/200). The being is no longer evident as ready-to-hand but turns into something or other present-at-hand. A mistake concerning the foundational relation here alluded to (presence-at-hand is a way of being founded in readiness-to-hand) will have negative consequences for the foundation of science. For an inversion of the relation overlooks the fact that the question concerning the relationship to "things themselves" (as the "present-at-hand" of a science) refers back to another question: in what is the original openness of beings grounded?

An important result has thus been attained: A being that is evident from itself openly allows that new and different determinations of it are "brought forth" and given to it (i.e., it gives them to itself) through a changed attitude. This result must be more carefully considered for the sciences, especially the natural sciences. The ground of accessibility for the objects of science must be sought on behalf of the sciences in the primordial openness of beings.

In *Being and Time*, Heidegger interprets "Dasein" as "being-in-the-world." Insofar as in modern thinking the origin has been lost from view, a gradual "loss of worldhood" can be discussed in the sense of Heidegger's concept of world (world as "entire context of significance"). Thus, it is necessary, especially in the current age, to recognize that by grasping things in terms of their sensibly perceived properties (e.g., taking a wooden chair as an object of a certain hardness, weight, and color etc., this chair is already no longer perceived as "a thing in the environment"). Even though the qualities mentioned need not yet be perceived as quantities of a certain value, they are already, as determinations, properties that are attributed or denied to the being as thing-descriptions that are used in subsequent assertions. Heidegger says in connection with a chair described in this way: "what we have just said of the perceived can be said of any piece of wood whatsoever" (GA 20, 49/38). Something is said about the chair, "not qua chair-thing, but rather as a thing of nature, as *natural thing*" (GA 20, 49/38)

A difference must be acknowledged when a being is determined as a thing of nature. When, in his famous second meditation, Descartes describes a piece of wax by its sensibly perceived properties (as colored, of a certain form, size, hardness and coldness), this is according to Heidegger already an apprehension of the thing as "nature-thing." It already constitutes a "theoretical apprehension" (GA 20, 247/182) that is no longer guided exclusively by a primordial understanding of significance. Descartes then takes a well-known further step. He considers the identity of the wax throughout its changing qualities, and forces the reader to admit that henceforth determinations by the intellect are relevant (the thing is thereby determined as "extended, flexible, and changeable"). A direct route follows from these intellectual determinations of a thing of nature to the Kantian determination of the "object of cognition." For Kant, it is assumed *a priori* that the "object of cognition" is always located in a well-defined space and time, and that other properties of the thing can be described in well-defined "degrees of intensity." This is the kind of "projection" that natural science seeks using mathematical relations between measurements. Insofar as natural science cannot abandon the thought of "*a* nature," it strives to overcome several essentially different descriptions of nature in favor of *one* determination.

So a being is experienced in the sciences in a new way of what-being and how-being. A change has taken place that is grounded in the "projection" of science. What the object of science can be is thereby determined from the outset. Heidegger says about this change in the way of seeing: "a different being is not related to and discovered, but rather the being of the already evident being is from the beginning seen, taken and determined differently" (GA 27, 186). The projection of science "drags the being clearly into the light, without changing it into anything else." Such knowledge can be understood as a "letting-be" [Seinlassen] of a being qua object that must "make use of the openness of the being" (GA 27, 180). The being, removed from its original context, reveals itself "as lying-before [als vorliegendes], as positum." In this sense, scientific knowledge is "positive knowledge" (GA 27, 197). Heidegger carefully considered such relations of lying-before [vorliegenden]. On one hand, science gives itself its object area, and therewith also the relevant ground of its experience; but on the other hand, its ground is already given in the primordial lifeworld. We see in more detail how this latter ground is already given, although science as science can have no insight into it.

There is, however, another difficulty that first must be overcome: the realm of natural things [Naturdinge] evidently extends further than the realm of things of use [Gebrauchsdinge]. The latter are things of nature, but conversely, we make a wealth of statements about natural things to which we have no access—especially in modern natural sciences. The "systems" set up in these sciences are markedly similar on a general level, despite substantial individual differences. Concerning "what" the objects are, one can always reply they are physical "systems," and the "how-being" of these systems is determined by physical values that have a universal applicability. If the values and laws (e.g., the relations between lawlike values of place, velocity, and mass in physics) are examined more closely, then it becomes clear that science applies to more than objects of use in a novel but legitimate extension. Through a totally changed constitution of the being [Seinsverfassung] of beings, it becomes possible to say something about objects that are in no way "affected by that constitution" [Verhaltung] (as ready-to-hand objects of use), to which we "do not need to reach with our methods, and upon which we do not need to touch" (GA 27, 182). The entire field is "outlined in advance" by the projection.

Heidegger's subsequent insight is also significant for scientific theory. The "active" approach of the scientist toward nature, which expresses itself in experimental arrangements, is seen by him in the right light. I have already hinted that according to Heidegger science can prompt a being to reveal what was previously not known, that it can "get the latter out" of the being. It is important here to understand properly how the activity of the scientific approach does not lead to a subjective falsification of objects. This activity (so reads his analysis) has the character, when properly seen, of "withdrawing before the being" (GA 27, 183). What experiment reveals about an object is revealed by the way in which the experiment is arranged.

These arrangements are tailored right from the start towards the object under investigation in such a way that it is prompted to produce a particular kind of answer about its how-being. The entire process is carried out through measurement of what is meaningfully questionable and answerable by physics according to its projection (at a derivative level). The primordially perceptual object is, however, he reminds us, "the specific environmental thing, even if it remains hidden."

The ground of knowledge has therefore been overlooked from the very beginning, according to Heidegger, if the enquiry takes as its starting-point present-at-hand things that somehow encounter us and concerning which science works out an optimal description. He makes his point precisely: "the apprehensibility and the objectivity of a thing is grounded in the encounter of the world, but objectivity is not a presupposition for the encounter" (GA 20, 258/190). Because this relation is usually understood the other way around, "problems of correspondence" inevitably develop (see the next section). There also is an inversion of this relationship if things in nature, which present themselves in one way or another as suitable, unsuitable, or similarly, are subsequently attributed certain value-predicates such as "good, bad, plain, beautiful, suitable, unsuitable and the like" (GA 20, 247/183).

Up to this point, natural science has been used as the main orientation. As we have seen, it secures its object area *a priori* through a mathematically determined projection. But nature is already present "by itself," and the objectification of nature (to a "second [kind of] nature")[6] remains dependent on nature thus present. "Objectivity" is only "*one* way . . . in which nature exhibits itself" (VA 58/174). Natural science cannot "embrace" the fullness of the coming to presence of nature. In two senses, nature does not let itself be "gotten around" by science. On one hand, natural science depends on something more fundamental, a prior openness, that it cannot avoid. On the other hand, it cannot "get around" nature in the sense that it is "unable to encompass" nature.

What about other sciences than natural sciences? Heidegger begins by remarking that as soon as we have "once caught sight in one science" of the already present "that which is not to be gotten around [*das Unumgängliche*] and have also considered it somewhat," we can easily see it "in every other" (VA 59/174). In the other sciences, there also are particular aspects of inquiry that can only be drawn out of a being by a specific science. The sciences demarcate themselves by means of such "entrapping securing" [*Sicherstellen*] (VA 59/175) of themselves against one another. In his lecture "Science and Reflection," Heidegger addresses some of these disciplines explicitly. Starting with natural science, he elaborates: Psychiatry represents aspects of the life of the human soul "in terms of the objectness of the bodily-psychical-spiritual unity of the whole man" (VA 59/174). Historiography secures itself in an area "that offers itself to its theory as history" (VA 59/175). Finally, philology "makes the literature of nations and peoples into the objects of its explanation and interpretation" (VA 60/175). But (just as with nature) people, history, and language can be discussed in a more original sense (as that which forms the basis). For these sciences have the same "that which cannot be gotten around" in both its senses (VA 62/177). So when Heidegger adds that the natural sciences cannot as such decide "whether nature, through its objectness, does not rather withdraw itself than bring to appearance the hidden fullness of its coming to presence" (VA 59/174), this inability to decide applies beyond the natural sciences. It is therefore important to keep Heidegger's following comments in mind when evaluating the social sciences (VA 63/178ff):

1. One such "that which cannot be gotten around" is Dasein, "wherein human being as human being ek-sists."

2. The final ground for all "historiology" is the "destiny in historicizing" [*Geschick im Geschehen*].

3. All human language is grounded in the event of truth as an event of "clearing" [*Lichtung*] and safe-guarding self-concealment [*bergendem Sich-verbergen*]. "Language speaks" already "before" the person speaks. So language "speaks" already "without becoming literature and entirely independently of whether literature for its part attains to the objectness with which the determinations of a literary science correspond" (VA 60/175).

The question, raised at the beginning of this section, concerning the "accessibility" of the objects of science, is accordingly brought back to a prior "openness of the being itself." It refers to a primordially presencing being that can "reveal" itself from itself in a special way as object. The "activity" of knowing must be seen as a way of "stepping back before the being." The search for conditions for the possibility of demonstration leads to conditions of being in which the "already evident" lets itself be revealed as standing "in the open region."

Whoever asks at this point what possibility exists for the sciences to bring the inaccessible "that which cannot be gotten around" better into focus, is asking how Heidegger in his subsequent philosophy constructed his phenomenological hermeneutics to pursue the question of being. This theme is beyond the scope of this section. But—again using the example of nature—a hint has been given which will be further developed in the section, " 'Challenging Revealing' and the Increasing Danger of the Gigantic in 'Machination'." If we enquire with Heidegger into "nature" in a more original sense, we are led to his later concept of "earth." What needs to be considered here is connected to a question he himself formulated in this context: "Who enkindles that strife in which the earth finds its open, in which the earth encloses itself and is earth?" (GA 65, 278/195)

THE FUNDAMENTAL PRESUPPOSITION FOR OVERCOMING DUALISTIC POSITIONS

What has been said so far acquires additional significance, and deeper clarification becomes essential, upon recognition of the necessity in scientific research to exclude from the outset global error concerning the object of research. Descartes' answer, that our own representations, as something immediately given, are especially excluded from any doubt, will hardly satisfy anyone who pays special attention to his recovery of the corporeal world. Here it is clear, one might argue, that Descartes was obliged to rely on supporting hypotheses (such as the role of the pineal gland) that are no longer credible. To anyone who does not ignore Descartes' especially influential philosophy, remark concerning the possibility of global error makes one thing clear: if the theoretical descriptions of physics reconstruct nature in subjective thinking, then a correspondence problem affects the question of demonstration of each physical theory, and the truth of its statements. It may seem that such possible doubt applies only to distant or multiply mediated natural processes, but the immediacy with which an object is grasped is not the issue. If comparison of the knowledge of nature with its attempted reconstruction in subjective thinking causes problems, then these problems affect the legitimacy of the whole business of knowing nature.

Of course the problem just touched on is not particularly new. Skeptical objection to theories that depend on the assumption of a "transcendent reality" is frequently voiced, and connected to long-standing, heated controversy. The assumption of knowledge of so-called "transcendent objects" is obviously infected with the difficult problem of the "reaching out" to them. Modern scientific theory, which does not overlook this problem, offers attempts at solution beyond the obviously flawed "correspondence theory" (although there is no room to go through such attempts here).[7] Heidegger offers in my opinion a deep clarification of the problem by demonstrating the mistake of approaches that start from the "*Adaequatio rei et intellectus*" [adequacy of the thing and the intellect].

Heidegger radically rejects all claims that the truth of knowledge reveals itself in the relationship of the "representing subject to an object." In 1941, he says that "the coming together of subject and object (and vice versa) is rather only possible in an already essentially open region, the openness of which has its own essential origin that has not yet been questioned by all previous philosophy" (GA 49, 56). To understand this insight, what has already been said must be kept in mind: Knowledge of an object can only be reliable if the "object itself" is accessible. Moreover, I have shown the extent to which Heidegger put special emphasis on primordial behavioral and experiential processes in human life. In this context, his fundamental idea is relevant: "So far as it exists, the Dasein is always already dwelling with some being or other, which is uncovered [entdeckt] in some way or other and in some degree or other" (GA 24, 296/208). If the primordial "openness" of the being (here called "uncoveredness" [Entdecktheit]) is "passed over," then one cannot see how further ways of experiencing a being can "encounter" their object or contain the possibility of grasping it correctly.

The drawing away [Entrücktsein] of Da-sein into an "openness of being" is rightly the dominant focal point, but this must be shown in more detail. The starting point is still the question of what we experience in "faulty vision." That our primordial sense-perception directs us more or less correctly concerning the perceived sensible qualities of objects and their properties, turns out to be an unsatisfactory answer because it overlooks something essential. Starting with "sense perception" can only be justified if we acknowledge that what is here called perception is informed by a primordial, guiding understanding of significance (see the previous section). What does it mean that a being is revealed in its way of being as ready-to-hand, and simultaneously in its what-being, in disclosure of relations of significance? According to Heidegger, disclosure of relations of significance is grounded in Dasein as being-in-the-world. This must be thought in terms of what makes possible the accessibility of beings, whether as ready-to-hand or present-at-hand. Here "truth" [Wahrheit] is grounded in an original sense (i.e., in the sense of "openness"). Because "Dasein" is always already included in a "world" of significance, it is according to Heidegger already drawn away [entrückt] into an "openness" [Offenheit] that grounds the availability [Offenbarkeit] of beings. This insight is especially important for the sciences. Insofar as their knowledge is founded on a prior understanding, they are freed by this insight from the deadend that threatens their relation to the "things themselves." The sciences take advantage of the openness, into which Dasein is drawn away, in their specific way. In "On the Essence of Truth," published in 1943, Heidegger says that the possibility of a directedness of the self toward present-at-hand beings—a self-directedness that asserts itself as a binding standard—arises precisely because we have already entered freely into an "open region" for something "opened up which prevails there [waltenden Offenbaren]" (W 185/123)

In *Being and Time*, unconcealment (as "openness of being") is to be understood as a drawing away [Entrücktsein] that shows itself as a "being temporally enraptured [*sich zeitigendes Entrücktsein*]." Dasein exists as "thrown projection" in the disclosure of the world. Projection, thrownness and being-alongside are shown to be ecstases of temporality. §21 of *Basic Problems of Phenomenology* explains further how "horizonal temporality" (GA 24, 378/267) is brought about in the ecstatic temporality of Dasein. In horizonal time, the "farthest horizon" of "being in general" is seen, which vision also includes the openness of beings which are not Dasein. Later Heidegger sees only a "first name" for the "truth of being" in the "open region of time." The focus of the investigation shifts to an original dimension of time from which the "destiny of being" can be suitably brought into view as the "domains of the ways of time." Movement into the openness, i.e. the "being-able-to-reveal [*Entbergen-können*]" that is particularly "granted" to Dasein, relates back to what opens itself "from itself" and the "already open" unconcealment. The projection, and even Dasein itself, is experienced "as happening through being." It is no longer sufficient to talk about "projection" [*Entwurf*]. Thrownness is thought as taking place through a "throw" [*Zuwurf*]. This is a consequence that for the very first time does justice to the dependence of human being on the destiny of being.

That the sciences can, as said above, take advantage of "openness" in their specific way, can be appropriately explained in more detail using the example of physics. The being studied in physics is the present-at-hand "object." What has been determined as present-at-hand is no longer grasped under the structure of involvement—readiness-to-hand—as detailed above. The properties belonging to the present-at-hand are "got out" of the being itself. This getting out originates in a letting-be, which is a letting-be of a being, and prior to that, a letting of being. The so-called "activity" of knowledge was explained above in the previous section as this "allowed" letting of being. The letting of being entails the possibility of any determination [*Setzen*], which is in physics the "determination [*Stelle*] of challenging forth." Fully mathematized physics aims at this determination. If one asks "what" is the being of physics, the answer would have to be "a system of measurements." Measurements are mathematically determined, and mathematical relations between measurements are established in possible proofs, which strengthen the reliability of measurements in the relevant empirical realm. In such a stabilizing, systematizing context, the being stands "over against" the knower; it literally becomes an object, that which stands against [*Gegen-stand*].

There is a temptation to perceive the "opposition" as merely a "represented" relation. If one falls for this superficial idea, then it is no longer possible for physics to establish the legitimacy of its relation to its objects. This temptation, which has in large part arisen from physics itself, has widely influenced the general way of thinking. As a result, the dualistic representation of thinking as "lying over against" and describing the "world of objects" (with all the difficulties of agreement that can therein develop) have become standard. Thus "correspondence problems" persist just below the surface to the present day.

Heidegger's position can be clearly inferred from a short paragraph in §44(b) of *Being and Time*. He stresses that an entanglement with problems of correspondence arises only with the "switching over" of the already primordially open being (its "discovery") to relations of presence-at-hand. The being, which as present-at-hand lies "outside" the mind, is placed in a *represented relation* in the present-at-hand representations of a science (especially the objective representations of physics). But then the status of the terms of the relation, as well as of their relationship, becomes an insurmountable problem. (GA 2, 297; SZ 224-25/267–68).

HEIDEGGER'S DISTINCTION BETWEEN "CORRECTNESS" AND "TRUTH"

Heidegger's comments quoted in the previous section are of special significance for "representative assertions" and the distinction between "true" and "false." This significance lies primarily in potential clarification of the words "correctness" and "truth," which are used atypically by Heidegger.

Something is said about things according to Heidegger in a "demonstrative assertion" that attributes or denies them properties. Assertion is distinguished from interpretation (with its hermeneutic as-structure) by change in the as-structure (to an apophantic as-structure). The fundamental relation of assertion to interpretation is herein expressed. Scientific assertions about represented objects as such depend on the delimiting projection of a science's specific field.

The problem of truth is usually discussed in the context of the difference between existent and nonexistent matters of fact. Existent and nonexistent matters of fact can be represented and appear in assertions. Verification of such representations entails a problem insofar as the matter of fact seems also to play here a role as not represented—the representation must be compared with the matter of fact to establish correspondence. Husserl had the groundbreaking insight that in sensible intuition, the intentional object can "prove identical"—the representation and the matter of fact are indistinguishable. Thus, a matter of fact can be said to be self-presencing. Husserl was thus brought back through the problem of "categorial intuition," which already permeates sensible intuition, to an essentially pure philosophy of consciousness. In contrast to Heidegger, he was unable to start from a solid basis in the seeing intellect. Still, the "openness of the being itself," which was from the beginning assessed by Heidegger as fundamental and thus grounding, is a consequence of the great influence, in some respects noted explicitly by Heidegger, that Husserl had on him.

When a matter of fact that "actually" exists is represented or asserted as existing, or when a matter of fact that does not "actually" exist is represented or asserted as not existing, then the representation, or likewise the assertion, is usually called "true." When, on the contrary, a matter of fact that "actually" exists is represented or asserted as not existing, and vice versa, then the representation or the assertion is accordingly called "false." This is supposed by many to agree with Aristotle's analysis at Chapter 10, Book IX of the *Metaphysics* (at 1051 b), wherein he explained "truth" in the sense of the "correctness" of the representation or assertion. Heidegger brings significant counter-arguments against this interpretation of Aristotle, and in particular disputes the thesis that for Aristotle the judgment is the "primordial 'locus' of truth" (GA 2, 298; SZ 226/268). Leaving this dispute aside and going instead to the heart of the problem, in the usual explanation of "true" and "false," as already mentioned, the question immediately arises as to what exactly is meant by an "existent" or "nonexistent" matter of fact. What is essential here can best be shown by examining a false assertion.

A false assertion usually occurs when one is mistaken about the how-being of a being; for example, I say my neighbor's cat is gray, when "actually" it is black. My neighbor's cat is for me something in my disclosed world that lies in the environment, and to which I already have access. Thus, I arrive, through something I see, at the conviction: "My neighbor's cat is gray." This means that when I see a being that appears to me gray, I do not confront an illusion; on the contrary, I "encounter" a determinate cat (as the being itself). I thus pick out a "being itself" with respect to what I see. I see a determinate gray cat. If I perhaps see the cat again sometime later, I will notice that I made a mistake insofar as the cat is black. I did not have a representation that did "not correspond to reality" in any way at all, but on the contrary I saw a being (about which I was mistaken) from the very beginning.

It is helpful to cite one of Heidegger's more radical examples. When one mistakes a tree for a person, it is assumed that one is mistaken as to what this being is (GA 24, 88/63). The important point about error here is that it is not committed in relation to an illusory representation, but to the being itself; that means that a man himself is picked out in the example "person instead of tree." One falsely takes the observed being at a glance to be a man. Heidegger explains that "it would be wrong to say that this perception is directed toward a tree but takes it to be a man, that the human being is a mere representation" (GA 24, 88/63), for "in this mistake in perception the man himself is given to me and not, say, a representation of the man" (GA 24, 89/63).

There can be no "serious" doubt that when I say something meaningful, I make a true or false assertion, and that each being about which I speak truly or falsely is "discovered" by me. It is better to ask instead what this alleged "not showing itself" implies about the being, whether this is unthinkingly supposed or taken as a question about "transcendent reality." Could it not be that this situation needs to be discussed in an entirely novel way in which the data must be necessarily accounted for, despite the fact that it is presented in more or less erroneous form? Heidegger formulates the issue as follows: the scandal of philosophy is not that proof of the "reality of the outside world" has not yet been given, but that "such proofs are expected and attempted again and again" (GA 2, 272; SZ 205/249), instead of recognizing the extent to which we already "are outside."

This is clearer when what was said about the accessibility of beings in the previous sections is reconsidered. It was there concluded that allegedly "being outside" only becomes an issue when not just the being itself and its representation, but also the being of the being comes into view. As a brief reminder of what was said, without the "openness of being," the "being-outside" of beings would not be granted, because the possibility of orienting oneself toward present-at-hand beings presupposes a "standing in the openness of being." This openness (later called the "clearing") is called "truth" by Heidegger, whereas the previously discussed assertions are called "correct" or "incorrect." Sensible criteria for "correctness" and "incorrectness" can be given in the sciences, but regardless of the criteria by which the correctness of an assertion is determined, the possibility of correctness is indebted to truth (openness of being), and thus to a "realm from which man and Being have already reached each other in their active nature" (ID 25/33).

There are frequently accusations against Heidegger that he simplifies the problem of demonstrating truth in such a way as to arouse the suspicion that he takes every demonstration to be analogous to the immediate, intuitive grasp of what is concretely given. If so, it would be clear from the outset that he can contribute nothing useful to demonstrations of the complex relations of scientific theories. This objection arises from a misunderstanding that overlooks his essential insights, and moves the discussion back to a level to which he was rightly opposed. Things are different when his opposition to views about the unavoidable, subjectively determining character of the sciences is considered. He is rightly attributed with such opposition, but it is imperative here to deal in more detail with what he says.

Science deals with "correct" and "incorrect" assertions. But it does not need to relate to a being that is separated over and against it, and about which it can only have knowledge in the sense of a secure "correspondence." As explained already, an accessible [*offenbares*] being lies before the scientist, and the object to which science relates can give itself to the science as the "thing itself," whereby the accessible being is determined in a new way by scientific projection, and the radius of comprehended things is greatly extended. The comparison to what is intuitively given is only justified insofar as it is often easier to see the extent to which the intuitively given can be comprehended "in itself." But it is absurd to allege that Heidegger would not see that in physics, for example, things come to knowledge through a universal, regulated network of understanding.

Heidegger temporarily lost sight of the fact that early in Greek thinking the "unconcealment" of being has already turned into a concealing of the original event, so that soon a concept of truth in the sense of "correctness," and thus also "adequation in the sense of the correspondence of representing with what is present" (SD 78/71), moved into the foreground. He corrects this temporary "error" concerning the deterioration that has already taken place with the added remark that in this deterioration, "man's ecstatic sojourn in the openness of presencing is turned only toward what is present and the existent presenting of what is present" (SD 78/71).

"CHALLENGING REVEALING" AND THE INCREASING DANGER OF THE GIGANTIC IN "MACHINATION"

If Heidegger's critique is not aimed at the sciences as such, then—it may be asked—at what is it aimed? This question makes it necessary to name that which can (not "must") be overlooked, but which certainly is overlooked when the sciences are no longer judged by what "is their own," so that a specific concept of "reality" attains a dangerous monopoly.

That which is overlooked in this dangerous way is elaborated in the following three different (but closely connected as regards content) ways:

1. The already discussed "that which cannot be gotten around" for the sciences in their ownmost possibility.

2. The question "Who is human being?" To what extent is human being "called" to itself?

3. The safeguarding of truth in beings (which is threatened by the "destruction of the earth").

"That which cannot be gotten around" for the sciences was discussed in detail in the second section of this chapter, and should be understood in this further investigation in the way it was explained there.

Starting now with the next question, "Who am I?," Heidegger writes that the utmost distress (of lack of distress) expresses itself where "everything is held to be calculable and, above all, where it is decided, without a preceding question, who we are and what we are to do" (GA 65, 125/87). Dasein, so runs one of the most famous phrases of *Being and Time*, is "always mine." At first glance, this will appear to some as self-evident. One might say that, whether human or not, every object is always this and not some other object. In the case of human being, one can of course reflect on oneself, and then voice the trivial sentence: "I am this and not some other being." Whoever talks so obviously suffers no distress with the question, "who am I?," yet fails especially to see that Heidegger is not talking about a secondary reflection on one's own "consciousness," or an "objective system." Rather, he is talking about something that for all reflection comes "too late." I "know" "always already" in a certain sense (and entirely without reflection) that Dasein is "always mine," and at the same time I do not "know" it. In order better to understand "mine-ness," we need to pay attention to the second of Heidegger's remarks ("to what are we called?"), which is also easy to misunderstand. This second point does not (at least not primarily) ask, "what we should we do as moral agents?" The primary concern here is not obligation in the modern moralistic sense. Rather, it is the fact that Dasein is called to itself. That the "itself" is at the same time not, is implied here when being "called" and thus also the call are mentioned. That is, a decisiveness is necessary in order that we really are—in order to be the "there [Da]" of Da-sein.

What has just been said remains unintelligible to one who takes the "I" as something present-at-hand among many present-at-hand entities. Each I that is present-at-hand in this way perishes after a certain time, as experience shows is the case with human bodies (including their so-called "consciousness"). For Heidegger, it is essential to consider things entirely differently. He is concerned with respect to Dasein about the being carried away [*Entrückheit*] of human being into the "truth of the being." He then asks, what is the "other of this carrying away?" His concern is not the presence or absence of something previously present-at-hand, but rather "the totally other of the t/here [*Da*], totally concealed from us, but *in this* shelteredness-concealedness [*Verborgenheit*] belonging essentially to the t/here and needing to be sustained along-with the inabiding of Da-*sein*" (GA 65, 324/228). The shelteredness-concealedness of the t/here mentioned here can be open for Dasein. There is for Dasein the possibility of bringing the shelteredness-concealedness into a relation to truth; that is, there is a possibility of attaining its own self and therewith at the same time of arriving at the "sustaining" of what is utmost for Dasein, the "traversal of the widest removals-unto" [*Entrückungen*] (GA 65, 324/227).

According to Heidegger, the possibility of being in an actual sense "oneself" is therefore given to human being along with the openness of being—the openness of the shelteredness-concealedness of the there [Da]. This selfhood, according to the later Heidegger, is grounded in mastery of "owning in enowning" [Eignung im Ereignis] (GA 65, 320/224). Simultaneously, owning-to [Zu-eignung] and owning-over-to [Über-eignung] belong here. Said differently, Da-sein can come to "itself" in owning-to only if this owning to "itself" happens through being and is an "owning-over-to in enowning." The entire weight of this question, which includes the "secret," shifts itself for each "self" (actually for us "mortals" together) to the essence of the truth of being. This question is not to be treated as a theme of modern science, since a "secret" of this kind is different from anything that can be thematized in any such science. The question belongs rather to that which can be threatened by the sciences, that is, by an illegitimately generalized concept of reality.

The third threat named above relates to the safeguarding of the truth of being in beings. Safeguarding through Da-sein belongs to the standing-in of the "there" in truth. Safeguarding presupposes that the earth hides itself. When the earth is "accessible" [offenbar] in the openness of being, truth is sheltered in beings. This applies both to things of nature and to what is made by people. Thus, the "earth" as such is not an object of knowledge for the sciences, even though it belongs to their foundation that "cannot be gotten around." As what then do we primordially experience nature and natural entities when they reveal themselves to us in their significance? In the lecture "Building, Dwelling, Thinking," Heidegger says that by "serving and bearing" the earth makes possible our "building and dwelling"; it is "spread out in rock and water" (VA 143/149). "Rising up into plant and animal," it at the same time makes possible "nourishing" of our bodily being. Moreover, the question arises, as what do we *see* the things of nature? Can we explain scientifically the primordial happening of a "bright colorfulness," of a "dull nuisance," or of a "carefree giving-itself," without destroying it? Not at all. But also the distinction between "primordial" and "derivative" is not sufficient here. Above all must be asked how it is possible that what for science is a completely legitimate concept of reality, generalizes itself and makes this distinction disappear.

It certainly does not belong to the internal tasks of a science to question, with respect to the authority of its projection, what is bound by it and what not. But if such clarification is missing, and if our "fascination" with well-established practical achievements solidifies into a comprehensive concept of reality, then the task of thinking itself is to counteract such danger. Here we find in Heidegger much timely advice for the present age.

The revealing [Enthüllung] of being, that is, the way of revealing [Entbergen], has changed substantially according to Heidegger since the era of the Greeks. Heidegger calls the "rising-from-itself" of nature for them, as well as the revealing of what is manufactured and artistic, a revealing that "brings-forth" [*hervorbringendes Entbergen*] (VA 15/10). In contrast, the essence of modern technology lies in a "challenging gathering-together [*herausfordernden Versammeln*] into ordering revealing [*bestellende Entbergen*]" that "holds sway already in physics. But in [physics] that gathering does not yet come expressly to appearance" (VA 25/22). Physics challenges and "sets upon [*stellt*]" nature, which it makes present itself as calculable relations. Challenging revealing has, in contrast to revealing that brings-forth, the character of entrapment [*Nachstellen*], and at the same time, securing [*Sicherzustellen*]. Science entraps "objects belonging to the real, in their objectness, in order to secure them in the unity of objectness" (VA 56/172). Tremendous success and many possibilities for application are therein fascinating. Thus it is understandable that reflection on the authority of science (and on the way the corresponding concept of reality strengthens an unjustified monopoly of science) is particularly lacking.

What is given up if the concept of reality belonging to the natural sciences is the only one? This "reality" encompasses only differences in the how-being of systems, and it confines changes in how-being to mathematical formulae. Assuming that we have understood physically what a "healthy mind" is for a person, in contrast to a "mind impaired" by some defect, then we comprehend with the certainty of physical science only the difference between two kinds of physical system. The classifications "impaired" and "healthy" rely on nonphysical criteria, so the science of physics contributes nothing directly to this distinction. Accordingly, claiming that the physical means for transformation to the second kind of system (the "healthy mind") are available, and that physics is in principle in a position to correct the defect and thus bring about a "healthy mind," exemplifies unjustified extension of the scientific concept of reality. Similarly, it will soon be possible to prolong for an extended time the life of a present-at-hand human body, from which a functioning brain is "absent;" but such a physical advance says nothing about happiness (or perhaps unhappiness). And the rest of the scientific community? Can one really be convinced that "happiness" is a theme concerning which the cooperation of all the sciences can deliver information with "certainty"?

Yet the gigantic [*Riesenhafte*] of machination [*Machenschaft*] marches on unimpeded according to Heidegger. Insight concerning scientific authority, the concept of reality, and what cannot be gotten around [*Unumgängliche*] fall by the wayside with growing danger. Heidegger raises questions concerning this situation in the 1930s by attending to the "safeguarding of truth:" "Who enkindles that strife in which the earth finds its open, in which the earth encloses itself and is earth?" (GA 65, 278/195; cited above, p. 7).

HOW DO THE PARTICULAR PROBLEMS OF MODERN SCIENCE STAND IN RELATION TO HEIDEGGER'S THINKING?

A question often figures in interpretations of Heidegger of the extent to which it is possible for him to understand the philosophically relevant problems of modern science with respect to their scientific content, and thus account for their relation to his own constellation of problems. This question is especially important for contemporary natural science, which dominates modern-day thinking. This question will therefore be illuminated in what follows using characteristic examples.

When Heidegger talks about the mathematical projection of natural science in the sense explicated above, he often provides philosophical notes concerning the measurements of classical physics. In classical physics, measurements of, for example, location, time, velocity, acceleration, mass, and force are each in their turn introduced. They all appear in Heidegger in the context of comments on Newtonian physics. That a particular problem arises in the measurement of mass (H. Weyl stresses the importance of the law of momentum for the possibility of making mass relations accessible through relations of velocity)[8] indicates a thoroughly important, specific problem concerning the attribution of properties to objects. But for Heidegger's assessment of Newtonian physics, knowledge of these special problems is not necessary. He pays close attention to what is relevant to his own inquiry; for example, the demand that the sciences remain within the framework determined by their projection. Thus, Heidegger rightly stresses concerning Newton that space functions as the "locus primarius," and that a corresponding significance is attributed to physical time. Heidegger mentions the question (answered by Newton in the negative) whether physical time itself can be located in a different kind of time, and whether likewise place itself can be located in a different kind of place. Both questions are, as Heidegger says in agreement with Newton, meaningless questions (GA 27, 194).

One might still ask whether the twentieth century cannot call decisive scientific discoveries its own? Is not for instance "Gödel's incompleteness theorem" awarded philosophical significance of the highest rank (according to H. Scholz as a "20th century critique of pure reason"), so that Heidegger can hardly be considered responsible for raising questions about the grounding of knowledge? And has quantum theory not also posed philosophically revolutionary questions that presuppose a knowledge of physics that Heidegger understandably could not have at his disposal? Such questions cannot remain unanswered. But how do things stand if these questions are illuminated concretely?

Gödel's famous incompleteness theorem says something about certain statements in closed systems, especially in arithmetic. It shows that these statements are "formally insoluble" yet "true in content." The truth content applies to the "correctness" of the statement, whereas the "formal insolubility" refers to the limited means of every strict "formalization" through calculation. A sufficiently conceived "constructive mathematics" circumvents the defect of the "formalization of arithmetic"; that is, the statement that is true in content is now constructively provable. Gödel's proof stands also in close relation to the existence of "an insoluble calculus" for which—this is what "insoluble" means here—there is in principle no algorithm for general solution of the derivation of statements. This does not exclude searching for a derivation for a particular statement not yet derived (because the possibility of a success regarding this cannot be excluded), and it likewise does not exclude that the underivability of this statement is proved using means "outside" calculus (because the possibility of a constructive proof of underivability also cannot be denied). It is therefore not nonsensical to seek constructive proofs for particular statements.

Overall, we are dealing with internal problems regarding the *correctness* of mathematical statements with regard to the appropriate methods of proof for such correctness. The difficulties relate especially to the "tertium non datur," by means of which the problem moves primarily to the question of permissible, constructive methods of proof for "correct statements" that can be constructed with logical parts. Because the goal in sight for all these observations is truth in the sense of correctness, there is no reason to remark critically that Heidegger should consider the relevant research in particular detail. Furthermore, if it is said that the "tertium non datur" refers to the opposition between constructivists and Platonists, then in this context the term "Platonic" means nothing other than the assumption that the "alternative" expressed in the "tertium non datur" must—as long as it has not been refuted—be true "in itself" (a refutation is difficult to free from suspicion of a proof of contradiction).

The discussion can be clarified considerably in relation to I. E. J. Brouwer and his work on the "continuum." Brouwer points to cases where in his opinion a "tertium non datur" is shown to be invalid. To understand what this means, Brouwer's most important insight must be explained, according to which it cannot be claimed, for any finite interval of real numbers, that any two real numbers are either different or not. The salient point is that defining real numbers as intervals of rational numbers allows for choice between intervals, and that the "tertium non datur" belongs in a higher realm of discourse (claimed for all real numbers in the interval) and holds the position of an existence operator. Thus, it cannot be claimed for any two real numbers either that there is an interval for which the interval containing both falls short, or that there is no such interval.[9] That it is possible to prove, by considering choice between intervals, this astonishing statement concerning the continuum intensifies the peculiarities that occur with reflection on the continuum. If the continuum is a quantity in the usual sense, then this quantity must be "nondenumerable" (as Cantor proved). But in contrast, it is also possible to assert the meaningful concept of an "open quantity." An "open quantity" can by definition be expanded, whereby Cantor's proof is invalidated. Surprisingly, nothing useless is asserted with an "open quantity." Statements shown for an open quantity take their provability from the proven irrelevance of their proof to their extension. The latter concept of openness suggests, in the interest of mathematics, taking a closer look at how this openness relates to "now-time" [*Jetzt-Zeit*] (in Heidegger's sense of clock-time, SZ 370/422 ff; 413/466), and how it relates to the possibility of action in the framework of now-time.

Reflection on this becomes more urgent, if the "surprising" consequences forced on us by the results of the quantum theory are considered. I emphasize from the outset that I have no intention here of aligning myself with those who exploit the mystifications of quantum theory that focus on the conscious intervention of the human subject into natural processes. I caution against such a sensationalized twist on the subject–object problem, and add that Heidegger, with keen sense, attributed no ontological significance to the so-called "complementarity" of certain values in quantum theory. He found himself in no way forced to recant concerning the status of the objectivity that determines knowledge (and challenges human being). Heidegger gives a correct analysis that in quantum theory, a particle's "state of motion may on principle only be determined either as to position or as to velocity" (VA 56/172). Objectivity shows in quantum theory "*fundamental characteristics completely different*" insofar as it "admits only of the guaranteeing [*Sicherstellung*] of an objective coherence that has a statistical character" (VA 56/172). In this sense, the theory entraps the objects of reality in order to "guarantee" them. And that means for Heidegger that human being is itself challenged to challenging orderings [*herausfordernden Stellen*].

But what does it mean, if in light of the quantum theory the "nonobjectivity" of events is discussed? This is in fact a common but also misleading conception. Briefly put, it means that a quantum system is no longer describable in general as "separable" from Newtonian "apparatus" and "measurements," and that the values of its measurable quantities cannot be unequivocally attributed to a separate "reality." Since in particular the values of "complementary quantities" (i.e., values described mathematically as inherent values, not interchangeable operators) can never simultaneously occur "in themselves" in the quantum system, one also cannot measure them simultaneously. But one can, in a mathematical description, assign a "state" to a particle, which allows for the calculation of a probability distribution for the value of the complementary property. These are values that determine the quantity of a measurement with a specified quantum mechanical probability.

If an emitted particle cannot in general be represented as following a "trajectory," can we ever arrive at an adequate representation of reality? In Young's widely discussed double-slit experiment, the intensity of a beam of electrons is reduced so that one particle at a time is passed through a screen with two slits, and detected behind on a photographic plate. The assumption that the particle is on a trajectory that goes through either one or the other of the slits (if it is detected afterwards on the plate) proves in the general case to be false. One can set up the experiment to take place in two equal time intervals, during the first of which the first slit is closed, and during the second of which the second slit is closed. This experimental set-up leads to a very different result (explicable in classical terms). But when both holes are open, the result is obtained of an image on the photographic plate that corresponds to the calculation made using a state-function, which does not describe each particle as passing unequivocally through one hole; on the contrary, a quantum mechanical probability distribution for every particle must be assigned. Such experiments have been conducted in many variations; for example, so that only two routes (different, but equally accessible)—like tunnels—are possible for each particle. Here too the result contradicts the assumption that each particle (photons were used in the best-known experiment) takes either one or the other route.

The assignment of one state to a quantum system is therefore not possible such that the system could be attributed with unequivocal values for quantities (that could then reveal themselves in single measurements), so that it would be possible to *separate* the values as an objective determination. A radical interpretation says that a quantum system only becomes "real" in its interaction with a Newtonian system (in the above example, via its impact on the photographic plate as a Newtonian measuring device). But since it is however possible to calculate the probability distribution of the mathematical state function for the values of quantities in potential measurements, one can also say that the quantum system exhibits a state of *futuristic relevance*.[10]

A concept purely futural to understanding again is at play in the context of—as Heidegger calls it—"now-time" [*Jetzt-Zeit*] (GA 2, SZ 421/474 ff). This touches on a philosophical problem insofar as now-time is not self-evidently characterized by a special "openness" in relation to the future. It owes its so-called "direction" to a "more original time" (Heidegger), which it is not the task of physics to explain, although the peculiarities of "original time" very much find expression in physics.

Heidegger sees how strong opposition to philosophical reflection in his sense must be. His questions go back to a place that remains foreign. They are therefore all too easily mistaken for unfounded systems of thinking, which according to an analogy of Kant's resemble a "fragile dove" that wants to rise above on supporting air, without recognizing the risk of a sudden fall. Nevertheless, it is absolutely clear that Heidegger felt himself obliged to maintain an exceptional discipline in thinking; and that he thus has much to say to modern science, which he does not oppose as such.[11] Today it is important to focus increasingly on the misuse of science (see the previous section). Such focus is at the same time productive with respect to the problem of laying a foundation for the sciences, and their future development, which can occur without the danger of the "gigantic in machination" only on the ground of what doubly "cannot be gotten around" (see the second section of this chapter).

NOTES

1. Cf. Ewald Richter, "Heideggers Seminar in Wellingsbüttel" *Heidegger Studies* 16 (2000), 221 ff.

2. Cf. Trish Glazebrook, *Heidegger's Philosophy of Science* (New York: Fordham University Press, 2000). The research in this book makes absolutely clear that essentially philosophical insights about the relationship between "the essence of modern technology" and an original understanding of "nature" can be taken from Heidegger's decades-long confrontation with modern science.

3. Friedrich-Wilhelm von Herrmann, *Hermeneutik und Reflexion* (Frankfurt am Main:, 2000), 39 ff.

4. Von Herrmann (2000), 26.

5. Paul Lorenzen, *Methodisches Denken* (Frankfurt am Main:, 1968), 26 ff.

6. The comment regarding "second nature" appears in a footnote at GA 7, 56, but not in VA.

7. On these attempts, cf. Ewald Richter, "Truth and Logic" in Ewald Richter, ed. *The Question of Truth* (Frankfurt am Main:, 1997), 125 ff.

8. Cf. Hermann Weyl, *Philosophy of Mathematics and Natural Science* (München:, 1927), Section Two, II, 20, (concept formation).

9. Cf. Weyl (1927), Section One, II, 9.

10. Carl Friedrich von Weizsäcker—departing from the peculiarities of the quantum theory revealed in the double-slit experiment—suggested the concept of a "temporary logic." Cf. Carl Friedrich von Weizsäcker, *Time and Knowledge* (München: Carl Hanser Verlag, 1992), 192 ff. and 743 ff.

11. In the following publications, the author has attempted to stress the significance of Heidegger's work for modern science, with special consideration to the "belonging together" of thinking and being: Ewald Richter, *Heideggers Frage nach dem Gewährenden und die exakten Wissenschaften/Heidegger's Question about the Offering and the Exact Sciences* (Berlin: Duncker & Humblot, 1992), *Ursprüngliche und physikalische Zeit/Original and physical time* (Berlin, a. a. O., 1996); "Heideggers Kritik am Konzept einer Phänomenologie des Bewußtseins"/"Heidegger's critique of the concept of a phenomenology of consciousness" in P.-L. Coriando, ed. *Vom Rätsel des Begriffs/On the Mystery of the Concept*, Festschrift in honor of Friedrich-Wilhelm von Herrmann, Berlin, a. a. O., 1999), 7 ff.

III

SCIENCE AND THE HUMAN EXPERIENCE

FROM ANIMAL TO DASEIN

Heidegger and Evolutionary Biology

Lawrence J. Hatab

A significant element of Heidegger's thought is its attention to modern science and its concurrent expression in modern technology. Here the emphasis is primarily on the physical sciences. In this chapter, I ask if modern biology raises comparable and perhaps different issues of philosophical interest. No discussion of modern biology can ignore its complete reliance on evolution theory, which of course raises fundamental questions about the relation between animal life and human life, between nature and culture. In *The Fundamental Concepts of Metaphysics* (GA 29/30), Heidegger offers a surprisingly extensive and detailed discussion of biology in trying to differentiate animal life and the phenomenological concept of Dasein's world. Taking heed of this discussion, I want to explore possible convergences between Heidegger's early phenomenology and evolutionary biology, and how Heidegger's phenomenology can readily address certain puzzles and reductive excesses that have commonly marked evolution theory. In general terms, I believe that Heidegger's thought can help open up and clarify central philosophical questions that arise at the intersection of biology and culture.

I

At the outset it must be stipulated that Heidegger challenges scientism, not science, that he critiques metaphysical realism (the world is the way it is *independent* of our approach to it), not objectivity. For Heidegger, objectivity is a real possibility that emerges out of a background of meaning structures that themselves cannot be reduced to objective "independence." Objectivity is born out of the *vorhanden* perspective, which itself arises out of the *zuhanden* perspective, when disturbances to familiar competencies and practices open up a disengaged look on things and their properties simply there before our attentive gaze. For Heidegger, both *vorhanden* and *zuhanden* perspectives are genuine modes of disclosure (cf. SZ, 55/82; 88/122; 138/177). The problem is that modern philosophy has taken scientific objectivity—which is a highly sophisticated extension of *vorhanden* thinking (SZ, 361–62/412–13)—as the primary, even exclusive, measure of being. Heidegger challenges this reduction by showing how objective categories arise out of a prior horizon of practical, temporal, meaning-laden care structures.

For Heidegger, all understanding is born out of Dasein's existential possibilities, the concrete movements of Dasein's tasks of concern. Interpretation arises out of understanding, as the articulation of Dasein's possibilities and concerns. Different fore-structures of understanding emit different as-structures of interpretation. Even science—which takes the world *as* a set of objects divorced from human involvement in order to uncover causal explanations—presupposes a host of meaningful possibilities and engagements: for example, an *interest* in causal explanations, the habits and competencies of scientific practice, the adventure of science as a quest to uncover the secrets of nature, which delivers the mind from ignorance and allows technological manipulation of nature. For Heidegger, even the most "exact" sciences are not divorced from existential concerns; they simply reflect the *narrowing* of the scope of an existential base (SZ, 153/195). Differences in disciplines reflect the *extent* of existential concerns and how the different regional ontologies open up the world in the light of these concerns. Even the "hard sciences," then, are interest-laden, and therefore not purely "objective" in the manner of metaphysical realism.

Heidegger's phenomenology does imply a certain ordering of disciplines. Because scientific objectivity and causal thinking do not as such articulate the meaning structures imbedded in their enterprises, then a philosophical language laying out those structures will have a certain priority, owing to its comprehensiveness. This is simply a phenomenological arrangement that is *not* a denial of, or replacement for, causal thinking. Causality can have its place. As Heidegger says, being is shown in many ways, according to different kinds of access we have to it (SZ, 28/51). Different possibilities prompt different questions that open up different modes of interpretive access (Why does my stomach hurt? Why do I love philosophy?). Heidegger's hermeneutics does not suggest that knowledge is arbitrary or up for grabs. Interpretation should be *appropriate* to phenomena in question (SZ, 148–50/188–91). We can say that modern causal explanations are fully appropriate to causal questions in the realm of natural processes. The problem concerns a reductionism that presumes to apply causal explanations to any and all questions. With Heidegger we can say that the disclosure of meanings and purposes, which have been specifically exiled in a modern metaphysics of nature, can be given an ontological status that avoids the modern tendency to "subjectivize" such phenomena. Art, science, religion, history, and ethics can coexist as *appropriate* disclosures of different world regions. In this respect, Heidegger can be called a kind of pluralistic, phenomenological realist.

The important point is that Heidegger's phenomenology amounts to neither a subjectivism nor a denial of all senses of objectivity. As Heidegger says, the "objectivity" of a science is given in terms of how it *uncovers* beings in its sphere of questioning, and thus it is nothing "subjective" (SZ, 395/447). The same is said with respect to truth (SZ, 227/270). Disclosure does not "create" truth, as though it were nonexistent prior to its discovery: "Once entities have been uncovered, they show themselves precisely as entities which beforehand already were" (SZ, 227/269). The "beforehand" suggests a kind of realism, but the "uncovering" that occurs by way of Dasein's concernful being-in-the-world undermines the strict sense of "independence" implied in metaphysical realism (GA 26, 194/153). Heidegger even says that the existence of Dasein is dependent on the extant presence of physical nature (GA 26, 199/156). From a philosophical standpoint, however, the meaning of being cannot be grounded in an objective ontology of nature because the very disclosure of such a setting presupposes the meaning structures of Dasein's being-in-the-world. If the *question* concerns Dasein's physical placement in nature and relevant causal investigations, then natural science can do its job. Yet,

> the existence of the material things of nature is not the only existence; there are also history and art works. Nature has diverse modes: space and number, life, human existence itself. There is a multiplicity of *modi existendi*, and each of these is a mode belonging to a being with a specific content, a definite quiddity. The term "being" is meant to include the span of all possible regions. (GA 26, 191–92/151)

> Plants are something other than geometrical objects, while the latter are completely different from, say, a literary work. . . . The determinative thinking which is to measure up to the particular being in question must also take into account a corresponding diversity regarding what and how the being in each case is. The thought determination, i.e., the concept formation, will differ in different domains. . . . The logic of these disciplines is related to a subject-matter. It is a *material logic*. (GA 26, 2/2–3)

The objective categories of modern science, then, are genuinely disclosive, but they tend to miss or suppress how Dasein, even in its scientific orientation, inhabits its world, an engagement captured in the term "dwelling" (SZ, 54/80). Moreover, Dasein's practical immersion in its lifeworld is animated by a temporal openness, by way of the dimensions of past and future that exceed conditions of immediate actuality and that extend Dasein's world immeasurably beyond extant conditions. Dasein's being-in-the-world, then, can be called an engaged, situated openness, which is both at home in the world and open to possibility. The meaning structures intrinsic to Dasein's engaged openness are lost when a reductive naturalism overextends scientific categories to cover questions that exceed the appropriate scope of natural science.

II

An important question concerns whether and how we can correlate scientific and nonscientific orientations because we should not want to say that they are utterly separate tracks of understanding. Of the natural sciences, it is biology that most intimately connects with human concerns, since the science of life addresses our capacities as living beings. In fact, biology exhibits a hermeneutic circle, as Heidegger says, because as living beings our inquiry into life presupposes a certain preconception of life (GA, 29/30, 266/180). In *Being and Time*, Heidegger indicates that Dasein can be understood purely as a form of life (SZ, 246/290), but including Dasein with other life forms constricts our understanding. Indeed, Heidegger claims that life as such can only be understood by way of Dasein (which would be consistent with the hermeneutic circle), and then "privatively" extended to other life forms (SZ, 49–50/75; 58/85), in other words how they are differentiated from Dasein.

Shortly after the publication of *Being and Time*, Heidegger, in *The Fundamental Concepts of Metaphysics*, offered extended reflections on biology and the phenomenon of life, in the context of trying to understand the meaning of "world." Heidegger indicates that animal life and findings in biology are "instrumental," even "indispensable" in addressing the philosophical question of world. Accordingly, he departs from his earlier notion that animal life can only be understood as a privation of Dasein's mode of life; zoology will help in coming to understand animal life in its particular essence (GA 29/30, 310/211–12; 389/268). He also interjects that the question of human descent from apes must be held off until the very nature of animality and humanity can be clarified and distinguished (GA 29/30, 265/179).

Heidegger begins by drawing distinctions between material objects as "worldless," animals as "poor in world," and humans as "world-forming"

(GA 29/30, 263/177). The world-poverty of animals is not a biological concept based in biological research, which is an appropriate investigation of facts about animals (GA 29/30, 274/186); it is a concept that speaks of the "essence" of animality, which is the presupposition of biological investigation (GA 29/30, 275/186). Essential philosophical concepts open up the possibility of a science as such and so cannot be drawn from that science (GA 29/30, 275/187). Yet even though the world-poverty of animals cannot be derived from biology, it cannot be understood *independently* of biology either (GA 29/30, 275/187). The relationship between philosophy and positive scientific research is "ambiguous" and in any case it cannot be broken up into separate spheres (GA 29/30, 277–79/188–89). The relationship between philosophy and science requires an "inner readiness for communal cooperation" (GA 29/30, 280/190). Such cooperation is hindered by 1) a hyper-sophistication in philosophy that sees itself as superior to the sciences owing to its universality, and that leads it to lecture science "in a supercilious manner," and 2) an intransigence in the sciences owing to a stubborn appeal to "facts," forgetting that all facts presuppose interpretation. In this way both philosophy and science talk past each other and adopt the "spurious freedom" of leaving each to its own devices, a freedom actually based in insecurity (GA 29/30, 280–81/190). Heidegger calls such segregation a "danger." Both philosophy and science can be understood as part of Dasein's existence and historicity (GA 29/30, 282/191).

For Heidegger, each science is historical in that the fundamental way it conceives itself goes through transformations (GA 29/30, 277/188). What Heidegger finds particularly favorable in the biology of his day is the tendency to rescue the understanding of "life" from the "tyranny" of chemistry and physics (GA 29/30, 277–78/188), particularly in the direction of nonreductive conceptions of an organism (GA 29/30, 311/212). Heidegger thinks that recent biological research can be relevant in exploring the philosophical question of world and the world-poor condition of animals (GA 29/30, 284/192). The difference between animal world-poverty and human world-forming is one of *degree*, as in rich and poor (GA 29/30, 285/193). The world is accessible to animals, but in a limited and circumscribed manner. The bee is familiar with blossoms, their color and scent, but knows nothing of the nature of plants. The world of humans is rich in its greater range, particularly in being "constantly extendable," hence the notion of world-*forming* (GA 29/30, 285/193).

Heidegger interjects that poor and rich should not be understood as a hierarchical evaluation in terms of higher and lower. Some animal capacities are far superior to humans, and humans can "sink lower than any animal" in depravity. Even the amoeba is "complete" in itself, and hence not "imperfect" (GA 29/30, 286–87/194). Poverty means being deprived of something, rather than lacking something that can or should be present (GA 29/30, 287/195). Heidegger's text is not clear on the difference between the animal and human world. In comparison with a stone, an animal has a world, but in comparison with humans, animals lack world. Indeed, the world of animals is *not* a degree or species of the human world (GA 29/30, 293–94/199–200). I think that the radical openness of the human world can support this point, but the denial of degree conflicts with a previous claim in the text. Heidegger mentions our cohabitation with domestic animals, saying that we live *with* them but they do not share our sense of world (GA 29/30, 308/210). In some respects, I think this is true, in other respects it is highly debatable; just ask any pet lover. Later in the *Zollikon Seminars*, Heidegger puts the point more clearly:

> . . . an animal only *is* insofar as it moves within an environment open to it in some way and is guided by this environment which itself remains circumscribed by the nature of the animal. The animal's relationship to this environment, which is never addressed [by the animal itself], shows a certain correspondence to the human being's ek-sistent relationship toward the world. Thus, in a certain way the human being in his ek-sistent Dasein can immediately participate in and live-with the animal's environmental relationship without ever coming to a self-disguising being-with the animal, or vice versa. . . . According to its own proper and essential relationship to the environment, the animal's situation makes it possible for us to enter into this relationship, to go along with it, and, as it were, to tarry with it. But it is not enough to consider only that. It remains far more essential to see that an animal (as opposed to a rock) shows itself to us only then *as* an animal insofar as we humans as ek-sistent have *engaged in advance* the relationship to the environment proper to the animal.[1]

Returning to GA 29/30: A stone is worldless. It lies on the earth but does not "touch" it. For a lizard basking in the sun on a rock, the rock is "given" to it in a certain way (remove the lizard and it will seek to return to the rock). A stone is worldless in having no access to beings. The lizard does not know the rock *as* a rock in any factual sense, but the manner of its being with the rock shows some kind of access (GA 29/30, 291/197). What seems to distinguish animals from humans is as follows: The animal has a "specific relationship to a circumscribed domain," with respect to nourishment, prey, enemies, and sexual mates. So the animal has access to the world in which it moves, an access that is specific, appropriate, and not arbitrary. But the animal's environmental world is confined, a "fixed sphere that is incapable of further expansion or contraction" (GA 29/30, 292/198).

Heidegger maintains that a living being must be understood as an organism in a holistic sense rather than as a collection of parts (GA 29/30, 311/212). An organism is neither an instrument (*organon*) nor a mechanism (GA 29/30, 313/214); it is essentially a movement (GA 29/30, 317–18/217). Organisms are different from machines in being self-produced, self-regulating, and self-renewing (GA 29/30, 325/222). Heidegger adds that there *can* be mechanical explanations of an organism's movements, but such explanations are insufficient for capturing the essence of an organism's movements as such (GA 29/30, 318/217). Heidegger praises developments in the biology of his day that recognize the holistic character of organisms (GA 29/30, 379–80/261), something already recognized by Aristotle.[2] The self-organizing and self-directing nature of an organism will not abide strictly mechanical or atomistic explanations. A kind of "reverse engineering" can be explicative of various functions and processes, but it misses the phenomenon of the organism as such. If one did not know what an organism is ahead of time, atomic explanations could not get you there (if you did not know what a cake is, no molecular or chemical "assembly" could get you there). In the language of *Being and Time*, the fore-structure of understanding an organism, and of *being* an organism, determines the as-structure of how the parts of an organism are interpreted.

Heidegger continues that an organism's "organs" cannot be understood as "instruments," but as potentialities and capacities incorporated into the organism (GA 29/30, 319–20/218–19). An organ's capacity is different from an instrument's potentiality. A pen, for instance, has readiness for writing, but not capacity (GA 29/30, 322/220). Heidegger then says that it would be wrong to see an organism as simply an aggregate of capacities. Organs, rather, arise out of the organism's capacity (GA 29/30, 323/221); they need to be understood in terms of the "central direction" of the organism (GA 29/30, 326/223). To make his case, Heidegger draws on something compatible with evolution theory. Single-cell organisms like the amoeba show that a capacity precedes the development of fixed "organs" (GA 29/30, 326–27/223). The development of a mouth and intestines would follow *from* a basic capacity to take in nourishment, which in the amoeba is a generalized fluid capacity (GA 29/30, 327/224). The well-developed and fixed organs of higher animals lead us to think of organs as *vorhanden* "instruments." Simple organisms show us that organs and specific functions must be seen as emerging out of the "way of being" of an animal called *life* (GA 29/30, 329/225). Organs belong to the capacity of an organism to live and are "incorporated" into the organism in a subservient manner (GA 29/30, 331–32/227). The capacity of organs is, again, not simply readiness; a hammer is ready for hammering but it is not an "urge toward" hammering (GA 29/30, 331/226). Organic capacity is only found where there is *Trieb*, or drive (GA 29/30, 334/228).

The self-driving of an organism is primal, preceding an analytical breakdown into parts. Drive has a "dimensional" structure, understood as a kind of traversing, a "self-driving toward its own wherefore" (GA 29/30, 334/229). That is why an organism cannot be taken as a mechanism, because there can be no mathematical expression for this dimensional structure of drive, especially when a drive is "anticipatory" (GA 29/30, 334–35/229). Drive is never *vorhanden* because it is a "toward" (GA 29/30, 335/230). And because of this toward-structure, an organism's environment is essential in understanding animal capacities. Developed organs in a way are "between" the organism and its environment (GA 29/30, 337/231). The life of an organism is a self-moving comportment-toward its environment (GA 29/30, 345–46/237). Heidegger says that the self-movement of an animal is its "selfhood" in terms of "being proper to itself," and not as a mode of "consciousness" or self-reflection (GA 29/30, 339–40/233).

Heidegger then distinguishes animal and human comportment by saying that animal behavior is an "absorption in itself" called "captivation" (GA 29/30, 347/238–39). Animals are captivated or "given over to" their capacities and environments (GA 29/30, 350–58/241–46). The animal does not apprehend things *as* present or absent (GA 29/30, 354/243), and so it does not stand within the manifestness of beings as beings (GA 29/30, 361/248). And yet, as a relatedness-toward its environment, the animal must be understood as a kind of openness-for its milieu (GA 29/30, 361/248).

Heidegger describes animal behavior as a set of instinctual drives within a "disinhibiting ring," an environment that instigates instinctual behavior (GA 29/30, 362–74/249–57). This is a circular structure that cannot be divided into separate conceptions of stimulus and response; in order for a stimulus to work, an organism must be originally related to, predisposed toward, its environment. For this reason, Heidegger claims that the Darwinian notion of self-preservation, although understandable, is misleading, because it misses the encircling structure of animal behavior; the struggle to survive and reproduce presupposes an essential relatedness to the world (GA 29/30, 377/259).

Heidegger praises developments in biology that associate with ecology (GA 29/30, 383–84/263–64). Ecology derives from the Greek *oikos* or house, and it considers how animals are at home in the world, conceived as an *Umgebung*, an environment (GA 29/30, 382/263). Heidegger interjects that this is the strength of the Darwinian conception of adaptation, and yet it conceals the environmental structure of the organism by suggesting a *vorhanden* entity that adapts "to" its environment. It would be better to say that an organism adapts its environment "into" itself, because it is the fundamental openness-to-the-environment as such that permits not only responsiveness but also the "leeway" that allows for adaptation (GA 29/30, 384–85/264). So the "captivation" of the organism is not something rigid or static (GA 29/30, 385/265). It should be noted that this idea of a responsive openness is indeed recognized in evolutionary biology with the notion of an organism's "plasticity," without which evolution could never get off the ground.

Comparing animal and human life, Heidegger maintains that the animal's encircled captivation does not allow the opening up of beings as such, the letting-be of beings (GA 29/30, 367–68/253). In fact, animal life is mainly struggle and is "eliminative" in the sense of devouring and avoiding (GA 29/30, 363,374/250,257). Finally, animals in their captivation can be said to die in the manner of coming to an end, but unlike humans they are not aware of death as such. And awareness of the negativity of death is correlated with appreciating the value of life, a clue to how humans are open to the meaning of being as such by way of a negative dimension (GA 29/30, 387–88/266–67). Moreover, profound boredom (analyzed in great detail in this text) is counterposed to the captivation of animals (GA 29/30, 409/282). Boredom, like *Angst* and being-toward-death, shows how Dasein is not strictly bound to its world, but fundamentally open to otherness. And this finite openness to otherness is precisely what animates the context of meanings that characterize the human "world," which Heidegger summarizes in general terms as follows: the manifestness of beings as beings, letting-be, not letting-be, the "as," comportment-toward, and selfhood (GA 29/30, 397–98/274). Most specifically, it is language that accomplishes the human openness to beings as such (GA 29/30, 376/259). Indeed, sections 68 to 78 of this text offer an extensive analysis of language as the opening up of world and the world-forming character of human life.

Animals have access to beings, but not *as* beings (GA 29/30, 390–400/269–76), and in this sense animals have "no world" (GA 29/30, 391/269). Such not-having, however, is still in relation to a having; its "poverty" is different from the worldlessness of the stone. And yet animals do not have "less" of a world than humans, they do not have a world at all (GA 29/30, 392/270). This aspect of Heidegger's analysis, as I have indicated earlier, is not clear. How can an animal not have a world and yet not be worldless? In some respects, which I address shortly, the lack of world in animals can be defended. But in other respects I am not so sure that animals can be divided from human experience in the way that Heidegger suggests (especially because he had said that the difference between the animal and human world is one of degree). Is it possible to talk of animals "not having" a world as a kind of "concealment" opened up by Dasein's disclosedness? Consider how we can come to understand the world and history of animals in a way that they cannot; such understanding can even lead to our improving their lot, as in the case of veterinary science. This disclosive connection between humans and animals would allow us to address a number of issues: 1) the ease with which we tend to analogize animal behavior to our own (and vice versa); 2) the ways in which animal and human lives can come to affect each other intimately; and 3) the evolutionary links that are the presupposition of modern biology. At any rate, Heidegger does admit that the world-poverty of animals remains a provisional and problematic thesis (GA 29/30, 396/273).

III

No one can rightly ignore evolution theory and its intellectual ramifications. Evolution is the core of modern biology and it offers an alternative to supernatural or transcendent explanations in accounting for life, its variations, and the emergence of human life. Humans are continuous with, and arise from, a natural process of evolving forms, a process explained by natural selection, where a combination of random mutations, varying environmental conditions, and sexual selection produce new adaptations over vast stretches of time, succeeding in complexity up to the higher mammals and finally to humans. To be precise, evolution is not a theory; the fossil record, which indicates an invariant order in the movement of simple to complex organisms, shows that evolution has in fact happened. For Darwin, natural selection was the theory that explains evolution; and today, the so-called modern synthesis has joined genetics to Darwin's theory to account for evolution. Scientific debates and controversies concerning evolution (which creationists have tried to exploit) are not about whether evolution happened, but how. It seems clear that evolution theory stands as the "best explanation" for the natural facts at our disposal.[3] The real trouble starts when we realize that if evolution is true, then somehow human "culture" emerged out of "nature," and one does not have to be a religious fundamentalist to see stresses between these two spheres. I believe that Heideggerian phenomenology can be brought to bear on this question in productive ways. Heidegger's finite environmental structures can be compatible with, and even assist, the findings of evolutionary biology in trying to make sense out of organic life forms. At the same time, his phenomenology can show how culture "transcends" nature (narrowly construed) but without transcending the earth (broadly construed). By this I mean that cultural phenomena are not directly translatable from physical and biological conditions and yet that this "transcending" aspect of culture is nothing transcendent; in fact it is "natural" in the sense of being *intrinsic* to Dasein's being.[4] In other words, Heideggerian phenomenology can show what is wrong with a reductive naturalism without appealing to what Daniel Dennett calls "skyhooks," namely transcendent, dualistic, religious, or ad hoc explanations.

Heideggerian phenomenology and evolution can complement each other in a number of ways. Both can be called fully worldly and earth-bound orientations. We might call evolution an account of being-in-the-natural-world, and we should understand part of Dasein's "thrownness" in terms of its biological inheritance. Both Heidegger and evolution advance dynamic models of the world constituted by time, change, movement, tension, and finitude; in these respects both speak against metaphysical essentialism.[5] It seems to me that evolution theory should find much more affinity with some of Heidegger's phenomenological concepts than with the discrete, atomistic, mechanical categories inherited from modern science. The environmental structures of being-in-the-world, thrownness, zuhanden competence, affective attunement (Befindlichkeit), are all more appropriate for understanding biological adaptation and development than are the more crudely delineated push–pull, stimulus–response models common to modern science. Evolution presupposes that organisms be conceived as "dwellers," predisposed to respond to their milieu, and adaptive to the practical needs and problems of navigating a tensional environment. Heidegger's work in drawing attention to prereflective competencies in Dasein's existence can be extended to the rest of the lifeworld, too, with the advantage of not having to rely on mentalistic metaphors that confuse more than they illuminate. Evolutionary "capacities" and "adaptations" can be construed as modes of nonreflective "know-how" that in their way are "intelligent" and thus not utterly blind.[6] At the same time, such capacities need not be analogized to conscious intentions and concept formations inherited from the modern epistemological subject–object binary; so we can be suspicious when we hear of evolution "creating" this or that, organisms "aiming" for this or that, genes "directing" their organisms, an ape's "concept" of this or that, and even Dennett's presumed alternative to strict cognition, namely "engineering."[7]

An ecological, organic holism (suggested by Heidegger) can avoid crude reductions that constrain evolutionary biology. It is not that atomic or mechanical explanations are false, but that they are insufficient in accounting for organisms as self-organizing wholes.[8] This applies particularly to genetic reductions, as in the infamous notion of the "selfish gene."[9] Here the exclusive unit of natural selection is the gene, so that an organism's behavior is simply in the service of reproduction. Since life began at the level of genetic DNA, all subsequent manifestations of life, including cultural constructs, presumably must be at bottom nothing more than epiphenomena in the mindless process of gene replication. Such a model can be apt in analyzing statistical tendencies and long-term population patterns. It also seems intuitive that whatever life form presently exists must have been made possible by reproductive fitness patterns in the past. But reducing the behavior of an organism to the dictates of the gene pool misses the complex nature of organisms in relation to their environment. Some genetic traits are stable, but others are flexible in being modifiable by environmental variations. The plasticity of a phenotype organism in fact contributes to the furthering of the organism and hence to the long-term development of the genotype. So the behavior of a life form and its evolutionary trajectory must be understood in terms of the complex intersection of genetic and environmental factors, as well as new factors that emerge by way of the interaction of genetic and environmental factors, particularly when the organism modifies its environment in such a way as to further its fitness.[10]

Even if behavior were rigidly determined by genetic traits, we would want to correct conceptual mistakes and bad metaphors that often prevail in genetic reductionism: that genes are "using" their organisms for their own "interests," that behavior "serves" reproduction, "in order to" further the genetic line. It would seem more accurate to say that an organism's present behavior is following genetic traits that *happen to have been selective* in the past. If there were a radical change in the present environment, so that the traits were now unsuitable, we can see the mistake in talking of the primal "interest" of the genes "for which" the organism is acting. What we would have now are "stupid" genes or "suicidal" genes. The point is not to deny the role and force of gene replication; such explanations can have value as extensive statistical surveys of long-term populations. Rather, the issue is that such accounts bypass the actual behavior of the organism as such, as well as its plasticity. What is gained by saying of an organism that this maneuver or that desire or this function or that activity are "really" only surface strategies for its genes' need to replicate? Or what would be lost by saying that the behavior is in part a function of traits that happen to have been successful in the past and leave it at that, giving attention now to the complex interactions of the organism's inherited capacities, behaviors, ventures, and environment, with all these features taken together as a more appropriate account of the "being" of this life form? Since such an account would not ignore genetic factors, objections would seem to arise only out of an analytical fetish for atomic explanations.

IV

A crucial legacy of modern science has been, in Weber's phrase, the "disenchantment" of nature. Where Aristotle had conceived cultural meanings, purposes, and values as intrinsic to reality, modern science's mathematical base and disengaged objectivity stripped the world bare of such meanings and demoted them to mere subjective occasions.[11] Surely things are worse with sociobiology and the selfish gene. The reduction of life and culture to the blind process of genetic replication that surges forth simply for its own sake would leave us with a nihilism that only a Schopenhauerian pessimism could address with any authenticity.

Daniel Dennett has tried to mollify this pressing problem while keeping in line with the basic orientation of evolutionary biology. I agree with Dennett that evolutionary naturalism rules out essentialism, global teleology, and mind-first cosmologies (what Heidegger would call ontotheology).[12] And Dennett wants to make room for cultural meanings within an evolutionary viewpoint. But he does not persuade in posing the following (false) choice: Either we account for culture in a manner consistent with evolutionary science or we have to resort to some *deus ex machina* skyhook.[13] Dennett calls for a reductionism that is not "greedy" in canceling out culture in favor of blind forces.[14] Yet it is still reductive in remaining consistent with the basic "algorithm" of natural selection that is "substrate neutral" in being applicable to all phenomena, from nature to culture.[15] Following Dawkins, Dennett accounts for culture as an array of "memes," cultural constructs such as ethical norms that emerge through language out of biological processes and then "evolve" in a way analogous to genetic dynamics in natural selection. So memes compete for adaptive fitness and their organisms (humans) serve as carriers for their replicative strategies.[16] In this way we can have both nature and culture on a continuum without the need for skyhooks.

One of Dennett's main strategies is to align evolution with Artificial Intelligence (AI), to demonstrate a strictly naturalistic alternative to skyhook explanations of mentality.[17] But a significant challenge to strong AI is the idea, going back to Aristotle, that what we know as intelligence is a function of living, desiring beings. Heidegger shows how phenomena such as curiosity and affective dispositions operate in "knowing" the world, and how these phenomena are better rendered in the dimensional structure of a striving being attuned to its environment. It is hard to understand how an algorithmic computer program can capture the "absences" (wonder, denial) and the strivings (curiosity) that seem so indigenous to world-disclosure. Of course computers are "intelligent" in their operations, but they would not be intelligent in the full existential sense opened up by Heidegger. If intelligence is intrinsically a function of life, it is hard to understand how life's structure as a holistic self-emerging-drive-toward-and-amidst-its-environment could be replicated by artificial constructions that must *begin* with piecemeal components that lack such a structure.[18]

Dennett's position on memes runs into a number of problems. Even though a memetic theory might have some interesting correlations with structuralism, nevertheless the insistence on a close analogy with genetic natural selection is suspect.[19] First of all, cultural transmission of memes needs no connection with genetic transmission. And memes are "directed mutations" that emerge out of the interests of human organisms, thus disrupting the analogy with blind genetic mutation and replication. Finally, memes pass along acquired traits, and so they commit the Lamarckian fallacy so anathema to Darwinism.[20] But the most vexing problem, I think, is this: Dennett understands that to avoid greedy reductionism one must give cultural constructs their due in their own terms in some way. Ethical memes, for instance, cannot be explained simply in terms of gene replication. Although genetics would be the *ultimate* historical source of present moral values, that does not mean that gene replication is the *proximate* source or *beneficiary* of ethical practices.[21] Dennett grants that the emergence of memes in humans allows us to "transcend," "overpower," and "escape" biological constraints in important ways.[22] In other words, culture allows us to *exceed* nature in some ways, and yet it seems to me that Dennett simply stipulates this as an emergence out of a biological base, without much attention to *how* it is so and without thematizing the "excess" as such.

This is where Heidegger's thought makes an important contribution. The phenomenology of Dasein's immediate lived experience is able to show that so-called cultural constructs come first, before the uncovering of so-called objective conditions in nature. With Dasein's ek-static structure of being-in-the-world, where meanings constitute Dasein's existence all the way down, where Dasein *finds itself* in a world saturated with existential significance, we can say that cultural meanings are *there* in Dasein's world (rather than being mere subjective projections) and *given* to Dasein in immediate, concrete ways (rather than through transcendent sources). Most importantly, Heidegger thematizes the temporal finitude of Dasein's world, especially in terms of how "negative" experiences (practical disruptions, anxiety) figure in the generation of meaning and sustain an openness beyond extant conditions. In other words, Heidegger tries to articulate how our understanding of the world involves correlations of absence and presence, otherness and being. Analytic philosophers often ridicule Heidegger for his linguistic ventures, but in fact he is trying to bring into words the existential background for much of the very things analytic philosophers simply take for granted uncritically: for instance, Dennett's stipulation that culture "exceeds" nature, or the common notion that ordinary human rationality involves deliberation that requires recollections of the past and anticipations of the future, in other words, giving presence to absences and exceeding the present. Heidegger's portrayal of Dasein's situated openness is able to offer a concrete, viable account of what Dennett wants out of culture without a reduction to extant conditions of nature and without an ascension to transcendent sources.

V

From a physical and biological standpoint, I have no doubt that the human animal emerged out of an evolutionary continuum, that the human *animal* evolved from other animals. Yet if we consider humans as cultural animals, there are conceptual problems haunting an evolutionary "continuum" of nature and culture, which presupposes empirically discernible causal connections. Although I have some problems with Heidegger's division between human and animal "worlds" (or the lack of world in the latter), I think his distinction between the world-poor condition of animals and the world-forming condition of humans is phenomenologically evident, and at certain levels the *openness* of the human world displays a kind of "break" that eludes empirical descriptions or causal explanations.

Culture is what (apparently) distinguishes humans from other animals and is indicated in such phenomena as art, religion, morality, technology, science, and especially language. I take cultural phenomena and natural descriptions to be incommensurable at certain levels. So although a biological continuum is evident and makes sense, when considering the emergence or presence of culture, the notion of a continuum runs into trouble. Here I am not referring to specific products of culture, but (in a Heideggerian fashion) to the background meanings and conditions of cultural phenomena. Religion speaks of alternative dimensions beyond ordinary experience. Philosophy arises out of wonder, perplexity, and questions. Art fashions new or altered images beyond brute encounters with given things. Morality involves the modification of brute behaviors. Technology transforms physical nature. And abstract thinking, one consequence of which is science, involves concepts and structures that reach beyond immediate experience and perception.

It is evident that the atmosphere of such cultural phenomena (e.g., questioning, evaluating, creating, conceiving) presupposes a "transcendence" of the brute given or the immediate present. Culture as such exceeds extant (*vorhanden*) conditions. But then explaining the emergence of culture by means of extant conditions (e.g., as a special organization of material, biological, environmental, or behavioral states) makes no sense (phenomenologically). Put it this way: Culture cannot ultimately be accounted for because "giving an account" is a cultural phenomenon. Science too is a cultural phenomenon that exceeds the brute given. Seeing the world *as* a set of empirical conditions calling for causal explanation cannot itself be derived from empirical or causal conditions. This is why Heidegger insisted on differentiating causal explanations and phenomenological concepts. The hermeneutic circle shows that there is a prediscursive atmosphere of understanding (a *Vorhaben*) that makes possible various orientations toward objects, but which itself, therefore, cannot be an "object" of discovery. This circle forever frustrates attempts to advance strictly objective explanations of Dasein's world. The existential background of knowing cannot be put in the foreground because "putting" is a background operation. The background can *show itself* but it cannot be "explained" in the sense of a causal or logical trace to extant conditions or concepts.

There certainly are evolutionary, biological links between humans and other animals, especially our primate cousins. I think we can make sense out of certain quasi-cultural links as well, for instance, concerning certain social patterns and behaviors. It is easy to picture a clear continuum of ape and human posture, of ape and human locomotion, of ape and human manual dexterity, with the requisite approximations of end results. But it is difficult to understand how, for example, perceiving or remembering something can approximate or "lead to" painting it on a cave wall, or how witnessing the expiration of life can approximate or lead to burial customs, with their implicit sense of a "beyond."

The notion of a "culture gap" between humans and primates is not beyond dispute, of course. When we see a gorilla deftly trimming and using a twig to draw out insects from inside a log, we can clearly imagine a continuum in the direction of human tool making and use. But there is still a difference, I think, between such an example (which could have begun with a fortuitous accident followed by repeating, learning, remembering, even teaching the use) and tool *making* as a cultural phenomenon, a creative orientation beyond extant condition, that sees the world *as* "transformable." In Heideggerian terms, the human world is marked by a radical openness and differentiated otherness that does not seem to be evident in the animal realm. Human culture is more than life, it is living-for, coming-from, heading-toward, living in the midst of death, limits, and possibilities, all of which is a surpassing of actuality. For Heidegger, Dasein *is* a surpassing (GA 9, 138–39). This surpassing element is what distinguishes the human world from what Heidegger calls the encircling ring of animal horizons.

VI

I close with a brief consideration of perhaps the most important topic of all in this matter, the question of language, which unfortunately I cannot address adequately within the limits of this chapter. It is common to see language as the distinguishing mark of the human species. In fact, without language there would be no human culture at all. This is the spirit of Heidegger's remark that language is the "house of being" and that humans dwell in this house (GA 9, 313). Language is the environment in which the world opens up for human beings, and it too exhibits a circularity that eludes explication. Any attempt to "explain" language or connect it with prelinguistic elements must employ language to do so. Even "nonverbal" comprehensions or experiences bank on having been oriented into a language-laden environment from the first moments of life.

I would like to suggest ways in which language can be implicated in the openness that Heidegger insists marks the human world. To begin, as suggested above, language is the very shaping of the human world from the start. And in everyday linguistic practices and exchanges, we take language to be spontaneously disclosive of things. In direct conversations I simply understand immediately the disclosive effects of speech without marking a difference between the speech and its reference. Or if someone is verbally helping me learn a practical task, guiding my actions and pointing out aspects of the practice as we go along, I am immersed in this disclosive field without noticing "words" as distinct from "referents" (or puzzling as to how words relate to their referents). In this respect there is a "fit" between language and the world. And yet, there is a *difference* between verbal utterances per se and the subject of utterances, and it is this differentiated relation that makes possible the openness of language, particularly in terms of temporality. Words give a presence to things that can be retained in the absence of things. It is this presencing of absence that makes possible the stretches of temporal understanding that far exceed any primitive time sense that animals might possess. With words I can retain the past and project the future in vivid detail, and I can be released from the actual by envisioning the possible, I can scan temporal dimensions to compare present experience with past experience and uncover alternative futures based on the comparison. All of this is made possible by the *differential fitness* of language. Without fitness, speech about the past or future would not register. Without difference, speech would be trapped in actuality.

Differential space is what opens up language (and the world) and makes possible the dynamic openness of human language that does not seem evident in animal calls and their functions. In addition to temporal extension, consider examples of openness such as metaphor, comparisons, distinctions, negations, new or extended uses, misuses, deceit, asking questions, and meaningful silences.[23] All this indicates that the human linguistic environment is animated by traversals of otherness exceeding immediate states, which strictly empirical descriptions cannot convey. Interestingly, linguistic research shows that one of the few universals across different human languages is the capacity for negation, and that expressing negation is essential to language.[24] Negative dimensions, of course, are the core of Heidegger's phenomenology, which articulates how nonbeing is not the opposite of being, since absence is intrinsic to the rich scope of world-disclosure (cf. GA 24, 443/311–12; GA 9, 113–20). In light of the phenomenological negativity of language, from an evolutionary standpoint it is difficult to conceive how animal sounds and calls approximated or inched their way toward this surpassing dimension of language. It may be that the differential fitness of language itself opened up the dimensions of Dasein's radical finitude, as Heidegger understands it. But it is a puzzle to envision how mere sounds "evolved" to the point where language *as* fitting and *as* different emerged in some contiguous sense. The puzzle can be aggravated by noticing that my posing this question *as* a problem presupposes my already having been outfitted *by* language *as* a differential dynamic. And what about the *as* as such? What "is" the "as"? Can we take the "as" as, well, what? At this point one appreciates the aptness of Wittgenstein's appeal for silence, or more positively, Heidegger's talk of the self-showing marvel of being that is simply bounded by concealment and thereby not susceptible to explication. However human life has emerged from the earth, the very powers that let us explore this question cannot themselves be tracked in the earth.

NOTES

1. *Zollikon Seminars*, trans. Franz Mayr and Richard Askay (Evanston, IL: Northwestern University Press, 2001), 244. (Translation modified.)

2. GA 29/30, 384/264. It is surprising that Heidegger does not discuss Aristotle at all in this section of the text, given the significant influence of Aristotle on Heidegger and given the importance of biology in Aristotle's thinking.

3. For a stimulating overview of evolution and its philosophical ramifications, see Daniel C. Dennett, *Darwin's Dangerous Idea: Evolution and the Meanings of Life* (New York: Touchstone, 1995). Hereafter *DDI*.

4. For a treatment of ethics along these lines, see my *Ethics and Finitude: Heideggerian Contributions to Moral Philosophy* (Lanham, MD: Rowman & Littlefield, 2000), especially 110–12.

5. Darwin, of course, was a mortal blow to Aristotle's notion of fixed species in nature. Note, however, that Aristotle's three kinds of soul or life (rational, sensitive, vegetative) can be compatible with evolution, since the higher forms include the functions of the lower. Yet the idea of three clearly delineated, exclusive "kinds" would not likely hold up.

6. Richard Gregory calls this "kinetic intelligence." See *Mind in Science: A History of Explanations in Psychology and Physics* (Cambridge: Cambridge University Press, 1981). See also Marc D. Hauser, *Wild Minds* (New York: Henry Holt & Co., 2000). Hauser sees animal intelligence as a set of mental "tools," or various capacities for solving environmental problems that may be basic across species, such capacities as recognizing objects, counting, and navigating.

7. *DDI*, Ch. 8.

8. See Stuart Kauffman, *The Origins of Order: Self-Organizations and Selections in Evolution* (New York: Oxford University Press, 1993).

9. See Richard Dawkins, *The Selfish Gene*, 2nd ed. (Oxford: Oxford University Press, 1989), and E.O. Wilson, *Sociobiology: The New Synthesis* (Cambridge, MA: Harvard University Press, 1975).

10. See Larry Arnhart's discussion in *Darwinian Natural Right: The Biological Ethics of Human Nature* (Albany, NY: SUNY Press, 1998), 36–44; and Dennett's discussion of the "Baldwin effect," where the exploratory behavior of the phenotype creates fitness characteristics that are then favorable for selection, in *DDI*, 77–80. See also J. Scott Turner, *The Extended Organism* (Cambridge, MA: Harvard University Press, 2000).

11. It should be noted that an Aristotelian sense of teleology is still relevant in modern biology. See James G. Lennox, "Teleology," in *Keywords in Evolutionary Biology*, eds. Evelyn Fox Keller and Elizabeth A. Lloyd (Cambridge, MA: Harvard University Press, 1992), 324–33.

12. *DDI*, 39, 56, 66.

13. *DDI*, 144, 368.

14. *DDI*, 80–83.

15. *DDI*, 50–60.

16. *DDI*, 143, 342–52, 362–69. Dennett seems to grant that a scientific tracking of memes and their "descent" is unlikely, owing to the complexity and fluidity of culture; but he attributes this to *epistemological* limits rather than to any inadequacy in the evolutionary algorithm as such (356).

17. See *DDI*, Ch. 14.

18. For the classic phenomenological critique of AI, see Hubert Dreyfus, *What Computers Can't Do: The Limits of Artificial Intelligence*, revised edition (New York: Harper and Row, 1979). For confirmation from a scientific perspective of Heidegger's insights concerning the role of affects in knowledge, see Antonio R. Damasio, *Descartes' Error: Emotion, Reason, and the Human Brain* (New York: Avon Books, 1995).

19. For a recent attempt at a sociobiological reduction of culture, see E.O. Wilson, *Consilience: The Unity of Knowledge* (New York: Random House, 1999). For critiques of such approaches see Philip Kitcher, *Vaulting Ambition* (Cambridge, MA: MIT Press, 1985), and two essays by Stephen Jay Gould, "Sociobiology and the Theory of Natural Selection," and "Darwinism and the Expansion of Evolutionary Theory," in *Philosophy of Biology*, ed. Michael Ruse (London: Macmillan, 1989).

20. On these points, see H. Allen Orr, "Dennett's Strange Idea," *Boston Review*, 21/3.

21. *DDI*, 470.

22. *DDI*, 471, 491.

23. In critiquing the idea of human language "evolving" from prelinguistic animal capacities, one is faced with controversies surrounding intriguing animal research. Apes have been taught sign language and other nonvocal functions, and researchers point to an ape's capacity for innovative use and the complex grammatical structure implicit in various performances. Critics concede that apes are clever, bright, and communicative, but they are skeptical that apes *understand* language use beyond mimicry and experimental prompts. Chimps can be taught naming, but they seem to lack an *interest* in naming for its own sake, which is clearly evident in human children. See Jean Aitchison, *The Seeds of Speech: Language Origin and Evolution* (Cambridge: Cambridge University Press, 2000), 96–97.

24. Aitchison, *The Seeds of Speech*, 21, 177.

CARNAP AND HEIDEGGER

Parting Ways in the Philosophy of Science

Patrick A. Heelan

Both Edmund Husserl (1859–1938) and Martin Heidegger (1889–1976) were deeply and continuously involved, although in different ways, with the philosophical nature of mathematics and modern science (i.e., post-Galilean, theoretical/experimental science).[1] Both tended to see mathematics as the study of implicitly defined idealities and, in the spirit of the Göttingen School of theoretical physics, they saw modern science as the projection of such idealities on the world of experience, that is, as interpretations of experience projected in terms of mathematical symmetries or invariants related to group-theoretical operations or variations. In this way, the empirical operations on the manifolds of empirical data were represented as mathematical operators on ideal numerical manifolds.[2]

The emergence of phenomenology reflects a philosophical division about modern science in the German Neokantian School in the early part of the twentieth century. As Michael Friedman has shown in his book, *The Parting of the Ways*, Neokantianism in the early part of the twentieth century came to a parting of the ways. The Marburg School, among whose members were Rudolf Carnap, Ernst Cassirer, and Hermann Cohen, was focused on the natural sciences and was animated by a political philosophy that saw the future of human society as tied to one universal cultural language modeled on the natural sciences. The South-West German School that included Heidegger and Heinrich Rickert, was focused on the social and human sciences and was animated politically by a respect for a hierarchic plurality of national cultures. Husserl, trained in mathematics at Berlin under Weierstrass and a member of the Faculty of Philosophy (that included Natural Philosophy) at Göttingen from 1901 to 1916, tended to follow the natural science interests of the Marburg School while not belonging to

113

it politically. Under the political circumstances of prewar Nazi Germany, members of the Marburg School were forced to emigrate and went to the United States, Britain, and other countries. Members of the latter group, however, remained in Germany and suffered the consequences of Germany's defeat. What began as a philosophical and political "parting of the ways" in prewar Germany led to a more literal parting of the ways as more and more members of the Marburg School crossed the Atlantic, settled in universities in the United States, and awakened there a sense of the intellectual, political, and technological importance of the natural sciences as models of the kind of knowledge that proved its effectiveness in winning military victory and economic success in its Cold War aftermath.[3] It is then not surprising that after World War II phenomenology, which by that time had come to be associated almost exclusively with the human sciences, arrived late to the shores of the United States and that it bore the burden of being tainted by Heidegger's past expressions of loyalty to the national politics of defeated Germany. This history explains in great part why within the English-speaking world today there is a radical gulf of communication and purpose between phenomenological thinkers and the analytic/empiricist descendants of the Marburg School. Phenomenology and analytic philosophy live in different cultural and linguistic worlds and work from radically different platforms, the former emphasizing the subject–object role of intentionality in knowing and the importance of conscious self-appropriation, the latter focusing almost exclusively on the objective role of knowing. The prerequisites for a fruitful dialogue unfortunately require more than goodwill and mutual respect, they include the ability to share with some understanding and sympathy the incommensurable platforms they occupy as respective privileged philosophical spaces.[4] Nevertheless, there were some in the postwar era who found in phenomenological thinking elements of a significant, new and highly critical philosophical attitude toward the mainstream of British and American philosophy of science; the latter is often referred to in the United States as the "received view" of the philosophy of science.[5]

On the one hand, Husserl's influence on phenomenology was to regard scientific theory and the resulting theoretical invariants as the historical expression of the transcendental *eidea* intended by the infinite process of pursuing better and better empirical approximations in physical measurements. This is a view commonly held by theoretical physicists and scientists in many fields, as well as by a school of historians of science influenced by Alexandre Koyré, a close colleague of Husserl. These ideas had an important influence on generations of logicians such as H. Gödel; mathematicians such as O. Becker and G-C. Rota; psychologists such as M. Merleau-Ponty, K. Pribram, A. Damasio, and F. Varela; molecular biologists such as M. Polanyi, I. Progogine, and A-T. Tymieniecka; medical clinicians and researchers such as Heelan, D. Lader, and R. Zaner; theoretical physicists such as L. Brisson, W. Meyerstein, and H. Weyl; historians of science such as Koyré and P. Kertzberg; and philosophers of natural science such as Heelan, T. Kisiel, J. Kockelmans, E. Ströker, and A-T. Tymieniecka.

On the other hand and in contrast with Husserl, Heidegger saw these same mathematical models, not as transcendental *eidea* or metaphysical entities, but as historical inventions of the human spirit intending to reduce the world of experience to manipulable ontic entities with the suppression of what is particular, historical, contingent, creative, poetic, and ontological in Dasein's "being-in-the-world."[6] From the start, Heidegger saw the scientific culture of modernity as the "Age of the World Picture" in which the "real" is constituted by the theoretical representations of modern science rather than a revelation of what is constitutive of the foundational structure of what is, that is, what Heidegger called "ontology" (GA 5, 75–113/115–154). In later criticism, Heidegger came to see modern science as essentially entangled with technology in the constitution of objective frameworks of "standing reserves" (*Bestand*) or "mere value-neutral resources for human action" (*Gestell*) (VA, 9–40/3–35).

Heidegger has not had a strong following among those interested in the positive development of, and understanding of, mathematics and the natural sciences. Heidegger's influence in this area tends to be interpreted in the United States as a call to return to American Pragmatism. Heidegger's radical ideas have had a significant influence on the phenomenology of research, technology, and culture today. Among those influenced by these ideas are, for example, B. Babich, A. Clark, R. Crease, I. Fehér, D. Ginev, J. Haugeland, Heelan, D. Ihde, Kockelmans, Kisiel, G. Markus, Polanyi, J. Rouse, and R. Scharff. Polanyi, a molecular biochemist and also an important philosopher of science with a continental background, who is often seen as standing outside of contemporary philosophical traditions claims, however, that his "tacit knowing" comes to the same thing as Heidegger's being-in-the-world.[7]

In his critique of modern science, Heidegger argues that theories and mathematical models are "inauthentic" representations of Dasein's being-in-the-world and that they fail to establish modern science's or modern culture's relation to "aletheic" truth,[8] that is, truth based on "historicity," "authenticity," "openness," and "freedom" (GA 9, 177–202/115–138; SZ, 33–39/56–63; VA 9–40/ 3–35;).

"Authenticity" addresses the practice of a special phenomenological skill, that has to be learned, of avoiding that which in phenomenological analysis are the biases to "understanding" introduced by objective uses of abstract (theoretical) concepts and models. This "world" of Heidegger, this Lifeworld[9] of Husserl, is, perhaps, best seen for our purposes as the everyday world after the removal of all theoretical representational elements objectified as "real." However, in contrast with Husserl for whom the intention of truth is guided by unchanging "transcendental *eidea*," Heidegger holds that abstract concepts coerce being-in-the-world, into narrower or more determinate channels than those of the free, open, and historical life of the embodied inquirer thinking philosophically. Ironically, authentic phenomenological thinking can never be fully mastered, because our everyday "praxical"[10] lifeworld is never successfully purified of objects defined by abstract concepts, but is continuously challenged by them in a field of open inquiry. For Heidegger's phenomenology, however, the demand for authenticity in the Lifeworld is local, historical, contextual, emergent, cultural, and the subjective embodiment of imaginative, scientific, and poetic mind in the Lifeworld that is lived for its own sake as a terminal goal with respect to which problem solving serves only as a means to that end. The intention of truth is aletheic truth, free and open to emergent possibilities.

A developed phenomenology of nature would include phenomena revealed by natural science.[11] Phenomena are local, particular, historical, cultural, and contingent.[12] Scientific phenomena then are the contingent appearances of scientific entities in the subject's scientific—or better, postscientific—Lifeworld. In contrast with this view, the classical tradition of Western scientific culture focuses on objective abstract concepts and seeks them out by "epagoge" (induction), dialectics, or theory formation; for they are the objective fulfillment of every classical inquiry and reveal, as it is believed, the essences of what is. Classical science sees nothing limiting and deficient in their objective use. Phenomenology uses the same theoretical concepts but they function differently in a phenomenological philosophy of nature from how they function in classical science. This is discussed later. But it is only in the perspective of subjectivity, particularity, and history that the deficiencies of the objective use of theoretical concepts become manifest, for, in defining objective essences, they conceal the subjective, particular, and historical character of what is (i.e., the intentional unity of subject and object in the Lifeworld). For this reason, phenomenology sees theoretical concepts, not as pictures of a classical 'world,'but as a factor, usually in tandem with new technologies, in adding new scientific furniture to a changing the Lifeworld . . . but more of this later.

A phenomenology of natural science that seeks beyond theory for the grounds of aletheic truth is founded on the existential hermeneutical spiral (or circle[13]) that has a beginning in an original Lifeworld in which a problem is posed and ends in a transformed Lifeworld in which the problem is resolved.[14] The transformation referred to is effected by the emergence and naturalization in the Lifeworld of new theory-based technologies of understanding among which in particular are standardized measuring instruments and standardized measurement-based technologies.[15] These have been called *readable technologies*, that is, technologies capable of being integrated with the embodied self in the Lifeworld as lived phenomenologically for its own sake.[16] What then would be the minimum core of the aletheic truth of mathematical/experimental science within the phenomenological spiral of scientific research? It should at least comprise answers to the following questions:

1. How is modern science phenomenologically grounded in the local historical Lifeworld?

2. How do theories mediate computationally between pre- and post-theoretical phenomenological "data"?

3. What part do the "theoretical entities" of science play in the dynamic of change in the phenomenological Lifeworld?

First, the trajectory of a scientific inquiry is not a closed loop beginning and ending in the Lifeworld as unchanged; it is rather an existential hermeneutical spiral that involves Lifeworld change. Each successful advance in a scientific inquiry via the hermeneutic spiral changes the original Lifeworld in which the problem is initially located and embodied, into a transformed Lifeworld in which the problem is existentially resolved. By depositing in the Lifeworld new theory-designed processes (technological and/or institutional) that link emergent (perceiving) subjects and emergent (perceived) objects, new phenomena are, to use Fleck's words, "generated and developed."[17] This mutual adaptation of subject and object in the new Lifeworld is due in scientific studies to their entanglement with a readable technology. Included among readable technologies are legal instruments or institutions that can also produce new phenomena both social such as, say, universities, and physical such as, say, electrons, by redefining the political, social, and environmental structure of human agents and of the Lifeworlds they live in.

In these new and transformed Lifeworlds, new local scientific presences or phenomena (e.g., electrons) are revealed directly, not as theoretical entities (or literally as theory-laden entities) since these are abstract, but as new phenomena revealed by the manifolds of data produced by measurement processes. These processes use theory-designed technologies of the kind that enter praxically into the research life of the community.[18] Although the production process is theory-laden, such data and the scientific phenomena they refer to are not theory-laden but are or become praxis-laden in the new Lifeworld. The crucial fact is that the existence of scientific phenomena (and data about them) is not grounded on argumentative or theoretical grounds but on the success of such data and phenomena in transforming the Lifeworld; that is to say, in contributing to the goals of the research community within the praxical culture of research. The goal of this culture is to contribute to a new, different, and hopefully more desirable, quality of life "indwelling" through science within the Lifeworld. Regrettably, there is no guarantee that these new local scientific presences will be benign presences, for how they enter into the scientific Lifeworld is in the hands of those who manage the new technologies or institutions.

Second, how does theory computationally mediate between the original and the transformed Lifeworlds?

Consider medical research: scientific technologies can serve merely as indeterminate resources (or Ge-stell) or they can be specifically functional as praxes, dwelt in for their own sake. As specifically functional, some are invasive of the integrity of what is studied as in the killing by staining of living cells. And some are noninvasive of the integrity of what is studied, say, by imaging processes, such as sonograms or magnetic resonance imaging. And some can be Lifeworld-changing, quality-of-life-changing such as prostheses for the disabled or everyday user-friendly technologies because of the way readable technologies mutually transform both (perceiving) subjects and the (perceived) environment as described earlier. All of these cases can be studied in various scientific fields. For example, in the field of scientific medicine, Fleck (1979) gives a wonderful account of the way his Lifeworld as a medical researcher was changed by the progress of his research. He notes how under the influence of serological tests, the background and language under which he saw the patient's symptoms gradually changed until there was slowly revealed to him the presence of a new entity that had not hitherto been recognized as an entity. It was the disease entity to which the name syphilis applied. This process of discovery has the structure of an existential hermeneutical spiral.

It is not in medical research alone that measurements play complex hermeneutical roles. They do so also in natural science, with special attention to the very large and the very small, to cosmology and quantum physics.

To illustrate what a phenomenological hermeneutic of natural science can do, let me consider the role of theory in the received view of science. It is important to understand how in this view theory serves to explain a phenomenon and how an explanation is used to solve a scientific problem by computation or calculation. It does both usually by predicting the occurrence (or nonoccurrence) of a phenomenon. A theory explains by providing an ideal computational model that purports to represent the causes relevant to the occurrence (or nonoccurrence) of the phenomenon under standard circumstances. The model (as an ideal objective reconstruction) aims to replace the confused Lifeworld subject matter of inquiry, not for any and every purpose but for the (often tacit) standard purposes of the scientific inquiry. These purposes are achieved by using the model computationally to predict real Lifeworld outcomes in a Lifeworld transformed or to be transformed by equipment standardized for this purpose. The fact that the beliefs surrounding the received view work successfully for prediction and control in standard ranges of scientific phenomena, does not, however, imply that the received view is correct in its assumption that it is theory that constitutes the phenomena of science for the implications of standardization are not part of this analysis. It is well then to reflect, say, on what Heidegger says in *Being and Time* (SZ, 69–70/98–100) of what is implied by the meaning of theory in science as in ordinary life.

He begins with a worker engaged in a building project, using a hammer. The hammer unexpectedly breaks. Let us suppose that a replacement can't be found and that he has to have one made. He asks: what are the specifications of a hammer (of the kind he needs to finish the job)? The answer to this question will be a theory (about hammers) that explains a hammer's ability to do the hammer's job. What is a hammer's job? It is the meaning of a hammer. In this case it is a *cultural praxisladen meaning* dwelling within the context, let us say, of the building trade. Note that without a specification of context, the question is relatively indeterminate. In the context of the received view, however, the hammer is a physical entity specified by its specifications, it is a *theory-laden meaning* that lays out the physical conditions under which *it can become the host of the cultural meaning of a hammer. But whether or not it is assigned this cultural function is a separate and contingent matter.* The two meanings are not independent. The *theory-laden* meaning makes sense only if a local contingent existential condition is fulfilled, namely that the hammer referent is *praxis-laden* in the conventional sense.

If this condition is fulfilled, the hammer is a public reality constituted by a cultural meaning. But what if the existential condition is not fulfilled? 'It' would not be a hammer. It would not make sense even to call "it" theory-laden. Finally, "it" would have no more title to being listed in the hammer category among any categorial listing of the furniture of the Lifeworld than any old boot that could be used to drive a nail. "It" would become (in Heidegger's words) "a mere resource" (*Vorhanden* or *Bestand*) for hammering or some other indeterminate function, or just nothing in particular.[19]

Despite the fact then that (hammer) theory explains (hammering) praxis, the language of theory and the language of praxis belong to different although locally and contingently coordinated perspectives. Coordination does not imply, however, isomorphism between the two perspectives,[20] for someone working on a carpentry project could, perhaps, be served on this occasion by an old boot or something other than a hammer. Because theory and praxis are merely coordinated but not isomorphic, they can be taken as axes for a kind of cultural phase space within which there are zones of uncertainty between explanatory theory and Lifeworld praxis. This even suggests a general principle, a kind of Heisenbergian indeterminacy principle in theory–praxis phase space.

When we reflect on the fact that individual things in our Lifeworld experience are never without a role to play in the routines of human life, indeed that each may play multiple roles and be open to ever new roles, we realize that these routines belong to the Lifeworld and share the character of that world as historical, contingent, changing, praxical, and lived for its own sake. Everything in our experience then, including scientific entities, bears some resemblance to a hammer, or other tool or equipment and has (at least) two perspectives: (1) a *praxis-laden* cultural perspective (possibly with multiple *praxes*) that is constitutive of the Lifeworld, and (2) a *theory-laden* perspective—possibly multiply theory-laden—that explains this cultural perspective, makes it possible under appropriate conditions, *but does not constitute it*. If being-in-the-Lifeworld is—as I believe it is—the revelation to and for humans of what is constitutive of the foundational structure of *what is*, then scientific activity has a place in it and constitutes within it a powerful engine of change. But it is an engine of change within the dynamic of being-in-the-Lifeworld and not a substitute for it. And so we come to the third question: What part do theoretical entities play in this dynamic?

Every theory is a network of relationships among its terms. Usually the relationships are mathematical and the terms are quantified by measurement. In the received view, those terms and relationships constitute a model for, by being a representation of, some standardized aspect of the real world. They are called *theoretical entities*, meaning entities defined by a theory. Through the network of relationships, they explain the state that existed prior to the inquiry but was not previously understood as explained. What are we looking at? Let us name some of them: atoms such as carbon and hydrogen, molecules such as H_2O and CO_2, genes and proteins in living bodies, also their constituent protons, neutrons, electrons, quarks, and so on, and the properties of matter that they share, namely, local and temporal spatiality, energy, momentum, wave length, spin, and so on. These terms belong in the first place to the imaginative representational and constructed world of mathematical scientific models. Some of those terms in addition, however, manifest themselves as entities in public fora. In the received view they are taken to have metaphysical or real status like trees and tables. How are these terms assessed on Heidegger's philosophical platform? What do the terms stand for? How are they to be understood?

To belong to the Lifeworld has its own criterion: It is the possibility of realizing the named entity as a Lifeworld phenomenon.[21] This is achieved in the first instance by standard processes of measurement in a basic laboratory, for when a quantified variable is measured, the named entity (object) to which it belongs shows itself as present locally in the Lifeworld space of the laboratory as shaped by the practices associated with the standard measurement setup. In addition to being a public phenomenon in that forum, there are other public fora in which the named entity also has a presence with the status of a cultural phenomenon. These public fora feature, for example, everyday technologies, clinical medical procedures, pharmaceuticals, agricultural chemicals, finance, politics, religion, art, media, and others. All of these fora—like the basic laboratory—are local fora in which a named scientific entity, always in some familiar technological context, can play the role of a dedicated cultural resource (for everyday life, for clinical medicine, for pharmaceutical therapy, for agriculture, finance, politics, religion, art, media, etc.) and in this context can become part of the local furniture of the Lifeworld.

Although the distinction between invasive and noninvasive measurement processes is clear in biological research, the same distinction holds also for nonbiological objects of research. For example, electrons within atoms (bound electrons) are studied mostly by noninvasive methods, such as spectral analysis of emitted and absorbed radiation. With these methods, the holistic being of the atom is preserved and the bound electrons have to be treated just as so many functions of the controlling atom. In contrast, invasive methods destroy the internal environment of the atom and thereby free the electrons and the nucleus from their mutual bonds to make them independent phenomena. In the former (noninvasive) case the Lifeworld phenomenon is the atom. In the latter (invasive) case the electrons and the nucleus are freed from each other and they become phenomena each with their own constitutive properties in the Lifeworld and independently of the atom. However, antecedent to the emergence of atoms in cosmic history the properties of atoms as emergent holistic phenomena cannot be predicted just from the electronic properties of its constituents.

In all the local public fora just mentioned, the scientific entity and its Lifeworld data are meaningfully bivalent (possibly multiply so) and emulate the relationship between theory and praxis in the study of a hammer. If no distinction is made between invasive and noninvasive measurement processes, and no attempt is made to specify the kind of local forum in which the phenomenon will make its appearance, putative data cannot be assigned to phenomena in the Lifeworld.[22] Having no determinate phenomenological Lifeworld meaning, such data must be treated as indeterminate resources—or, on occasions, just noise.

By way of illustration of what has been said, consider how the Lifeworld of Europe was changed in the *quattrocento* with the invention of perspectival projection and the *camera obscura*. Terms of a new kind, scientific and theoretical, came to be introduced into everyday language with new practical—more precisely, *praxical*—measurement-based cultural meanings such as one universal mathematical Time, Space, and Calendar. The techniques of mathematical perspective revolutionized the Lifeworld of Italy and later of Europe, through art, architecture, urban planning, navigation, warfare, and much more. The geometry-filled productions of what were just craft skills in optics, astronomy, map making, painting, music, weapons' design, and other arts and skills prepared the way for the *elevation of practical artistic skills* to become known as applications of *scientific theory*. The products of the new arts came to be regarded no longer as works of art, but as works of geometrical reason or science.[23] Among other things, the works of geometrical reason changed the public urban space of Europe from a quilt of diverse local spaces and local times into a single universal space and uniform cosmic time based on measurement with standard rigid rulers based on common units and mechanical clocks synchronized with the stars. For those who looked for a unified cosmology, the way was prepared for Galileo and the Copernican revolution.[24] It was Galileo who helped convert works of thoughtful art—his deft experiments, such as, timing balls rolling down an inclined plane—into a world of science and reason, a move on which we now look back with more knowing and critical eyes.

It follows from what has been said that the furniture of the Lifeworld is not fixed with respect to categorial kinds, being dependent on the kinds of methods we use to realize them as recognizable free phenomena and to investigate them as such. Included among them are those like the Wassermann Test created by cultural institutions such as medical research laboratories. Some are categorized resources that are not functionally assigned (*Bestand* or *Vorhanden*), such as, chemicals, reagents, and appliances in central storage available for a variety of uses (SZ, 71–76/100–107; VA, 9–40/3–34). Others are functionally assigned (*Zuhanden*), such as those actually used in surgery (SZ, 70/99). Only the latter enjoy a meaning in the Lifeworld that is actually specified for definite tasks and, consequently, are part of the furniture of the Lifeworld.

Fleck, in his history of the scientific theory of syphilis, recognized the fact that in establishing the meaning of the new scientific term, *syphilis*, some special usage had to be negotiated between scientific terms normed by the thought collective of the research community and everyday terms normed by the thought collective of everyday life. To quote Barbara Duden:

> . . . as a practising bacteriologist, [Fleck] knew that his eyes were caught, not only in the norm imposed by the collective of the laboratory, but equally by the thought style characteristic of his everyday family life. It is this double anchorage—in the laboratory and at the table—that makes the scientist a conduit through which

scientific facts become confused with cultural interpretations. As a result, scientific facts . . . have a Janus-like face.[25]

What would result if the Janus-like face of a scientific fact were to go unnoticed or were to be flouted or ignored by convention in public fora of communication? This happens all too frequently, partly as a consequence of the widespread acceptance of the received view and partly because of the limitations on public discourse caused by the difficult subject matter. The result is distortions in communication. Two systematic errors become possible. Each, in its most innocent form, leads to the more or less conscious use of a figure of speech, something like a metaphor.

1. In one strategy the post-scientific praxis-laden perspective of the laboratory Lifeworld zone is simply redescribed metaphorically in terms of the theoretical scientific meanings.[26] Under such conditions, theoretical descriptions replace practical descriptions. But if the metaphorical character of the predication is not recognized, it is easy to take the replacement to be exclusive and metaphysical, and to think that the Lifeworld conditioning of phenomena no longer exists, and that only the theoretical scientific world exists. For example, perceptual space is assumed to be modeled by Euclidean geometry, colors by electromagnetic wavelengths, sounds by pitch and loudness, and syphilis by a positive Wassermann Test, when all such predications are no more than metaphors apart from the collaboration of the human senses, language, and cultural environment. Perceptual space, color, sound, and syphilis exist only as the product of interpretation through which they in their Lifeworld involvement become intelligible as phenomena of human experience. Modern scientific medicine then often has been charged with a weakness for reducing patients to a bundle of anatomical parts and physiologic processes, each having its scientific model at the level of chemistry, molecular biology, or physiology, and with little regard for the human life in which they are engaged and that uses such systems to cope, well or ill, satisfactorily or unsatisfactorily with the challenges of the patient's Lifeworld. All would agree that, ultimately, scientific medical models should not replace the Lifeworld of the patient and should be at the service of the patient's quality of life as lived in and tested by his or her Lifeworld.[27]

2. In a quite different strategy, the theory-laden perspective, which is privileged in the received view, is simply re-described metaphorically in terms of prescientific (naive or folk) Lifeworld meanings. In other words, because the scientific terms are not well understood in many public fora, they are simply filled with the old familiar prescientific Lifeworld meanings, possibly with a more or less conscious sense that this involves a metaphorical construction. From this awareness comes a warning, that should be heeded by ethicists, media pundits, and public policymakers who, confusing the

context of science with that of Lifeworld ontology, so easily and offhandedly fill scientific terms with prescientific Lifeworld meanings in their public discourse.[28] For example, in such discourse, scientific terms such as *cells, organs,* and *bacteria* are treated as (naive or folk) things like replaceable machine parts violating their natures as integral parts of a living organism, for unlike machine parts these organic terms are constituted by the continuous flow of chemical and energy exchanges across their interfaces with surrounding tissues.

What follows from the nonrecognition of these (at their innocent best) metaphorical transitions is, for instance, confusion in the public debate about such contentious practices as abortion, cloning, disease prevention, artificial intelligence, and much more, where scientific model terms, such as, *fetus, genotype, bacteria,* and *neural networks,* are filled in public discourse with meanings taken from related practical everyday (naive or folk) contexts, making them falsely synonymous with the everyday uses of the related everyday terms, *child, adult, cause of disease, intelligence,* and so on, respectively. This usage may be good politics, but it is itself a form of cultural disease.[29]

Despite the problems created by possible metaphorical usages due to the complementarity of explanatory scientific theory and preventive medical practices, scientific theories have been a very positive force in shaping the contemporary Lifeworld. There is no need to press this point. The bacterial theory of infection led to a host of new cultural practices dealing with food handling, personal hygiene, sewage and water systems, the urban environment, and the treatment of bacterial diseases. But these practices, of course, have to be carefully designed and prudently implemented. However, as scientific theories grow and change, a train of new and often contentious practical problems are beginning to emerge: for instance, genetic theory has led to noisy debates as to whether or under what conditions genetically modified foods should be admitted to the food chain. New cultural practices found to be effective also lead in their turn to new scientific theories, which in turn lead to better medical practices, which may lead to better scientific theories, and so on. Although often treated by public fora and sometimes even by the medical profession as stripping the mystery from Nature and as exposing what is constitutive of what really is, scientific theory is in fact no more than a tool for, or a way of coping with some living function of the human body constituted as meaningful by a lifestyle in the patient's Lifeworld. Because of the zone of uncertainty between theory making and cultural practices, and another between prescientific and postscientific Lifeworld terms, there is an inescapable tension is the public mind that can—and often does—result in changes, possibly also in confusion, concerning conditions for meaningful fillment and norms for public policy. Noting such changes, one captures something of the historicity and contingency of hermeneutic truth.

A critical example from medicine illustrates how the multivalence of scientific descriptions can create new moral perplexities in the Lifeworld. Duden, historian of the woman's body in clinical medicine, questions the scientific term *fetus* that belongs to contexts of scientific imaging and biology, and asks whether it is being abused in public fora of discussion when substituted for the term *child* that is used in the Lifeworld context of pregnancy and maternity.[30] She asks: Has the separateness of contexts between model-scientific, prescientific Lifeworld processes, and postscientific Lifeworld processes been illegitimately suppressed in our medical culture, in the media, and in public policy discourse? The terms *fetus* and *child* are, of course, correlative (each in its own context reveals something about what the other term refers to) but they are not isomorphic and interchangeable. A *fetus* is a term whose primary owner is the medical profession. A living fetus is recognized by sonographical and other imaging techniques apart from the mother's context in the everyday lifeworld. Even while inseparable from the living tissue of the mother, the fetus is generally described as a thing, as if, like a prescientific machine part, it had an existence separate from the mother. Duden notes with some concern that ethical rules and legislation in Western countries concerning pregnancies are presently being written in terms of the fetus where that term slurs the difference between the fetus as part of the scientific model, the fetus as an organic part of the postscientific Lifeworld, and the child as an element in the mother's (usually) prescientific pregnant life. Duden is unhappy with this and asks: Should the difference between the two cultural perspectives be recognized and an accommodation found that defers to the special cultural role of the mother in decision making about the child?

SUMMARY

In the assessment of scientific theory and practice, the critique of the analytic/empiricist view of science made via the phenomenological orientation of Husserl, Heidegger, and Merleau-Ponty toward the Lifeworld and Heidegger's hermeneutics (or interpretation) of experience has made it possible to assign different roles to theory and praxis. Theory is assigned to technological design for the purposes of environmental control, while praxis is assigned to ontological understanding for the purpose of human culture. Scientific theories then have a Janus-like face, one side looks in the direction of computational and technological control that is not constitutive of scientific knowledge but is merely a resource or tool for multiple praxes, the other looks in the direction of human culture which is ultimately constitutive of ontological scientific knowledge.

This bivalence underscores the prevalence of metaphor in scientific discourse and, in particular, in medical science and clinical practice under conditions where modern culture and the analytic/empiricist view tend to mask the presence of metaphor in such discourse.[31] It was shown, however, that under the broader analysis of phenomenology, metaphor is as fundamental for true scientific discourse as literality is for the analytic/empiricist view. Because the theoretical is mathematical and both the practical and the praxical are empirical, it makes no sense to predicate mathematical models literally of the phenomenological Lifeworld; at best, the two must come together consciously in some unambiguous but metaphorical way guided by professional experts in the spirit of (what Aristotle called) *phronesis* (prudent action), aware that they are seeking no more (and no less) than a praxical consensus about a set of relevant soluble Lifeworld issues.

NOTES

1. Husserl's influence on science and its critique; see, for example, E. Husserl, *The Crisis of European Sciences and Transcendental Phenomenology* (Evanston, IL: Northwestern University Press, 1970), *Ideas II. Collected Works*, Volume 3, tr. R. Rojcewicz and A. Schwer (Dordrecht and Boston: Kluwer, 1989), and *On the Phenomenology of the Consciousness of Internal Time*, Collected Works Volume 4, tr. J. Brough (Dordrecht and Boston: Kluwer, 1991).

2. See P.A. Heelan, "Hermeneutic Phenomenology and the Philosophy of Science," *Gadamer and Hermeneutics: Science, Culture and Literature*, ed. H. Silverman (New York: Routledge, 1991), 213–28.

3. See D. Hollinger, *Science, Jews, and Secular Culture* (Princeton, NJ: Princeton University Press, 1996), 17–41.

4. See W. Richardson, "Heidegger's Critique of Science" in this volume, originally appearing in *New Scholasticism* 42 (1968), 511–536.

5. The received view is the name often given to the mainline analytic/empiricist tradition in the anglo-American philosophy of science that stems from the influence of Descartes, Kant, and particularly, the Marburg School of Neokantianism (cf. M. Friedman, *The Parting of the Ways*, New York: Open Court, 2000), whose principal exponents in American were R. Carnap, C. Hempel, H. Feigl, and the contributors to the *Minnesota Studies in the Philosophy of Science Series* and the *Boston Studies in the Philosophy of Science Series*. For a further account, see the "Introduction" to F. Suppe, *The Structure of Scientific Theories* (Urbana: University of Illinois Press, 1974).

6. See Trish Glazebrook, *Heidegger's Philosophy of Science* (New York: Fordham University Press, 2000).

7. Polanyi's distinction between explicit meaning and tacit meaning parallels the distinctions made as SZ 66–72/95–102. In the Preface to the 1964 Torchbook edition of *Personal Knowledge,* Polanyi writes "Things which we can tell, we know by observing them; those that we cannot tell, we know by dwelling in them. All understanding is based on our dwelling in the particulars of that which we comprehend. Such indwelling is a participation of ours in the existence of that which we comprehend; it is Heidegger's *being-in-the-world*" (M. Polanyi, *Personal Knowledge,* New York: Harper & Row, 1964).

8. As an aside on *aletheic* truth, see M. M. Bakhtin, "The Problems of the Text in Linguistics, Philology and the Human Sciences: An Experiment in Philosophical Analysis," *Speech Genres and Other Late Essays,* eds. C. Emerson and Michael Holquist (Austin: University of Texas Press, 1986), 103–31: 126.

9. For the notion of the Lifeworld, see Husserl (1970); also Heelan (1991) and P. A. Heelan, "Husserl's Philosophy of Science," *Husserl's Phenomenology: A Textbook,* ed. J. N. Mohanty and W. R. McKenna (Washington, DC: University Press of America, 1989), 388–427. I am using the terms *lifeworld* or *everyday world* in a nontechnical sense. The term Lifeworld, however, is used in the technical phenomenological sense.

10. *Praxis* is the name Aristotle gives to a human activity that is pursued as an end in itself, that is, as a terminal goal of human living rather than as practical problem solving (Aristotle's *techne*) that serves as means to that end. *Techne* is related to technical, computational, or Heidegger's mathematical. *Praxis* is more difficult to translate. It is not just unreflective absorption in the activity of life, but an absorption that is continuously aware of the availability under the right circumstances of theoretical concepts as tools for action. They are not objectified realities but more like what Lonergan calls differentiations of consciousness (B. Lonergan, *Insight: A Study of Human Understanding, Collected Works,* Volume 4, Toronto: University of Toronto Press, 1994), 85–99.

11. See A-T. Tymieniecka, "The Ontopoiesis of Life as a New Philosophical Paradigm," *Phenomenological Inquiry* 22 (1998), 12–59, for another expression, deeply informed and scholarly, of the task of developing a phenomenological philosophy of science with special reference to certain aspects of new scientific thinking, especially in molecular biology, neuroscience, and evolution.

12. See Husserl (1989), and Heelan (1991).

13. The existential hermeneutical spiral must be distinguished from the philological textual or deconstructive hermeneutical circle that begins and ends in meanings that may or may not be fulfilled in experience. The existential hermeneutical spiral begins and ends in meanings fulfilled in their respective Lifeworlds.

14. GA 5, 75–96/115–136. H-G. Gadamer, *Truth and Method,* 2nd revised edition (New York: Continuum Press, 1995), 265–380.

15. For the analysis of measurement, see P. A. Heelan, "After Experiment: Measurement and Reality," *Amer. Philos. Qrtly.* 26 (1989), 297–308.

16. P. A. Heelan, *Space-Perception and the Philosophy of Science* (Berkeley and Los Angeles, CA: University of California Press, 1983/1988), 206. Researchers in the areas of robotics, such as Andy Clark in *Being There: Putting, Brain, Body and World Together Again* (Cambridge, MA: MIT Press, 1997), have adopted a cognate attitude toward robotics, that is, toward robotics as a cognitive technology.

17. Taken from the title, L. Fleck, *Genesis and Development of a Scientific Fact* (Chicago, IL: University of Chicago Press, 1979). A similar view is expressed in a study on the concept of the gene in P. Buerton, R. Falk and H-J. Rheinberger, eds., *The Concept of the Gene in Development and Evolution: Historical and Epistemological Perspectives* (Cambridge: Cambridge University Press, 2000).

18. Heelan, "After Experiment . . . (1989).

19. (SZ, 71–76/100–107; VA, 9–40/3–35). Kant, arguing from the *necessity* of natural scientific knowledge, found that its necessity was grounded in the apriori theoretical structure of human understanding, which he took to define the intelligible structure of empirical scientific phenomena (*Critique of Pure Reason*, B xii–xiv). To the extent that scientific understanding—in the long run at least—is historical and contingent, theory and experiment have come to be seen as mutually adaptable in the simultaneous creation of new scientific phenomena and new scientific theories. Although each *ideally* depends on the other, *in the practical order* as well as eventually *in the praxical order* the relationship depends on a background condition, all things being equal, that is not theoretically definable but subject to the practical/praxical judgment of experimenters in actual situations. It was the breakdown of such situations that eventually led to the need for relativity physics and quantum physics.

20. By isomorphism is meant a one-to-one translatability of any statement in one language into a unique statement in the other language. The two context-dependent languages refer to the same things but from different, often interacting and mutually interfering, actual perspectives. I have argued that these languages are related among themselves within a lattice structure which includes a least upper bound and a greatest lower bound as well as complements. This thesis is presented in Heelan (1983/1988), chapters 10 and 13; see also P. A. Heelan, "Hermeneutics of Experimental Science in the Context of the Life World," *Interdisciplinary Phenomenology*, eds. D. Ihde and R. Zaner (The Hague: Nijhoff, 1975), 7–50.

21. For a fuller account of what constitutes a phenomenon in Husserlian phenomenology, see also Heelan (1983/1988 and 1991), and cf. Heidegger, FD.

22. I am assuming Husserl's analysis of a phenomenon as the object of a noetic-noematic intentionality-structure, where the structure is group-theoretic relative to the connected set of variations (e.g., perceptual profiles) that maintain the invariance of the interest the subject has in the phenomenon. See Husserl (1989), 163–85, and Heelan (1991).

23. A. C. Crombie, Styles of Scientific Thinking in the European Tradition: A History of Argument and Explanation in the Mathematical and Biomedical Sciences and the Arts, Vols. I–III (London: Duckworth, 1994), 499–680.

24. Heelan (1983/1988), Chapter 11; P. A. Heelan, "Galileo, Luther, and the Hermeneutics of Natural Science," *The Question of Hermeneutics: Festschrift for Joseph Kockelmans*, ed. T. Stapleton (Dordrecht and Boston: Kluwer, 1994), 363–75.

25. B. Duden, *Disembodying Women: Perspectives on Pregnancy and the Unborn* (Cambridge, MA: Harvard University Press, 1993), 69.

26. M. A. K. Halliday and J. A. Martin, "General Orientation," *Writing Science: Literacy and Discursive Power* (London: Falmer Press, 1993), 3–21; C. Bazermann, "How Language Realizes the Work of Science," *Shaping Written Knowledge: The Genre and Activity of the Experimental Article in Science* (Madison: University of Wisconsin Press, 1988).

27. "The worlds of theory and the worlds of common sense partly interpenetrate and partly merge. The results are ambivalent . . . [for] it will also happen that theory fuses more with common nonsense than with common sense, to make the nonsense pretentious and, because it is common, dangerous and even disastrous" (Lonergan, 1994).

28. This is a warning that is more often heeded by historians of science and medicine than others.

29. From the point of view of contemporary functional linguistice, see, for example, the work of Halliday and Martin (1993), Bakhtin (1986), and Bazermann (1988).

30. Duden (1993).

31. See Bazermann (1988); G. C. Fiumara, The Metaphoric Process: Connections Between Language and Life (London and New York: Routledge, 1995); P. A. Heelan, "The Scope of Hermeneutics in Natural Science," *Studies in the History and Philosophy of Science* 29 (1998), 273–98; M. Hesse and M. Arbib, *The Construction of Reality* (Cambridge: Cambridge University Press, 1986); and G. Lakoff and M. Johnson, *Philosophy in the Flesh* (New York: Basic Books, 1999).

LOST BELONGINGS

Heidegger, Naturalism, and Natural Science[1]

David R. Cerbone

A "scientific" interpretation of the world, as you understand it, might therefore still be one of the *most stupid* of all possible interpretations of the world, meaning that it would still be one of the poorest in meaning. This thought is intended for the ears and consciences of our mechanists who nowadays like to pass as philosophers and insist that mechanics is the doctrine of the first and last laws on which all existence must be based as on a ground floor. But an essentially mechanical world would be an essentially *meaningless* world. Assuming that one estimated the *value* of a piece of music according to how much of it could be counted, calculated, and expressed in formulas: How absurd would such a "scientific" estimation of music be! What would one have comprehended, understood, grasped of it? Nothing, really nothing of what is "music" in it!

—Nietzsche, The Gay Science[2]

I myself believe, of course, that the religious hypothesis gives to the world an expression which specifically determines our reactions, and makes them in large part unlike what they might be on a purely naturalistic scheme of belief.

—William James, "The Will to Believe"[3]

There are more things in heaven and earth, Horatio,
Than are dreamt of in your philosophy.

—Shakespeare, Hamlet[4]

DISAPPEARING ACT

The first sentence of Thomas Nagel's now canonical (although, of course, not uncontroversial) "What Is It Like to Be a Bat?" declares the mind–body problem to be "intractable," owing, Nagel asserts, to the phenomenon of consciousness.[5] As Nagel argues, the phenomenon of consciousness confounds every attempt to formulate adequate identity statements linking states of consciousness and physical states of the brain (or the body more generally). Conscious states, as intrinsically subjective, seem thoroughly different in kind from physical states of the brain (body), scientific descriptions of which are, or at least strive to be, completely objective: This radical dissimilarity of the two kinds of states renders it unclear, to say the least, just how any such identity statements *could* be true. Physicalism, even if it (somehow) turns out to be true, is something Nagel thinks we do not currently, and perhaps never will, understand.

Nagel's argument might thus be summarized in the following way: Given that science strives for objectivity, it appears to be necessary that states that are by their very nature subjective will disappear from view. Put this way, Nagel's position can be seen, perhaps surprisingly, to share a presupposition with what is otherwise its spectral opposite in the philosophy of mind (i.e., eliminative materialism). Common to both views is a conception of science, of what scientific description seeks to achieve, which renders problematic the accommodation of mental or psychological concepts. Where they differ is in the lesson each draws from this failure of accommodation: Nagel sees the phenomenon of consciousness as so palpably real that we cannot simply dispense with our conception of the mental; accordingly, he recommends that we endeavor to produce an adequate phenomenology of consciousness, to stand alongside, for now, our endeavors to construct an objective account of reality. By contrast, the eliminative materialist (e.g., Paul Churchland) sees this failure as a kind of death knell for many long-familiar mental and psychological concepts.[6] Thus, we might be able to say that both agree in terms of what science is like, what its methods and aims are, while disagreeing about its *authority*: For Churchland, if something fails to appear within a mature (or, even better, a complete) scientific theory, then it fails to warrant further consideration as a denizen of reality and should instead be consigned to the scrap heap of outmoded (folk) theories, superstitions, and mythologies. For Nagel, on the other hand, the failure to appear of certain kinds of phenomena only shows the limitations of science, rather than something inherently defective about the putative phenomena so excluded. In considering this divergence, we may find ourselves feeling that an impasse has been reached, in that it is unclear what kind of argument or set of considerations would be sufficient to persuade either party to concede, to convince Nagel, say, of the ultimately illusory nature of that toward which he takes himself to be gesturing, or Churchland that something has indeed been left out, something that is not an illusion, and so that science is, and perhaps must be forever, incomplete. Intractable indeed.

Although in this paper I am concerned with the details of the debate thus far described, I have started with Nagel and the eliminative materialist because I see a kind of analogy between their debate and the one I do want to consider. That is, what interests me in (let us call it) the Nagel–Churchland debate is the shared conviction that something gets left out when scientific description gets going, that something, we might say, disappears. Martin Heidegger, in his later essays, likewise gives voice to such a conviction, and through language more dramatic than that found in contemporary philosophy of mind. Rather than disappearance, Heidegger writes instead of "annihilation [*Vernichtung*]." At stake for Heidegger is not a range of concepts associated with psychology or the philosophy of mind (Heidegger is not interested, as far as I can see, in "epiphenomenal qualia"), but instead the mere *thing*, whose humility perhaps makes it easier to overlook or reinterpret than consciousness. What Heidegger wants to assert is that with the total commitment to science and scientific description, things (e.g., the jug and the bridge) are, again, annihilated, that is, they fail to appear any longer *as* the things they are. This failure is not merely a matter of their not showing up within the vocabulary and concepts of "mature" scientific theories, but their failing to appear in our *lives* as well: At stake, then, is not merely mere things, but the possibility of what Heidegger calls "dwelling."

Heidegger's worries concerning the consequences of a total commitment to science and scientific description find their counterpart in the views of W. V. O. Quine. A precursor to the brand of naturalism advocated by Churchland and other contemporary eliminativists, Quine's naturalism maintains "that it is within science itself . . . that reality is to be identified and described,"[7] and, like Churchland's position, it has its own eliminativist tendencies. Thus, in examining Heidegger's sustained reflections on the thing in his later works, I consider the bearing of those reflections on the character and legitimacy of scientific naturalism. In doing so, I try to connect Heidegger's remarks on the distance or absence of things to Quine's eliminativism, which marks the disappearance of the thing in favor of, for example, the material content of a space-time region. Rather than merely the application of clear thinking and an example of ineluctable scientific progress, the rise of this way of thinking is seen by Heidegger as an occasion for loss; part of what his reflections do is open up the possibility of acknowledging and measuring that loss, and, perhaps, of redeeming it. In the latter part of this paper, I explore this last, redemptive feature of Heidegger's writing, and suggest that redemption in part is meant to be achieved simply by *letting* Heidegger's descriptions count as legitimate, as getting to "the things themselves." That is, one aim of Heidegger's writing, of the particular form it takes, is to get us to realize and reflect on our resistance to his descriptions of, for example, the bridge and the jug, as candidates for truth. What Heidegger persists in questioning is our dismissal of his language as *merely* poetic, and in doing so, he encourages a more complex, multifaceted view of the world, which allows both the poetic and the scientific to inhere.

SCIENCE AND THE FATE OF THINGS: FROM HEIDEGGER TO QUINE

Near the conclusion of "The Origin of the Work of Art," in commenting on the transformations in "the realm of beings" which have occurred throughout the history of the West, Heidegger offers this characterization of the modern age: "Beings became objects that could be controlled and seen through by calculation" (GA 5, 65/77). By tying the transformation of beings into objects to calculation and control, Heidegger is thereby linking that transformation to the rise of scientific thinking. Although not the central theme of his essay, Heidegger already here links the rise of this kind of thinking to a kind of loss. In developing the theme of how a work of art "sets forth the earth," he compares the manner in which that happens in artworks with what happens in scientific inquiry, where the earth is not so much set forth as obscured from view. Heidegger writes:

> If we try to lay hold of the stone's heaviness in another way, by placing the stone on a balance, we merely bring the heaviness into the form of a calculated weight. This perhaps very precise determination of the stone remains a number, but the weight's burden has escaped us. Color shines and wants only to shine. When we analyze it in rational terms by measuring its wavelengths, it is gone. It shows itself only when it remains undisclosed and unexplained. Earth thus shatters every attempt to penetrate into it. It causes every merely calculating importunity upon it to turn into a destruction. This destruction may herald itself under the appearance of mastery and progress in the form of the technical-scientific objectivation of nature. (GA 5, 33/47)

In this passage, Heidegger announces a theme that will recur in subsequent writings; indeed, it will become a central theme of some of those essays: the quantitative analysis characteristic of scientific theorizing, while undeniably revelatory of something genuine, at the same time serves to occlude something of vital importance. In the artwork essay from the 1930s, what gets excluded is the earth, while later, not unrelatedly, Heidegger concentrates on the disappearance of the thing. Indeed, Heidegger's essay entitled "The Thing" is throughout a meditation on lass, a kind of requiem for the proximity to things destroyed by "the frantic abolition of all distances" (GA 7, 167/165).

But what is a thing, and how is it that things are effaced by "technical-scientific objectivation," and why does it matter whether or not that effacement occurs? In this section, I concentrate primarily on the second of these three questions, although something must first be said about the first (the third question is considered subsequently). Rather than attempt at this point a thorough account of what Heidegger means by a thing, some preliminary remarks will have to suffice. Heidegger's account of the thing proceeds primarily by means of examples (the jug in "The Thing," the bridge in "Building, Dwelling, Thinking"). In attending to the jug, Heidegger emphasizes the difference between the jug, understood as a thing, and an *object*. He writes:

> As a vessel the jug is something self-sustained, something that stands on its own [*das in sich steht*]. This standing on its own characterizes the jug as something that is self-supporting, or independent [*Selbständiges*]. As the self-supporting independence of something independent, the jug differs from an object [*Gegenstand*]. An independent, self-supporting thing may become an object if we place it before us, whether in immediate perception or by bringing it to mind in a recollective re-presentation. However, the thingly character of the thing does not consist in its being a represented object, nor can it be defined in nay way in terms of the objectness, the over-againstness, of the object. (GA 7, 168/166–67)

Heidegger's talk here of the thing becoming an object when "we place it before us" is reminiscent of his discussion in *Being and Time* of the transformation of the hammer from something ready-to-hand to something present-at-hand when contemplation takes the place of active using.[8] Just as the equipmental character of the item of equipment is effaced in the act of contemplation,[9] so too is the thingly character of the thing obscured by an objectifying representation: "no representation of what is present, in the sense of what stands forth and of what stands over against as an object, ever reaches to the thing *qua* thing" (GA 7, 170/168–69). And just as the hammer reveals itself most authentically when we take hold of it and hammer, so too the thingly character of the jug is revealed in its use: "The jug's thingness resides in its being used *qua* vessel. We become aware of the vessel's holding nature when we fill the jug" (GA 7, 170/169).

In *Being and Time*, the revelation of the present-at-hand effected by the change over from active use to detached contemplation marks the beginning of scientific inquiry. Science recontextualizes the decontextualized *objects* discovered through contemplation, incorporating them into a systematic theory.[10] Not surprisingly, in "The Thing," Heidegger's contrasting of the thing with the object, although first spelled out in terms of perception and memory, quickly leads to a consideration of science, and in the following way: In working toward an adequate understanding of the jug *qua* thing, Heidegger first calls attention to the jug's being a vessel. Understanding the jug as a vessel leads in turn to an understanding of the jug as a *void*: in making the jug, the potter "shapes the void." "From start to finish the potter takes hold of the impalpable void and brings it forth as the container in the shape of a containing vessel" (GA 7, 171/169). The specific thingness of the jug is not primarily a matter of the material that composes it, but the void the material encloses because it is in terms of this void that its being a vessel to be filled and emptied is to be understood. It is at this point that Heidegger allows the voice of science to intrude, first by questioning this notion of avoid: "And yet, is the jug really empty [*wirklich leer*]?" (GA 7, 171/169). The appearance of "really" here marks an insistence that Heidegger will further develop: an insistence that it is the sciences, physics in particular, that will tell us the true nature of the thing. Heidegger writes:

> Physical science assures us that the jug is filled with air and with everything that goes to make up the air's mixture. We allowed ourselves to be misled by a semipoetic [*halbpoetische*] way of looking at things when we pointed to the void of the jug in order to define its acting as a container. (GA 7, 171/169)

What I am calling here the "voice of science" is one that expresses impatience with Heidegger's language, with its "semipoetic way of looking at things." Heidegger, for his part, is similarly impatient. While acknowledging the legitimacy of scientific description, he at the same time declares such description incapable of reaching the thing *qua* thing:

> These statements of physics are correct. By means of them, science represents something real, by which it is objectively controlled. But—is this reality the jug? No. Science always encounters only what *its* kind of representation has admitted beforehand as an object possible for science. (GA 7, 171/170)

Here we see an echo of the remark cited previously from "The Origin of the Work of Art," to the effect that the quest for objective control marks the effacement of the thing. Heidegger goes on to say that "science makes the jug-thing a nonentity in not permitting things to be the standard for what is real" (GA 7, 172/170) and, more dramatically, that "science's knowledge, which is compelling within its own sphere, the sphere of objects, already has annihilated things as things long before the atom bomb exploded" (GA 7, 172/170).

I want at this point to develop more explicitly this notion of annihilation Heidegger deploys here because it is at this juncture of his thinking that critical engagement with scientific naturalism becomes possible. Although, for Heidegger, the initial slide from thing to object in many ways already marks the "annihilation" of the former, it is nonetheless instructive to trace the fate of objects within scientific naturalism in order to bring out more fully the course and character of this effacement. Quine's essay, "Whither Physical Objects?," may be taken as illustrative.[11] Published in 1976, well on in Quine's philosophical career, the essay ha a retrospective quality as it traces the trajectory of Quine's own thinking about ontology. In keeping with his naturalism, Quine sees this trajectory as determined primarily by developments in the sciences themselves, there being for him no other serious source of ontological insight. In recounting Quine's own recounting of the development of a physicalist ontology, I emphasize its eliminativism: By Quine's own account, in the pursuit a physicalist ontology the physical object ultimately "evaporates."

To see this, let us begin as Quine does, with the notion of a physical body, which, according to Quine, is what we think of first as a suitable candidate for ontological commitment.[12] After all, on Quine's own account, our acquisition of language, as well as our everyday thought and talk, is oriented around familiar bodies such as tables, chairs, sticks, and stones, and it seems natural to take them to be ontologically respectable entities. Whatever naturalism this first thought might have is quickly pushed aside, as the notion of a body is "too vague in that we are not told how separate and cohesive and well rounded a thing has to be in order to qualify as a body," and "too narrow, since for ontological purposes any consideration of separateness and cohesiveness and well-roundedness is beside the point."[13] Instead, Quine recommends that we "understand a physical object, for a while, as the aggregate material content of any portion of space-time, however ragged and discontinuous."[14]

It is important to emphasize Quine's qualification "for a while," as even the notion of "aggregate material content," already itself far from the thing as something "self-standing" and "independent," does not survive for long. Quine raises two questions at this juncture. The first has to do with the status of "portions" or "regions" of space-time: Are they too to be admitted into a physicalist ontology? At this point in the trajectory of his thinking, Quine advocates forgoing any commitment to portions or regions, commitment to the material contents being sufficient. The second question, however, has more far-reaching ramifications, as it centers precisely on the notion of material content itself. Quine notes that as physics has developed, and here he has in mind developments in microphysics, "the naive conception of matter . . . tends to dissolve."[15] Such a dissolution is owing to the vagaries which attend the notion of a *particle* at the microphysical level: Questions concerning when one has one, rather than two, electrons, say, or the same electron first at one location, and then at another, do not admit of clear answers. Thus, Quine concludes that "matter evidently goes by the board. We are left rather with a field theory, a theory of the distribution of states over space-time."[16]

With the transition from material content to the distribution of states over space-time, a curious reversal is effected, such that the portions or regions of space-time so recently banished are readmitted:

> A little while ago it seemed that space-time regions were pretty flimsy affairs, and that our ontology would be better without them. But now our physical objects have themselves gone so tenuous that we find ourselves turning to the space-time regions for something to cleave to.[17]

Quine thus states that "the career of some body can be identified with a fixed portion of space-time, specifiable in terms of any of various systems in space-time coordinates."[18] At this stage, the objects are the portions of space-time, but a question remains as to the status of the states; about them, Quine asks: "do they have to be reckoned to our ontology as objects of a further abstract sort?"[19] There are two possibilities here: If the number of states is limited, then the language of the theory can include a simple predicate for each state, and so "there will be no need to recognize the states themselves as objects."[20]

If, however, the number of states is infinite in the sense that there is an infinite range of intensities of states, then, although one need not admit states a further objects, one must, Quine reluctantly acknowledges, admit *numbers* as the measures of these intensities:

> The admission of numbers and other abstract mathematical objects is an eventuality that has to be faced, melancholy though it be. There is no clear way to make natural science work without mathematics, nor make mathematics work without its objects.[21]

With the admission of numbers comes the apparatus of set theory, as it is "a familiar way of integrating the whole universe of mathematical objects."[22] Although Quine expresses reluctance about this initial admission, as it appears to enlarge the numbers of one's ontological commitments, the net result is a reduction of them:

> Now once we have reluctantly admitted all the ontology of set theory, we may get some consolation from a curious bonus that comes through: we can thereupon dispense with the other part of our ontology, the space-time regions. . . . Predicates that formerly attributed states to points or regions will now apply rather to quadruples of numbers, or to sets of quadruples.[23]

All that remains is the question of the "ground elements," that is, the "individuals in some sense" which are the members of these sets: ". . . what are they to be? Not physical objects; they gave way to space-time regions. But space-time regions gave way in turn to sets of quadruples of numbers; so nothing offers."[24] Nothing offers, indeed: the null or empty set suffices as the ground element, and so "there is the empty set, there is the unit set of the empty set, there is the set of these two sets, and so on."[25]

By thinking through physicalism, Quine's conception of ontology moves from an intuitive notion of physical body to ultimately nothing but sets, what Quine calls a "triumph of hyper-Pythagoreanism." As he happily concludes: "Our physical objects have *evaporated* into mere sets of numerical coordinates. This was an outcome, we saw, of physics itself."[26]

This excursus through some of the details of Quine's developing views on ontology amply illustrates Heidegger's claim that science annihilates the thing. Quine's invocation of the imagery of "evaporation" says as much: Heidegger's exemplary jug and the like are certainly nowhere to be found in the Pythagorean labyrinths of Quine's ontology, their having "gone by the board" in favor of, ultimately, "mere sets of numerical coordinates." Moreover, this foray into Quine illustrates a peculiar kind of convergence between Quine and Heidegger in terms of what they each take science to be like and what sort of view of the world the scientific perspective affords. The convergence is of roughly the same kind as that between Nagel and the eliminative materialist: just as neither of these letter two figures expects science to deliver an account of our mental or psychological lives in the terms we customarily use to depict them (what Churchland calls our "folk psychology"), neither Heidegger nor Quine sees science as a place where the "nature of the thing . . . comes to light [*kommt . . . zum Vorschein*]" (GA 7, 172/170).

Given this kind of agreement, adequately characterizing the disagreement between Heidegger and Quine's brand of naturalism is a difficult matter. The principal difficulty might appear to be one of just convincing the naturalist that, by recognizing only the deliverances of the sciences, something is being left out: given Quine's hasty dismissal of the criteria of separateness and cohesiveness, it is hard to imagine his being receptive to the charge of omission. Even this way of framing the disagreement is, however, problematic, because what is "left out" is not something that Heidegger thinks science in any way should (or could) accommodate. Rather than showing that there is something science has failed to account for, Heidegger wants, we might say, to question the exclusivity of science's *way* of accounting.[27] Although a difficult challenge to sustain against the Quinean naturalist, I suggest that there are resources available for developing the challenge in a way that makes the shortcomings of naturalism apparent, perhaps even to the naturalist. Not surprisingly, then, and in keeping with Quine's metaphor of "working from within,"[28] it will be best to develop the challenge Heidegger poses to naturalism from within Quine's own position.

IDIOMS, AUSTERE AND OTHERWISE

Heidegger pronounces that science does not let the thing count as the standard for what is real, and, as I have tried to show, with this Quine would readily agree. This pronouncement on Heidegger's part is not exactly a complaint about science, or if it is, considerable care is required in understanding just what the complaint is meant to be. That is, it is not at all clear that Heidegger is urging that science change its standards, that the natural sciences in some way accommodate the thing, or, more drastically, actually let the thing be the standard for what is real. Where then is the disagreement between Heidegger and Quine's brand of naturalism? To what does Heidegger's complaint amount?

Although Heidegger has no quarrel with science per se in the sense that he is not in any way out to reform the sciences, the problem vis-à-vis Quine would seem to be in the further step involved in scientific naturalism concerning what the "evaporation" of the thing from within the scientific perspective shows, namely that there is nothing real beyond the commitments dictated by one's best scientific theory or theories, suitably regimented in canonical notation. That is, the problem is not one of the legitimacy of this latter standard, but of its exclusivity. As Heidegger puts it in "Science and Reflection," the objectness of nature is, antecedently, only *one* way in which nature exhibits itself" (GA 7, 56/174). What Heidegger insists upon is a multiplicity of modes of exhibition, science's particular way being but one, albeit one which is extremely powerful and important. Heidegger's complaint against the naturalist is that the latter position fails to acknowledge the possibility of such other modes.

In other words, Heidegger wants an acknowledgment of something "outside" of science, something beyond the scientific perspective, to be exhibited by other means of engaging with, and talking about, the world. But what are these other means? Heidegger's most developed example is the *poetic*. In contrast to the scientific, the poetic constitutes an alternative way in which nature exhibits itself, a way, Heidegger claims, which allows the thing to thing and for human beings to dwell. In keeping with these latter claims, Heidegger will want to insist that poetry has a revelatory status more fundamental than the sciences. I discuss this toward the end of this paper. For now, I explore how a confrontation with the naturalist on these grounds might go.

That Heidegger wants to make room for the poetic as a way of approaching what is real again does not seem to be something with which the naturalist will necessarily disagree: What there is, poetically speaking, is of little interest to him. Sticking with Quine as our arch-naturalist, he is perfectly willing to acknowledge that there are ways other than the sciences of talking about, and conceiving the world, but as a naturalist, he sees those ways as markedly inferior to the methods and outcomes of the science, at least when it comes to telling us what the world is like. There is what Quine calls "official scientific business," which consists of "a relatively simple and austere conceptual scheme, free of half-entities," in contrast to what goes on elsewhere in various kinds of "second-grade" systems.[29] What Quine calls ontology, the determination of what there (really) is, proceeds according to this "austere conceptual scheme," according to the language of science, suitably regimented. Ontological investigation consists primarily of "scrutiny of this uncritical acceptance of the realm of physical objects itself, or of classes, etc.: Ontology makes "explicit what had been tacit, and precise what had been vague; of exposing and resolving paradoxes, smoothing kinds, lopping off vestigial growths, clearing ontological slums."[30] All of this smoothing and lopping off is in the service of arriving at what Quine calls variously "a theory of the world" or "a comprehensive system of the world," one devoted to "limning the true and ultimate structure of reality."[31] Within such a system "our notions of meaning, idea, concept, [and] essence" are all deemed "undisciplined and undefined," and so "hopelessly flabby and unmanageable," all denizens of the "slums" to be cleared away as ontological investigation proceeds.[32] There are, then, rather poor alternatives to the scientific method, these variously "flabby" and "unmanageable" notions, along with whatever "second-grade" systems into which they are incorporated. Quine feels no difficulty in acknowledging the existence of such systems, and so, in *that* sense could acknowledge that science is but one way in which nature can be exhibited; where he balks is in conceding that these other, second-grade ways get to nature at all. Thus, his acknowledgment is grudging at best.

The most prominent contrast with the "austere idiom" in Quine's own writings is the intentional one, exemplified in the attribution of "propositional attitudes" such as belief, desire, and the like. Quine's own attitude toward such attitudes is complex and ultimately, I suggest, highly problematic: Indeed, Quine's wavering over the intentional idiom, over questions concerning its irreducibility and indispensability, provides a kind of entering wedge for making good on Heidegger's complaints against the claimed exclusivity of the natural sciences.

Quine's animus toward the intentional idiom is abundantly manifest in his central work, *Word and Object*, and it persists throughout his many writings. To use the intentional idiom is, for Quine, to depart from the rigorous standards of scientific inquiry:

> In the strictest scientific spirit we can report all the behavior, verbal and otherwise, that may underlie our imputations of propositional attitudes, and we may go on to speculate as we please upon the causes and effects of this behavior, but, so long as we d not switch muses, the essentially dramatic idiom of propositional attitudes will find no place.[33]

The reasons why Quine thinks no place can be found for this "essentially dramatic idiom" are multiple: The nonreferential occurrence of terms in such idioms, in contrast to the strictly extensional, referential constructions of the austere idiom of science; the possibility of imputing such attitudes in cases where there is not even the possibility of corroborative verbal behavior, that is, the imputation of beliefs and the like to non-human animals; and, most importantly, the intimate connection between the attribution of propositional attitudes and determinations of linguistic meaning. Indeed, the latter is by far the most prominent target of Quine's attacks in *Word and Object* and elsewhere, the indeterminacy of translation being one of the doctrines with which Quine is most closely associated.[34] Since, for Quine, there is no fact of the matter in deciding between translation manuals that are mutually incompatible, and yet equally compatible with all speech dispositions observed or unobserved, there is likewise no fact of the matter separating attributions of rival sets of beliefs, each of which is couched in terms of the two manuals.

The difference between what is recognized from the standpoint of "the strictest scientific spirit" and what is allowed for when using the intentional idiom is, for Quine, the difference "between literal theory and dramatic portrayal."[35] In so relegating the intentional idiom, Quine thereby registers a kind of agreement with Brentano concerning the relation of the intentional to the nonintentional: Just as the dramatic cannot be rendered without loss in literal terms, so too the intentional cannot be reduced to the nonintentional. Again, this is in keeping with Quine's central indeterminacy thesis: Because conflicting manuals of translation are "physically equivalent,"[36] there is no hope of reducing them to, or identifying them in terms of, properly scientific (i.e., physical) theories. As Quine notes, "Brentano's thesis of the irreducibility of intentional idioms is of a piece with the thesis of the indeterminacy of translation."[37] Quine's agreement with Brentano only goes so far, however, because Quine draws an altogether different conclusion from the basic point about irreducibility:

> One may accept the Brentano thesis either as showing the indispensability of intentional idioms and the importance of an autonomous science of intention, or as showing the baselessness of intentional idioms and the emptiness of a science of intention. My attitude, unlike Brentano's, is the second.[38]

Quine means baselessness quite literally here, because there is nothing in physics to settle the rivalry between competing manuals of translation: Such rivalries outstrip, and so cannot be based on, matters of physical fact. And because Quine ultimately traces all real differences in the world to differences in physical facts, such that "nothing happens in the world, not the flutter of an eyelid, not the flicker of a thought, without some redistribution of microphysical states," the differences recorded by these manuals cannot be *real* differences.[39]

This passage about the lessons of Brentano's thesis contains, however, something of a fudge on Quine's part, since each of the alternatives consists of a conjunction of two claims and Quine really only rejects one of the two conjuncts he attributes to Brentano. Although Quine certainly does not share Brentano's opinion on "the importance of an autonomous science of intention," what is less clear is whether or not he regards intentional idioms to be *dispensable*. That is, although he clearly sees no hope of bringing the intentional into line with the strict standards of scientific theorizing, at the same time, he concedes the importance of such idioms, and in various contexts. Two passages from *Word and Object* are relevant here:

> Not that I would forswear daily use of intentional idioms, or maintain they are practically dispensable. But they call, I think, for bifurcation in canonical notation. Which turning to take depends

on which of the various purposes of a canonical notation happens to be motivating us at the time. If we are limning the true and ultimate structure of reality, the canonical scheme for us is the austere scheme that knows no quotation but direct quotation and no propositional attitudes but only the physical constitution and behavior of organisms.[40]

Not that the idioms thus renounced are supposed to be unneeded in the market place or in the laboratory. Not that indicator words and subjunctive conditionals are supposed to be unneeded in teaching the very terms—"soluble," "Greenwich," "A. D.," "Polaris"—on which the canonical formulations may proceed. The doctrine is only that such a canonical idiom may be abstracted and then adhered to in the statement of one's scientific theory. The doctrine is that all traits of reality worthy of the name can be set down in an idiom of this austere form if in any idiom.[41]

It appears, then, that the dramatic dimension of life's drama is here to stay.

In these passages, we can see Quine's attempt to have things both ways: to concede the indispensability of intentional idioms, while maintaining that they play no part in "limning the true and ultimate structures of reality," and so that they do not belong among the "traits of reality worthy of the name." I suggest that a tension appears at this point in Quine's position: on the one hand, science seeks "a comprehensive system of the world," and yet there appears to be something that it fails to comprehend, namely intentionality. This point alone does not yet constitute a criticism of the sciences: If science fails to accommodate ghosts, say, that is not a mark against science, but, on the contrary, a mark against ghosts. However, and here arises the tension, with the concession that the intentional idiom is indispensable, that is, one that we cannot simply forgo in favor of adhering exclusively to the austere scheme, then by some of Quine's own lights anyway,[42] intentionality would appear to be something the naturalist must acknowledge as a *bona fide* aspect of reality.

Quine, for his part, wants to insist on the inferiority of the intentional idiom. It is not at all clear, however, just how to cash out the claim of inferiority here, given the idiom's indispensability. Sentences formulated in scientific terms, couched in Quine's favored idiom, cannot take the place of those formulated in intentional terms. We are certainly not better off without the latter, as we cannot adequately cope with one another save by means of the concepts comprehended by the intentional idiom. It would appear that Quine's sole basis for claiming inferiority is that the intentional idiom does not adhere to the standards of the one he favors, but given the irreducibility thesis, this is hardly surprising. Quine's favored idiom begins to look like little more than favoritism, a penchant for austerity perhaps, but no longer with the claim to comprehensiveness: Quine's idiom may indeed be sufficient for all *scientific* discourse about reality, and hence for capturing all that is *scientifically* real, and why shouldn't it be if it is indeed the idiom best suited to regimenting scientific theories? But that is as far as it goes: To license a claim to completeness, to covering "all traits of reality worthy of the name," Quine must show how it is that one can dispense altogether with talk that appears to traffic in other traits.

The point I stress here is that Quine's admission of the indispensability of the intentional idiom shows that there are ways of talking or thinking that give shape to our lives, but about which science as Quine understands it falls silent. Indeed, in one telling passage, Quine is willing to confer the title of "wisdom" on this domain beyond the ken of scientific philosophy:

> Inspirational and edifying writing is admirable, but the place for it is the novel, the poem, the sermon, or the literary essay. Philosophers in the professional sense have no peculiar fitness for it. Neither have they any peculiar fitness for helping to get society on an even keel, though we should all do what we can. What just might fill these perpetually crying needs is wisdom: *sophia* yes, *philosophia* not necessarily.[43]

It is ironic, to say the least, that on Quine's conception of philosophy, something deserving of the name "wisdom" should fall outside the scope of the discipline named for its loving pursuit.

Quine's citation of the poem as one source of inspiration and edification constitutes a kind of grudging concession to Heidegger concerning the limits of scientific inquiry and discourse, to the effect that there is some form of writing, which is not scientific but whose effects are nonetheless important. In this respect, at least, poetry is on a par with those intentional idioms discussed at length above: something important is certainly accomplished in the day to day practices of attributing intentional states to one another. But there is an important disanalogy as well: Although life without intentionality, without the practice of attributing intentional states to one another and to ourselves, is in some deep sense unthinkable, such that even Quine must acknowledge the idiom's indispensability, an unpoetic life is, Heidegger fears, all too possible, indeed quite likely given our current scientific–technological predilections. the sense of loss that pervades Heidegger's "The Thing" underscores this point. With this in mind, I return to "The Thing" and other writings from Heidegger's later period.

POETRY, THINGS, AND THE POSSIBILITY OF DWELLING

The principle aim of my discussion of Quine's ambivalence toward what he calls the intentional idiom was to point to a weakness in the naturalist position, at least as articulated by Quine, which may be summarized in the following way: Despite Quine's deployment of the metaphor of working from within as the slogan for philosophical naturalism, sense can still be made of something from without as well. To put the point another way, that something disappears from view from the standpoint of scientific theorizing is not sufficient to establish its nonreality: Indeed, the example of intentionality shows there to be something pervasive and fundamental to our lives which science, by Quine's own lights, leaves, and must leave, out of account. What before appeared to be on the part of the naturalist a kind of don't-care acknowledgment of nonscientific modes of discourse, of all those flabby denizens of ontological slums and the like, might now be seen as a much larger concession, one that undermines the claim to science's all-encompassing authority.

It is one thing to have established this general point, and quite another to work out the particular one made by Heidegger. Nonetheless, the general point does serve as a kind of entering wedge against naturalism; moreover, this general point serves to expose naturalism's inherent, and from Heidegger's perspective, alarming tendency to dismiss any nonscientific modes of discourse or thinking or relating to the world as non-serious, as second class, and as, at best, merely inspirational. That is, naturalism's championing of the sciences as having exclusive purchase on "all traits of reality worthy of the name" is symptomatic of precisely what Heidegger thinks is wrong with the present age. Beyond this diagnosis of our current predicament,[44] Heidegger's further contribution is to elaborate a particular alternative.

Near the outset of "'... Poetically Man Dwells ...,'" Heidegger remarks:

> But when there is still room left in today's dwelling for the poetic, and time is still set aside, what comes to pass is at best a preoccupation with aestheticizing, whether in writing or on the air. Poetry is either rejected as a frivolous mooning and vaporizing in the unknown, and a flight into dreamland, or is counted as a part of literature. (GA 7, 191/213)

In contrast to this prevailing attitude, Heidegger wants to claim for poetry and the poetic a kind of primacy, such that poetry is not merely a kind of literature, an ornamental embellishment, or source of comfort, inspiration, and edification. Thus, Heidegger would not be satisfied with the grudging concession on Quine's part concerning the possibility of nonscientific forms of writing, since Quine's naturalism still dictates that the nonscientific not be afforded complete legitimacy. In the wake of the discussion of intentionality in the previous section, all Quine's attitude comes to is that the poetic, as one nonscientific activity (or discourse, or form of thinking), is, well, not scientific, but why should we expect otherwise? Only if one is committed, as naturalism is, to the primacy and exclusivity of the sciences for "limning the true and ultimate structure of reality," does this observation count as criticism, and I have tried thus far to demonstrate how difficult maintaining this commitment turns out to be.

Still, the questions remain: What exactly is at stake in the acknowledgment of (or the failure to acknowledge) the poetic dimension of human existence; and how is that acknowledgment bound together with the recognition of things *as* things (rather than, say, physical objects)? The quick answers to these questions are as follows: at stake for Heidegger is nothing less than the full realization on our part of what it is to be human,[45] and this full realization, what Heidegger calls dwelling poetically, is a matter, at least in part, of living a life of intimate relations with things, a life we might call one of *belonging*. Let me try in the space remaining to unpack these answers a bit. In doing so, I want to suggest that overcoming resistance to Heidegger's writings, to his philosophical and yet poetic renderings of human existence, can be a first step in accomplishing that intimacy.

The passages that initiated the comparison of Heidegger and Quine, those from "The Thing" on science and annihilation, continue in what I find to be a curious way. Commenting on the lately introduced idea that science annihilates the thing, Heidegger remarks:

> The thingness of the thing remains concealed, forgotten. The nature of the thing never comes to light, that is, never gets a hearing. This is the meaning of our talk about the annihilation of the thing. That annihilation is so weird because it carries before it a twofold delusion: first, the notion that science is superior to all other experience in reaching the real in its reality, and second, the illusion that, notwithstanding the scientific investigations of reality, things could still be things, which would presuppose that they had once been in full possession of their thinghood. (GA 7, 172/170)

Although this passage begins with an appeal to *forgetfulness* as our current relation to the thing, the "twofold delusion" Heidegger goes on to explicate serves to complicate that idea. The first aspect of the delusion is unsurprising, as it reiterates Heidegger's hostility toward a hegemonic conception of the sciences of the kind enshrined in, for example, Quine's naturalism; the second aspect, however, gives one pause, as it suggests that things have yet to come into "possession of their thinghood." Heidegger continues:

> But if things ever had already shown themselves *qua* things in their thingness, then the thing's thingness would have become manifest and would have laid claim to thought. In truth, however, the thing as thing remains proscribed, nil, and in that sense annihilated. This has happened and continues to happen so essentially that not only are things no longer admitted as things, but they have never yet been able to appear to thinking as things. (GA 7, 172/170–71)

These passages have the net effect of reorienting our understanding of Heidegger's mourning because it now appears to have the form of a kind of pessimistic longing, a desire for something that might yet be, but faces overwhelming difficulties in coming to pass. These passages also help to deflect the charge that Heidegger, in his longing for the thing, is simply indulging in nostalgia, despite his predilection for rustic and rugged examples. As he himself acknowledges toward the end of the essay, "Nor do things as things ever come about if we merely avoid objects and recollect former objects which perhaps were once on the way to becoming things and even to actually presencing as things" (GA 7, 183/182). Indeed, if we take these passages seriously, it is not clear that there really are any examples of things (and certainly not of things thinging), but at best approximations.

That examples of things are not ready to hand is, for Heidegger, bound up with the idea that we dwell unpoetically. What Heidegger calls our "restless abolition of distances" (GA 7, 167/166) near the outset of "The Thing" is symptomatic of this failure. That our lives have become "frantic" and "restless," that we live in such a way that "everything gets lumped into uniform distancelessness," suggests a kind of pervasive inattentiveness, a failure to attend to things in their particularity (as opposed to their objectivity). Heidegger sees poetry, and the fundamentally poetic nature of language, as the antidote to this kind of frantic oblivion. In "'. . . Poetically Man Dwells,' . . . " Heidegger remarks:

> That we dwell unpoetically, and in what way, we can in any case learn only if we know the poetic. Whether, and when, we may come to a turning point in our unpoetic dwelling is something we may expect to happen only if we remain heedful of the poetic. How and to what extent our doings can share in this turn we alone can prove, if we take the poetic seriously. (GA 7, 207/228)

"The poetic," Heidegger claims, "is the basic capacity for human dwelling" (GA 7, 207/228). Whereas science annihilates the thing, effacing it through a kind of quantitative homogenization of space and time, poetry calls to things in their particularity, thereby establishing (or holding out the possibility of establishing) a kind of proximity to things:

> The naming calls. Calling brings closer what it calls. However this bringing closer does not fetch what is called only in order to set it down in closest proximity to what is present, to find a place for it there. The call does indeed call. Thus it brings the presence of what was previously uncalled into nearness. (GA 12, 18/198)

"The naming call bids things to come into such an arrival. Bidding is inviting. It invites things in, so that they may bear upon men as things" (GA 12, 19/199).

Poetry names, calls, and so invites things in their particularity through the particularity of poetic language. What I call here the particularity of poetic language is revealed in the ways in which poetic language has its own kind of regimentation, its own demands for order and exactitude, but of a kind other than that found in the sciences. That a poem employs *this* word, rather than a near synonym, with *this* stress, and in *this* relation to the words around it: Understanding these demands is essential to a proper understanding of poetry, of its peculiar kind of necessity. To substitute words and expressions wantonly, even for more or less synonymous words and expressions, is to fail to grasp the nature of the poem and of the poetic use of language.[46] Language, we might say, *matters* in poetry,[47] and so it is in poetry that something essential about language is revealed.

In saying that language matters in poetry, I do not want to suggest that Heidegger wants to call our attention to further, and otherwise neglected, features of language (e.g., its "aesthetic qualities") over and above or apart from, language's "cognitive" dimension, it meaning proper. Rather, Heidegger's conception of the poetic, of the poetic nature of language, is more radical, as it aims to reorient our entire understanding of the nature of language, and in a way which resists these sorts of distinctions between the cognitive and the aesthetic, between content and form. The reorientation Heidegger seeks is suggested by his distinction between "speaking language [*Sprache sprechen*]" and "employing language [*Sprache benützen*]." As he puts it in *What Is Called Thinking?*:

> To speak language is totally different from employing language. Common speech merely employs language. This relation to language is just what constitutes its commonness. But because thought, and

in a different way poesy, do not employ terms but speak words, therefore we are compelled, as soon as we set out upon a way of thought, to give specific attention to what the word says [*eigens auf das Sagen des Wortes zu achten*]. (WD, 87–88/128)

"To give specific attention to what the word says" involves a recognition of the particularity of the word, such that what we might otherwise be tempted to separate out into its aesthetic and cognitive features are bound together. To recognize the inseparability of these features is to recognize the word as a *word*, rather than as a *term*. As Heidegger acknowledges, such recognition may not be easy to achieve:

At first, words [*Worte*] may easily appear to be terms [*Wörter*]. Terms, in their turn, first appear spoken when they are given voice. Again, this is at first a sound. It is perceived by the senses. What is perceived by the senses is considered as immediately given. The word's signification attaches to its sound. . . . Terms thus become either full of sense or more meaningful. The terms are like buckets or kegs out of which we can scoop sense. (WD, 88/128–29)

Poetry, as Heidegger understands it, directs our attention to the words themselves, rather than terms that act as mere containers ("buckets" or "kegs") of their meaning. As Heidegger puts it, words "are not like buckets or kegs from which we scoop a content that is there" (WD, 89/130). Instead, "words are wellsprings that are found and dug up in the telling, wellsprings that must be found and dug up again and again, that easily cave in, but that at times also well up when least expected" (WD, 89/130).

The particularity of words in poetic language is of a piece with the particularity of the thing. "The word makes the thing into a thing—it 'bethings' the thing" (GA 12, 220/151). Coming to appreciate the particularity of poetic language is thus a step at least toward coming to appreciate the particularity of the things called forth by the poem: By attending to the words themselves, we thereby come to attend to the things themselves. To insist on the possibility of redescription of the thing without acknowledging the loss that redescription might exact (redescribing the jug, say, as the material content of a space-time region rather than as a vessel, indeed *this* vessel, for pouring) is to fail to acknowledge the particularity of the thing. Note what Heidegger says in connection with the jug:

In the scientific view, the wine became a liquid, and liquidity in turn became one of the states of aggregation of matter, *possible everywhere*. We failed to give thought to what the jug holds and how it holds. (GA 7, 173/171; emphasis added)

Heidegger's reference to "the states of aggregation of matter" is not far from Quine's Pythagorean ontology, where there are only sets, and sets of sets, with the null or empty set everywhere the ground element, and where, not unrelatedly, discourse about the world is ideally to be regimented into one uniform notation.

Coming to know the poetic, coming to take the poetic seriously, is Heidegger's way of talking about the possibility of redeeming the loss of things in our lives: Fostering a reverence for the forms of language is a contribution to the realization of that possibility. Although Heidegger cautions that realizing this possibility cannot be effected by "a mere shift of attitude" because that alone "is powerless to bring about the advent of the thing as thing," a change of attitude toward poetry and poetic language might still constitute a first step. Moreover, that change can in part be a matter of our coming to be receptive to Heidegger's own words, to the particular descriptions that constitute his progress toward the nearness of things. Earlier, I cited a passage from "The Thing" wherein Heidegger describes his own way of looking at things, at the jug in this instance, as "semipoetic." Not yet poetry, and yet on the way to poetry, this description suggests, or rather demands, something from the reader, namely a willingness on the reader's part to take Heidegger's descriptions seriously as they are. Some of Heidegger's remarks on the example of a bridge in "Building, Dwelling, Thinking" are instructive on this point.

About the bridge, Heidegger says our common tendency is to

> think of the bridge as primarily and really merely a bridge; after that, and occasionally, it might possibly express much else besides; and as such an expression it would then become a symbol, for instance a symbol of those things mentioned before [viz. earth and sky, divinities and mortals]. (GA 7, 155/153)

In contrast to this tendency, Heidegger wants instead to insist:

> But the bridge, if it is a true bridge, is never first of all a mere bridge and then afterward a symbol. And just as little is the bridge in the first place exclusively a symbol, in the sense that it expresses something that strictly speaking does not belong to it. If we take the bridge strictly as such, it never appears as an expression. The bridge is a thing and *only that.* Only? As this thing it gathers the fourfold. (GA 7, 155/153)

Heidegger's rejection here of the notion of the symbolic as a way of understanding his depiction of the bridge is of a piece with his demand that we come to take poetic language seriously, as getting at or getting to what things really are. Heidegger's talk of "gathering," his invocation of the "fourfold" as essential to characterizing the particularity of the thing adequately is apt to be dismissed as merely symbolic, as something superadded to how things really are. On the contrary, to be receptive to Heidegger, and so to be on the way to being receptive to poetic language more generally, one must come to recognize those "semipoetic" descriptions not as symbols, projections, or fanciful imaginings, but rather as picking out, to borrow Quine's expression, "traits of reality worthy of the name."

NOTES

1. A version of this paper was presented at the June 2001 meeting of the International Society for Phenomenological Studies in Asilomar, California. I would like to thank the members of the audience for their comments and criticism, especially Steven Affeldt, Steven Crowell, Hubert Dreyfus, Mark Okrent, John Richardson, Joseph Rouse, Charles Siewert, Iain Thomson, and Mark Wrathall. I would also like to thank Trish Glazebrook, Randall Havas, and especially Edward Minar for comments on written drafts, and Steven Crowell and Charles Guignon for encouragement and support. Financial support for the research which led to this essay was provided in part by a Senate Research Grant from West Virginia University.

2. Friedrich Nietzsche, *The Gay Science*, tr. Walter Kaufmann (New York: Vintage Books, 1974), §373.

3. William James, "The Will to Believe," *The Will to Believe and Other Essays in Popular Philosophy* (New York: Dover Publications Inc., 1956), 1–31: 30.

4. Howard Staunton, ed., *The Complete Illustrated Shakespeare* (New York: Park Lane, 1979), 325–408: 345 (Act I, scene v).

5. Thomas Nagel, "What Is It Like to Be a Bat," *Philosophical Review* 83 (October, 1974), 435–50.

6. See, for example, Paul Churchland, *Matter and Consciousness*, revised edition (Cambridge, MA: The MIT Press, 1988), especially Chapter 2, §5.

7. W. V. O. Quine, "Things and Their Place in Theories," *Theories and Things* (Cambridge, MA: Harvard University Press, 1981), 1–23: 21.

8. See SZ, Division I, Chapter 3, especially §§15–16.

9. That the equipmental character of equipment is effaced in the act of contemplation, and thus in the change-over from the ready-to-hand to the present-at-hand, provides a basis for criticizing physicalism, and hence those forms of naturalism, like Quine's, which tend toward physicalism. Thus, it is not just in Heidegger's later philosophy that resistance to naturalism can be found. I discuss the ways in which *Being and Time* contains resources for constructing a critique of physicalism in "Composition and Constitution: Heidegger's Hammer," *Philosophical Topics* 27, no. 2 (Fall, 1999), 309–29. In many ways, this paper is a sequel to that one.

10. See SZ, Division II, §69b. I discuss these processes of decontextualization and recontextualization, and their import for the charge that Heidegger is an

idealist in "World, World-Entry, and Realism in Early Heidegger," *Inquiry* 30, no. 4 (December, 1995), 401–22.

11. W. V. O. Quine, "Whither Physical Objects?," *Essays in Memory of Imre Lakatos*, eds. R. S. Cohen, P. K. Feyerabend, and M. W. Wartofsky, Vol. 39 of *Boston Studies in the Philosophy of Science* (Dordrecht: D. Reidel, 1976), 497–504. In the discussion to follow, I have benefited greatly from Barry Stroud, "Quine's Physicalism," *Perspectives on Quine*, eds. R. Barrett and R. Gibson (Cambridge: Blackwell, 1990), 321–33. In the first half of the essay, Stroud likewise attends closely to "Whither Physical Objects?" In the second half, he goes on to criticize Quine's physicalism, but from a perspective different than the one developed here. Of Quine's many writings, the following are also especially relevant in this context: "Ontological Relativity," *Ontological Relativity and Other Essays* (New York: Columbia University Press, 1969), 26–68; "Facts of the Matter," *Essays on the Philosophy of W. V. Quine*, eds. R. Shahan and C. Swoyer (Norman: University of Oklahoma Press, 1979), 155–70; and Quine (1981).

12. The claim that a "physical body" is what we first think of is itself, of course, contestable, and is certainly one that Heidegger would want to resist. Indeed, from Heidegger's perspective, the effacement of the thing is complete with this claim alone. It is nonetheless instructive to proceed through Quine's ontological reflections, as they make explicit just why this seemingly innocent first step really marks the effacement of the thing. That is, Quine's reflections show the path down which this first step leads.

13. Quine (1976), 497. Notice how quickly Quine pushes aside as "beside the point" the criteria of separateness, cohesiveness, and well-roundedness, precisely the kind of criteria Heidegger deploys in characterizing the thing (compare self-standing, self-supporting, independent).

14. Quine (1976), 497.
15. Ibid., 498.
16. Ibid., 499.
17. Ibid.
18. Ibid., 500.
19. Ibid., 500.
20. Ibid., 500.
21. Ibid., 500.
22. Ibid., 500.
23. Ibid., 500–501.
24. Ibid., 501.
25. Ibid., 501.
26. Ibid., 502 (emphasis added).

27. Consider Quine's criterion of ontological commitment: to be is to be the value of a variable. Cf. W. V. O. Quine, "On What There Is," *From a Logical Point of View* (Cambridge, MA: Harvard University Press, 1953) 1–19. Heidegger's criticism of scientific naturalism is not that there are more values for variables than the naturalist allows, but rather that this entire way of thinking about being in terms of the values of variables, quantification, and canonical notation at the very least stands in need of supplementation.

28. See W. V. O. Quine, *Word and Object* (Cambridge, MA: The MIT Press, 1960), Chapter 1.
29. W. V. O. Quine, "Speaking of Objects" in Quine (1969), 1–25: 24.
30. Quine (1960), 275.
31. Ibid., 221.
32. W. V. O. Quine, "Has Philosophy Lost Contact with the People?" in Quine (1981), 190–93: 192.
33. Quine (1960), 219.
34. The *locus classicus* for Quine's statement of, and argument for, the thesis of the indeterminacy of translation is the second chapter of *Word and Object*.
35. Quine (1960), 219.
36. Quine (1981), 23. Quine explains the idea of physical equivalence in the following way:

> suppose, to make things vivid, that we are settling still for a physics of elementary particles and recognizing a dozen or so basic states and relations in which they may stand. Then when I say that there is no fact of the matter, as regards, say, the two rival manuals of translation, what I mean is that both manuals are compatible with all the same distribution of states and relations over elementary particles. In a word, they are physically equivalent.

37. Quine (1960), 221.
38. Ibid.
39. W. V. O. Quine, "Goodman's *Ways of Worldmaking*," in Quine (1981), 96–99: 98.
40. Quine (1960), 221.
41. Ibid., 228.
42. What I mean here by "Quine's own lights" is well expressed by Nelson Goodman, in his introduction to Quine's Carus Lectures, published as *The Roots of Reference* (La Salle, IL: Open Court, 1973), Goodman praises Quine for his insistence on the "sterling principle": "If you do say something about something, don't think you can escape the consequences by saying you were only talking" (xi–xii). The point that Quine's insistence on this principle renders questionable his own attitude toward the intentional idiom has been made by Barry Stroud, in his "Quine on Exile and Acquiescence," *On Quine: New Essays*, eds. P. Leonardi and M. Santambrogio (Cambridge: Cambridge University Press, 1995), 37–52, a paper to which I am greatly indebted.
43. Quine (1981), 193.
44. In speaking of a diagnosis of our current predicament, I do not wish to suggest that our predicament is identical to, or exhausted by, a commitment to scientific naturalism. Although in this paper I have concentrated on the bearing of Heidegger's thinking on scientific naturalism, it should be clear that he aims in his writing to engage critically with a much more pervasive mindset, one that makes naturalism come to seem attractive and perhaps even inevitable.

45. In *What Is Called Thinking?*, Heidegger talks in terms of "achieving [*vollbringen*]" our humanity [*unser Menschein*]. See WD, 157/144.

46. This aspect of poetic language is alluded to by Wittgenstein late in the Philosophical Investigations, tr. G. E. M. Anscombe, 2nd ed. (New York: The Macmillan Company, 1958). At §531, Wittgenstein writes:

> We speak of understanding a sentence in the sense in which it can be replaced by another which says the same, but also in the sense in which it cannot be replaced by any other. (Any more than one musical theme can be replaced by another.)
>
> In the one case the thought in the sentence is something common to different sentences; in the other, something that is expressed only by these words in these positions. (Understanding a poem.)

47. For Heidegger, poems are not the only examples of poetic language. He notes that "pure prose is never 'prosaic.' It is as poetic and hence as rare as poetry" (GA 12, 28/208).

IV

TECHNOSCIENCE

HEIDEGGER'S PHILOSOPHY OF SCIENCE AND THE CRITIQUE OF CALCULATION

Reflective Questioning, Gelassenheit, and Life

Babette E. Babich

> Zeit ist das, was sich *wandelt* und *mannigfaltig*,
> Ewigkeit hält sich einfach.
>
> —Meister Eckhart

HEIDEGGER'S PHILOSOPHY OF SCIENCE: CALCULATION, CRISIS, AND LIFE

Like Nietzsche, Heidegger argues that beyond theoretical reflection or scientific analysis, philosophy is an explicitly active questioning, especially so in the case of the philosophy of modern science and modern technology. It is in terms of the importance of reflection in philosophy that Heidegger argues that "all science is perhaps only a servant with respect to philosophy" (GA 29/30, 7/5). The critical spirit of this early account of the specific difference of philosophical reflection and scientific theorizing finds its most famous expression in the later Heidegger's provocative dictum "science does not think,"[2] a claim that is already to be heard in his 1927 *Being and Time*: "ontological inquiry is more primordial or original than the ontic inquiry of the positive sciences" (SZ, 11/31).

By saying that "all scientific thought is merely a derived form of philosophical thinking" (EM, 20/26). Heidegger's claim is that, in consequence, philosophy "is prior in rank."[3] Heidegger thus opposes the creatively foundational activity of philosophic reflection to the then popular articulations of epistemological investigations into the sciences of his era as the kind of "logic" (Heidegger sets this off in quotes) following after science, "'limping along in its wake,' investigating the status of [any given] science as it chances to find it in order to discover its 'method'" (SZ, 10/30). By contrast, the *"productive* logic" (SZ, 10/30, emphasis added) of philosophy is ultimately capable of illuminating breakthroughs and even within science itself (but only to the extent that it is "philosophical" as Heidegger emphasizes). And more than one scholar writing on Heidegger has noted an anticipatory parallel with Kuhn as Heidegger continues to write in *Being and Time*, this creative logic leaps ahead "into some area of Being, discloses it for the first time, in the constitution of its Being, and, after thus arriving at the structures within it, makes these available to the positive sciences as transparent assignments for their inquiry."[4]

For Heidegger, writing in Husserl's critical foundational spirit with respect to the development of mathematical physics, "What is decisive" is the creativity of "the mathematical project of nature itself" inasmuch as the project "discovers in advance something constantly objectively present (matter) and opens the horizon for the scientific perspective on its quantitatively definable moments (motion, force, location, and time)" (SZ, 362/413–14). The "founding" of "factual science" is "possible only because the researcher understood that there are in principle no 'bare facts'" (SZ, 362/414), that, in other words, the material project of nature must be given in advance, *a priori*. Only then is it possible for a science to be "*capable* of a crisis in its basic concepts" (SZ, 9/29).

Heidegger, who remained committed to phenomenology throughout his life, emphasizes that beyond any superficially obvious call "to the things themselves,"[5] phenomenology "presupposed life."[6] To understand this presupposes a specific and hermeneutic attention, in the methodic sense Thomas Seebohm has underlined, like Heidegger indeed with reference to Dilthey.[7] In methodic focus for Heidegger himself and in addition to history and the philosophical question of time, was the transformation of the biological sciences that was well underway at the time and in addition to the more well-known transformations of physics as a mathematical science, and to a lesser degree, of chemistry[8] in the same vein.

Including and exceeding Claude Bernard's *milieu intérieur*, as the evolution beyond Cartesian mechanism, Heidegger's own reference would be critically *ecological* in Ernst Haeckel's original sense (*assuming* we might be able to translate as Pierre Duhem long ago suggested that we could not do, between the sensibilities of "German" and "French" science),[9] radically environmental, in today's rather specifically "English" (and not American or Canadian or Australian, etc.)[10] terminology: "Life is that kind of reality which is in the world and indeed in such a way that it has a world. Every living creature has its environing world not as something extant next to it but as something that is there [*da ist*] for it as disclosed, uncovered."[11] And in 1925, Heidegger emphasized that "for a primitive animal, the world can be very simple," explaining that we run the risk of missing "the essential thing here if we don't see that the animal has a world."[12] Heidegger's original continuum of complexity–simplicity must accordingly be added to contemporary readings of Heidegger's subsequent discussions of the world-poverty of the animal in terms of *indigence*.

In his 1929–1930 lecture course: *The Fundamental Concepts of Metaphysics*, Heidegger alludes to Hans Dreisch's account of chemical gradients in embryological development,[13] and he cites the Czech biologist Emmanuel Rádl on the significance of animal phototropism (GA 29/30, 356/244). In that same course, Heidegger also invokes the theoretical biologist Jakob von Uexküll's 1909 expression of the "*Umwelt*" (GA 29/30, 382-3/263-4), emphasizing the gulf [*Abgrund*] between human and animal (GA 29/30, 384/264), an anti-biologistic contrast to the Cartesian tendency of modern scientific biology to define both animals and human beings in mechanistic terms.

Our medical paradigms continue this mathematic and often effectively numerological preoccupation with calculation.[14] Manifestly, if now expressed increasingly in terms of information (genomics), this mechanistic or calculating tendency persists in modern experimental biology, a mechanistic orientation that underlies its reliance on "models," not only computer and other simulations but also the very practice of animal experimentation, as this is also a specifically Cartesian legacy.[15] Such experiments hold the promise for us that they do despite ethical concerns—concerns the scientists themselves long ago and in order to be scientists in the first place, learned to set aside.[16]

Heidegger's specific argument in this text thus reprises his hermeneutico-phenomenological case for the interpretive ontology of the human being as an animal bound to world-making or -invention (GA 29/30, 397/274). Despite several efforts, a Heideggerian philosophy of medicine has yet to be fully worked out.[17] Precisely for the sake of phenomenology, I urge that nursing research be considered. In this spirit, Heidegger invokes Nietzsche's perspectival sense of the human being as the "not yet determinate" or "still unfinished animal" [*das noch nicht festgestelltes Tier*]. It is in this projective, that is, yet-unfinished sense of the human as being-in-the-world that Heidegger reflects that the "world that is closest to us is one of practical concern. The environing world [*Umwelt*] and its objects are in space, but the space of the world is not that of geometry."[18] There are overtones of a battle already fought (and already won) by Bertrand Russell and the analytic tradition in philosophy and the reductivist tradition in the natural sciences, a battle only appropriated (not coined) by Carnap contra Heidegger, as a conflict between the spirits of finesse and geometry, a conflict expressed in the life sciences, and likewise already then decided as the battle against vitalism. Historically, it would be the mechanistic conception of life that was to triumph over the notion of "vital movement" nascent in Driesch (although it is an error to reduce Driesch's concerns to the mystical vagaries of "vitalism," as is evident in Driesch's own emphasis on electro-chemical gradients), whereas Heidegger for his own part took care to explore the living trajectory of life as opposed to its "calculable" course.[19]

Heidegger's focus on life also recurs in his reference to chemistry in 1929 in order to speak of the organic and biological sciences, emphasizing how little is science says about "the living being" when we "define it in terms of the organic as opposed to the inorganic" (GA 29/30, 311/212). As Nietzsche reminds us, the notion that there is "nothing unchanging in chemistry" is little more than "a scholastic prejudice. We have dragged in the unchanging, my physicist friends, deriving it from metaphysics as always. To assert that diamond, graphite, and coal are identical is to read off the facts naively from the surface. Why? Merely because no loss in substance can be shown on the scales?"[20] Echoing Nietzsche's contrastive differentiation of reductively identical kinds in chemistry, Heidegger considers the example of "organic and inorganic chemistry" just to the extent that "organic chemistry is anything but a science of the organic in the sense of the living being as such. It is called organic chemistry precisely because the organic in the sense of the living being remains inaccessible to it in principle."[21]

If Heidegger began *Being and Time* by referring to the crisis in the sciences and just in this context with an argument on behalf of *philosophical* over and against *scientific* reflection, he also argued that each particular science articulates its own regional ontology in terms of its basic constitution [*Grundverfassung*], (SZ, 9/29) beginning with the example of the foundational controversy of mathematics in his (and still in our own) day: "between the formalists and the intuitionists" (SZ, 9/29). Thus Heidegger adds, in good Husserlian fashion, that what is at stake in this debate turns upon "obtaining and securing the primary way of access to what are supposedly the objects" (SZ, 9/30) of mathematical science. As he invokes the theory of relativity, Heidegger articulates the same foundational revolution in physics. This means that science begins with its own fundamental concepts. Accordingly, inquiry into these foundations cannot itself be an object of or for scientific research just to the extent that scientific research per se is only possible on the basis of these same concepts. Hence, philosophical inquiry or what Heidegger calls "ontological inquiry" can *only* be "more primordial, as over against the ontical inquiry of the positive sciences" (SZ, 11/31).

For Heidegger, philosophy is, and can be, the science of science, in Husserl's terminology now, not because of a venerable tradition of so regarding philosophy but just because specifically philosophical research must and most ontically, most importantly, "*can*" as Heidegger claims, "run ahead of the positive sciences" (SZ, 10/30). Thus, as Heidegger clarifies this point, the contribution of Kant's *Critique of Pure Reason* "lies in what it has contributed toward the working out of what belongs to any Nature whatsoever" (SZ, 10-11/31). Rather than epistemology, Kant's "transcendental logic is an *a priori* logic for the subject matter of that region of Being called 'Nature'."[22]

In this productive, disclosing sense, which Heidegger also expresses as the constitutional eventuality of aletheic truth as discovery or "uncovering," the scientist, reflecting philosophically, effectively opens up the truth of nature. In the aletheic context of such a specifically scientific disclosure, Heidegger observes that *before* "Newton's laws were discovered, they were not 'true'" (SZ, 226/269). But to say this also is to say that through "Newton the laws became true: and with them entities became accessible in themselves to Dasein. Once entities have been uncovered, they show themselves as the entities which beforehand they already were. Such uncovering is the kind of Being which belongs to 'truth.'" (SZ, 227/269, modified). By way of Dasein's "being in the truth," (SZ, 226/269) the laws of Newtonian physics only first "became true. "Newton's laws, the principle of contradiction, any truth whatever—these are true only as long as Dasein *is*" (SZ, 226/269).

CALCULATING TECHNIQUE: THINKING THE QUESTION OF SCIENCE AND WORLDVIEW

Heidegger's concern with questioning grew into his thinking the technologically mediated understanding of truth as correctness [*Richtigkeit*] and what he called calculative [*rechnendes, planendes, forschendes*] thinking. In the same way, Heidegger poses what may be called the question of the question as he reflects on questioning itself in the wake of technology. Holding that technology is prior to science, the essence of technology is expressed as challenging-revealing, a very particular and precisely calculating deformation of questioning, far from its ultimate spirit as the "devotion," "dedication," or "piety of thinking" [*Frömmigkeit des Denkens*]. This same opposition between calculative (this includes the casual disposition that merely takes account of reports and information, all the news we consume with our morning newspapers and evening television shows) and dedicated or pious thinking recurs in his *Discourse on Thinking*.

What Heidegger means by questioning remains more elusive than even his more attentive readers tend to think. More is at stake in any question than is supposed at first glance and still more is involved in what, in his 1934 course on *Logic as the Question After the Essence of Language*, he names "authentic and genuine questioning" [*eigentliches und echtes Fragen*] (GA 38, 18), understanding such questioning as an invitation to reflection, that is, to the kind of thinking that *holds faith* with—and such we see again will be the meaning of piety, dedication, or devotion to—its own task. Delineating the formal structure of questioning in section two at the start of *Being and Time*, Heidegger begins by turning the question upon itself. From this reflexive perspective, he writes:

> Every questioning is a seeking. Every seeking is guided in advance by what is sought. Questioning is a cognizant [*erkennende*] seeking for an entity both with regard to the fact that it is and with regard to its being as it is. This cognizant seeking can take the form of an "investigating," in which one lays bare that which the question is about and ascertains its character. . . . Questioning itself is the behaviour of a questioner, and therefore of an entity, and as such has the character of Being.[23]

Thus, as Heidegger attempts to pose the question of Being as a question that has been forgotten, it is characteristic of his thinking that he will first find it necessary to reflect upon the Being of questioning (as such) and indeed as the Being of a seeking that is always guided in advance, in order to be a question at all, by that which is sought. In this way, every questioning includes: *das Gefragte*—that which, and to begin with, is asked about; *das Befragte*—that which is interrogated in the inquiry itself; *das Erfragte*—the aim of the inquiry: that which is to be discovered. A question thus spells out the range, object, or frame of what might be considered as a reply or answer but Heidegger will carefully distinguish between answer-bound or -determined questioning—the kind of question that seeks only a pregiven answer—from the kind of question that genuinely asks after what might come forth as an answering reply.

For Heidegger, philosophical thinking is an active questioning. Hence, "authentic" questioning is about the asking as such rather than the answer. With Heidegger, we are not only to reflect on the nature of questioning, but on the meaning of thinking, of thought itself, language, and indeed the embodied mortality of the inquirer. In opposition to calculative thinking, Heidegger opposed *sense-directed reflection* [*Besinnung*] for embodied, mortal beings such as ourselves. It is this sensitively, incarnate reflection that Heidegger contrasts as properly philosophical or indeed poetical thought to the rational, calculative (and effectively unquestioning because solution-informed and answer-driven) project of Western technologically articulated and advancing science. Such a poetically attuned task of reflective thinking would open its own way, just as "questioning always builds a way,"[24] by its questioning advance (ground, object, and aim). Thus questioning "is the unique habitat and *locus* of thinking" (WD, 113/185). However, and as opposed to the transparent and calculative inquiry that seeks solutions, authentic questioning ultimately turns out to be so rare that it is not clear that we can ever be otherwise than *underway* to questioning. One has, reflecting in this fashion, to take up a position in questioning, a disposition toward thought, an inclination. "To venture after sense or meaning [*Sinn*] is the essence of reflecting [*Besinnen*]. This means more than a mere making conscious of something. We do not have reflection [this is, as we shall see, also the meaning of what Heidegger names *Gelassenheit*] when we have only consciousness. Reflection is more. It is calm, self-possessed surrender to that which is worthy of questioning" (VA, 64/180; cf. WD, 116/189). What matters in questioning then will be thoughtful reflection (cf. VA, 64/180). But this brings us to a precipitously superficial insight—and Heidegger learned this best from Nietzsche—an insight reduced by trivializing convention. No sooner does one broach such a set of reflections, Heidegger warns, but "just as quickly, indeed, the next day, it is transmitted as the cliché: everything turns upon question-worthiness [*alles kommt auf die Fragwürdigkeit an*]" (WD, 113/185). Although, "with such an invocation one seems to belong amongst those who question" (WD, 113/185), such questioning is almost always other than authentic or genuine. We need then to be careful as we follow Heidegger in his reflections on thinking (and questioning) to avoid the lure of reductive convention.

As Heidegger distinguishes science (even as philosophy) and what he calls thinking (even as philosophy), when he says that "science does not think" [*die Wissenschaft denkt nicht*] (WD, 4/8; cf. WD, 57/33, 154/134, 155/135), he also observes that thinking per se does not figure in the calculative project of professional and university mathematicians and scientists. But, reciprocally, knowing as such turns out not to be the excellence of thinking ("thinking knows essentially less than the sciences" [WD 57/33]) and thinking is from its inception, distant from conceptualization as such: "the totality of the great thinking of Greek thought, Aristotle included, thinks non-conceptually [*begrifflos*]."[25] Heidegger's most critical claim here is that thinking is effectively impotent, inherently inefficient: "a doing that effects nothing" [*ein Tun, das nichts bewirkt*].[26]

Such "impotent," "ineffective" reflection remains, however, "more provisional, more forbearing, and poorer in relation to things" (VA, 65-6/181) and for this reason, thoughtful reflection is not to be reduced to analytic "problem solving."[27] In the attempt to learn such reflective or sense-attuned thought (as opposed to the technical culture of problem solving, manufactured knowledge, or even practicable wisdom, etc.), "we must [instead] allow ourselves to become involved in questions that seek what no inventiveness can find" (WD, 5/8). Heidegger specifies the calculative inquiry characteristic of modern technoscience (and the current environmental crisis, indeed, the current economic crisis, shows that little has changed), by contrast, as a solution-obsessed project: the "inquiry that aims straight for an answer. It rightly looks for the singular answer, and sees to it that the answer is found. The answer disposes of the question. By the answer, we rid ourselves of the question" (WD, 160/158). The circumstances of the current array of crises exemplify the limitations of such an approach while indeed underscoring its incorrigibility at the same time. Having "disposed" of the question, we are hard pressed to consider alternative solutions.

As a result and to conclude this section with an ecologically tuned variation, we emphasize wind power as an "alternative" energy source to invoke a technical fix, as it often is called by proponents and critics alike: ignoring the deadly consequences of the technology for the population of hawks and other birds of prey (and the boom in rodents, and the concomitant effects on plant life and soil, etc., etc., among the most immediate of consequences) but we also disattend to or fail to reflect on the long-range limitations of the wind as such. In other words, technically speaking, as wind technology currently stands, the wind towers currently manufactured and installed,[28] presuppose, like water power and like power derived from coal and gas, not only constant *but also* relatively intense prevailing winds (precisely as they are engineered to replicate the productive dynamics of water and carbon fuels). Furthermore, such alternative technology presupposes, giving the parallel limitations of the existing design of wind technology and electricity, well apart from any biotic concerns, that such large structural arrays as the fields erected in the western United States or Canada have no influence on those same winds. But what calls for thinking in any modest sense (i.e., not only in Heidegger's) is the question that asks where the effect begins and indeed, like the wind, whither it goes.

SCIENCE IN THE WAKE OF THE QUESTION CONCERNING TECHNOLOGY

In "Science and Reflection," Heidegger argues that science is more than a merely human or cultural, activity. In its technological articulation, contemporary research science increasingly defines and takes the measure of what counts as "real."

> Although science orders the object world with exacting precision, it fails to respond to the basic questions of being. More than a failure to respond, science as *the* modern epistemology erases the possibil-

ity of such questions in its refusal to accept their presence as that which cannot be arranged as objects. In striving for unmediated certainty, science overlooks the principle that is the precondition for its very possibility as a way in which truth is revealed: all scientific knowledge is *necessarily mediated* through a structure of representational knowing. (VA, 152/167–68)

Denying anything out of its ken as unreal, modern technological science dominates the globe in a singularly irresistible fashion that puts paid to the pluralistic rhetoric of globalization. If one could debate (as has been repeatedly done) the virtues of "folk" (or local) science—be it in terms of castigating Nazi science or the more respectable (because resolutely and hence conventionally science-celebratory) terms of a Bachelard, or a Foucault, or a Caanguilhelm, or a Braudel, or a Serres, or if one once considered the case for the insights of alternative traditions, as Feyerabend, in rather more incendiary fashion, argued with respect to astrology (notoriously going so far as to feature his own horoscope as frontispiece to one of his books),[29] or else to compare Western culture with Eastern acupuncture, these debates and considerations have become increasingly less not more cogent, and this is so despite Foucault, Deleuze, and certainly despite Lyotard.[30]

Globalization—another name for challenging-revealing—reigns even in the sciences themselves as the very nontechnological essence of modern technology. As it might be expressed in an Aristotelian sense, science comes forth as the sheer desire to know and is accordingly accounted a universal characteristic of humanity. If Heidegger uncovers the heart of modern technology as revelatory truth—a claim that turns out to be equally counterintuitive for both anti-technologists and pro-technologists (Luddite and anti-Luddite)—arguing that "something else reigns" in modern technology and accordingly in modern science, a "something other" that prevails "throughout all the sciences, but remains hidden to the sciences themselves" (VA, 42/156), he also argues that this "something other" is an inherently incorrigible state of affairs for the sciences. And we are, in Nietzsche's words now, "still pious" to the degree that we find this the least forgivable of Heidegger's many sins against the ideal of modern science. Heidegger argues against the technological fix and consequently against the promises of scientific salvation as the remedy for the global or environmental problems generated in the wake of modern science and modern technology.

When Heidegger explains that "physics as physics can make no assertions about physics" (VA, 61 176), that, as "the theory of language and literature, philology is never a possible object of philological observation" (VA, 61/176), his point, as the reflective priority of philosophy was noted at the outset, is to underscore that no matter whether its object be nature or poetry, science as such can never be "in a position to conceive and represent" its own essence (VA, 161/176). But, he continues, "if it is entirely denied to science scientifically to arrive at its own essence, then the sciences are ultimately incapable of gaining access to that which is not to be gotten around [that is nothing less than exactly their respective subject matter] holding sway in their essence" (VA, 62/177). This foundational "not to be gotten around" is the most elusive element of Heidegger's reflection on science (and reflection as such). Jacques Lacan names this the Real as we also hear this in Gaston Bachelard and recently and more popularly in Žižek.[31] This same elusive/obtrusive inconspicuousness is also for Heidegger, and this is no accident, that same nature that likes to "hide."

The *unscheinbar*, which can be translated as the inconspicuous, unimposing, insignificant, draws no attention to itself because it is, as inconspicuous, precisely not a problem: we do not attend to what is "intractable and inaccessible" (VA, 62/177). Here philosophic reflection on science comes specifically (this is rare) to expression in Heidegger's text. "Today," he writes, "we philosophize about the sciences from the most diverse standpoints." But "through such philosophical efforts, we fall in with the self-exhibition that is everywhere attempted by the sciences themselves in the form of sympathetic résumés and through the recounting of the history of science" (VA, 63/178).

If philosophy has the advantage and if one needs this philosophic advantage or priority for the reflective and indeed creative or revolutionary sake of science itself, why is the philosophy of science not more receptive to Heidegger's considerations? As Heidegger had already reflected in his "Letter on Humanism," the gatekeepers of the academy have already ruled on what is clear and what is unclear (Heidegger here speaks almost as Arendt speaks of the "peculiar dictatorship of the public realm," but in a different voice, perhaps echoing a sentiment he might have shared with Adorno: "language comes under the dictatorship of the public realm which decides in advance what is intelligible and what must be rejected as unintelligible") (GA 9, 317/221), and this ruling works against Heidegger's remonstration to go back before the very beginnings of philosophy, to "free ourselves from the technical interpretation of thinking" (GA 9, 314/218). This undertaking would be difficult in any case but it is rendered still more difficult given the then and still current policing of thought: "philosophy" by which Heidegger means the philosophy that claims the academic title or rank of the same, finds itself "in the constant predicament of having to justify its existence before the 'sciences.' It believes it can do that most effectively by elevating itself to the rank of a science. But such an effort is the abandonment of the essence of thinking" (GA 9, 314/218-9).

That means of course, and this remains endemic to the philosophy of science, that philosophers of science so far from emphasizing what Heidegger calls thinking (and a discussion of this is another topic altogether) endeavor instead to imitate the sciences. As philosophers of science are quite self-avowedly party to the scientific cause, this modeling of philosophy on science inevitably seems to exclude critique (or anything that looks like it). Said more routinely: to be thought anti-science in any fashion would be a nightmare for philosophy of science.[32] The philosophical problem with this anxiety is that it excludes critique.

In professional philosophy of science, neutrality or objectivity requires less questioning (Heideggerian or otherwise) than a conscientiously "sympathetic" and even emphatically pro-scientific approach. Philosophers of science stop well short of putting science in question and are consequently as suspicious of Heidegger as they are of Nietzsche but also and not less of both Feyerabend, at one end of the political spectrum, and dialectical critique and its several variations, on the other.[33] To these confessional limitations, Heidegger adds the specific limitations of science itself. Hence and almost inevitably, as philosophers of science we are inevitably inclined to miss, as the sciences for their own part likewise miss, what Heidegger named that which "is inaccessible and not to be gotten around" (VA, 62-4/177–79), a circumstance of nonadvertence which means and entails that this same inaccessibility "remains in inconspicuousness" (VA, 63/178).

But for Heidegger, and this is what confounds most readings of his position on science and the scientific worldview, here as in reflections on modern technology, and indeed in his related reflections in the *Spiegel*-interview, the problem is *not* the human inquirer, no matter whether philosopher or scientist or journalist. Like the essence of technology, the problem for Heidegger resides neither within us nor within our power to change.

Heidegger's deliberately gnomic suggestion that the sciences "lie" in this same very "inconspicuous state of affairs as a river lies in its source" (VA, 63/179) is drawn out of his extended hermeneutic reflection on what had been a formerly unremarkable or seemingly obvious (i.e., as we now see: *inconspicuous*) claim that "*science is the theory of the real*" (VA, 42/157). Saying that modern science "is the theory of the real" distinguishes it from its ancient and medieval counterparts. The distinction is not a matter of development or progress and hence it is more than a merely historical distinction.[34] A perfectly paradigmatic shift characterizes the exactly incommensurable differences between modern, medieval, and ancient Greek science and this shift has inspired several scholars who read Heidegger together with Kuhn on the nature of research science.[35] Heidegger's own account here is more historical than sociological in tenor.[36]

An observation that aims at empirical reports, or what counts as the same in the tradition of the medieval schoolman (we may think of Roger Bacon just as Heidegger also invokes him in this context), experience, "remains essentially different from the observation that belongs to science as research" (H, 81/121). For the experiment, as opposed to the contingent, empirical domain of experience, requires a pre-established rule, a stipulated law, and hence the very institutional and empirical framework of modern experimental science constitutes as such the very base-line of calculability and calculation: "to set up an experiment means to represent or conceive [*vorstellen*] the conditions under which a specific series of motions can be made susceptible of being followed in its necessary progression, i.e., of being controlled in advance by calculation" (H, 81/121). Because such an experiment is the expression of a projected law, one has both a criterion for and a limitation on possible results. This is, of course, the very possibility of measurement.[37] Heidegger will emphasize that experimental measurement is essentially not a matter of experience per se. The scientific observation made through experiment is not a matter of experience, be it personal or common. For Heidegger, it is "only because modern physics is a physics that is essentially *mathematical* that it *can be* experimental" (H, 80/121, emphasis added).

Beginning from a preliminary "ground plan of nature and sketched into it," literally calculated or built into it, Heidegger observes that "experiment is that methodology which in its planning and execution, is supported and guided on the basis of the fundamental law laid down, in order to adduce the facts that either verify and confirm the law or deny its confirmation."[38] This same *formulaic*, *methodological* rule generates the necessary *specialization* of modern science, which is, in turn, for Heidegger, less a result of the proliferation of results and progressive discoveries of modern science, than the presupposition *for* those same cumulative results and as such.

Almost more fundamental than mathematics and measurement, indeed even more than the scientific method per se, it is the ineliminably institutional nature of science that constitutes and only thus exemplifies modern science and it is this institutional character that distinguishes the collective dimension of modern science for Heidegger.[39] Merleau-Ponty, for his own part, also would attend to this same inherently collective group character of objectivity and scientific research. This scientific collectivity is at the same time amenable to a specific concentration or focus. Thus, the institutionally calculative project of modern science simultaneously, so Heidegger argues, yields its own isolation or field, that is, its disciplinary specialization. Here too and again an innocently simple connection can be made between Heidegger and Kuhn, but at this point it is perhaps more important to note the critical potential (sociocritical and historio-critical) of Heidegger's argument.[40] Everyday *busyness*, that is, the progressively accelerating enterprises of science in all its variants becomes the *business* of "science as usual," which Heidegger for his part (and it is only at this point that Heidegger introduces a critical reserve), analyses as the motor of the degradation of "ordinary science" (this would be what Nietzsche in a similar fashion named decadence):

> Ongoing activity becomes mere busyness whenever in the pursuing of its methodology, it no longer keeps itself open on the basis of an ever-new accomplishing of its projection-plan, but only leaves that plan behind itself as a given; never again confirms and verifies its own self-accumulating results and the calculation of them, but merely chases after such results and calculations. (H, 97/138)

For Heidegger, as for Kuhn (and, at least on my reading, even Feyerabend as well), a kind of decadent progress is not the default of science but much rather a sustainable degradation is here the key to the success of modern research programs.[41] Here, Heidegger's insight echoes the spirit of similar insights offered by Herbert Dingle (in connection with chemistry)[42] but also Rom Harré or Peter Medawar, and so on. And yet, the potential for a classically critical philosophy of science—and we have noted that this has numerous expressions, some continental, some not—continues to remain untapped.

In practice, as is well known today in the wake of social, political, and historical studies of both science and technology, institutional science belies the naive innocence of the "open" ideal of pure science.[43] The inherent stultification of regularized, conventionalized science must be resisted or "fought against at all times," so Heidegger insists, just "because research is, in its essence, ongoing activity" (H, 97/138). This focus on institutionalized creativity is one of Heidegger's key insights, an insight anticipating social studies of science and technology and one could take it in another sense to answer some of the more typical objections to Heidegger's understanding of science and technology as limited to (and therefore as only binding for) the science and technology of his own age.[44]

The institution that is also the *instrumentarium* of modern science (its technological being in the world) exemplifies modern science. Because scientific research is necessarily institutional, Heidegger argues (here in complete accord with Max Weber), that modern science is inherently systematic. "Ongoing activity in research is a specific bodying-forth and ordering of the systematic, in which, at the same time, the latter reciprocally determines the ordering. Where the world becomes picture, the system, and not only in thinking, comes to dominance" (H, 101/141). To put this in other words, namely in the language of "The Origin of the Work of Art": Science has its own fully institutional *world*, its *creators* and co-creators or *preservers*. But where, in the context of art, the world of the ancient work of art is lost, where the world of the past or contemporary artist is likewise temporary and so similarly vulnerable, the world of modern technological science is inherently secure. "Theory makes secure at any given time a region of the real as its object area. The area character of objectness is shown in the fact that it specifically maps out in advance the possibilities for the posing of questions" (VA, 53/169).

The world of modern technological science is built into the calculating inquiry of the scientific enterprise, with such overwhelming presence that, so Heidegger would argue, it has literally transformed the earth, and here—as is seen later in more detail—he makes a distinction. Today, the modern, techno-scientific project is poised to transform or absorb *life* itself. But before we return to the question of life already mentioned at the outset, it is important to consider the physical, already constructed or manufactured, scope of technological enframing.

The technological framework corresponds to the current array of scientific laboratory equipment, including linguistic and logical analysis as well as computer simulations and models, but also universities and government research agencies and the still broader field of technological industry and its influence in the culture of the everyday contemporary world at almost every cultural and social level on a global scale. Heidegger here goes beyond constructivism: Confirmable results (i.e., what will count as *viable* predictions) are possible (and indeed *replicable*) because from the start they are built into the project (and often the mechanics) of research science. Thus, in 1938 Heidegger could point to the extension of this ground plan from the natural to the social sciences, using historiography as illustration. At the same locus, Heidegger nicely describes (rather like David Lodge without the charm of the academic novel) the life of the modern academic in any field, particularly, to extend his observation on present-day terms, the academic researcher equipped with wireless technology or Internet access. In place of the traditionally polymathic scholar, Heidegger observes, we have "the research man engaged in research projects" (H, 85/125), and the life of such a new-age research academician is all about the cutting-edge of research: the latest discovery.

This Heidegger opposes to the old-fashioned image of "erudition" or book learning. And the whole point concerns the already at the time growing irrelevance of books. Today and in the current world of online and electronic research tools, Heidegger's observation holds more than ever: "the researcher no longer needs a library at home. Moreover he is constantly on the move. He negotiates at meetings and collects information at conferences. He contracts for commissions with publishers" (H, 85/125). The consequences of such an emancipation from the culture of reading to the culture of writing (Nietzsche reminds us that the two are mutually exclusive) to today's culture of "virtual" reading (what else are we doing when we download a text?) is still more volatile than in the era of the journalism Heidegger found so reprehensible because inherently nonreflective.

For such a media critique, we need more than the often forgotten insights of a McCluhan, we need, despite their forbidding Francophone complexity (a matter of academic context), a Baudrillard or a Virilio—that is, we need, as a hermeneutic and phenomenological supplement, a sociology of information and, perhaps still more crucially, of misinformation, a sociology of the communication/information market.[45] Beyond instant messages, blogs that crown themselves the arbiters of good and bad graduate schools (in philosophy, no less), email, cell phones, and so on, we have web-only journals, established overnight and as quickly out of operation, and there is no way here to point to the latest techno-manifestation. In this fashion, "established" scholarship is a self-perpetuating, self-reinforcing structure: One still "contracts" as Heidegger put it, for "commissions with publishers." But such contracts are the stuff of certification, credibility, legitimacy. Science as worldview, even philosophy as the same, is inevitably circular and this circularity is its virtue rather than a deficiency or a weakness one might rectify. Apart from this insular circle, everything else will be adjudged "unproven," everything else, in other words is denied the name of science, dubbed pseudo- or bad science, like "bad" philosophy, and discounted out of hand. Science (and academic scholarship in general) simply denies its attention to dissonant points of view (one does not cite these) or judges them (and this is a circular business that always comes out 'even' as it were) solely on its own terms. Thus, what the philosophy of science calls demarcation (what mainstream philosophy calls philosophy, and science recognizes as science) depends on just such self-absorbed, self-referring circularity.

For Heidegger (as for almost anyone writing on science), modern scientific research knowledge is all about calculation, as this calculative essence can be discerned in its methods, its corrective self-assessments, and its results. In its determinative prescription of the object of science as such, the calculative essence of modern science constitutes the scientific object: "whatever is called to account with regard to *the way in which* and *the extent to which* it lets itself be put at the disposal of representation" (H, 86/126, emphasis added). This is also at work in what Heidegger names *Ge-Stell* in his postwar essays on objectification and technology, which he also calls "insight" into what is, and elusively addresses as "the danger" in his lectures to the Club of Bremen.

In "The Age of the [Scientific] Worldview," which first appeared in *Holzwege* (1952), but is (instructively) contemporaneous with the composition of the notes that would become Heidegger's later posthumously published *Beiträge*, Heidegger had observed that the "objectifying of whatever is, is accomplished via a setting before, a re-presentation, that aims at bringing each particular being before it in such a way that the human being who calculates can be *sure*, and that means, *certain*, of that being" (H, 87/127, emphasis added).

Security and *certainty* in this case relates to the absolute centrality of the human being as the "measure" of all that is.[46] The human (this would be Nietzsche's "last" or "highest" man), takes him- or herself to be the center of the world, as Nietzsche expressed this human-focused perspective (a centrality, as Nietzsche reminds us, that is common to both the religious ascetic ideal and the scientific ideal). Because Heidegger's own reference to Nietzsche invokes the then-popular (indeed, *still* popular) reading of Nietzsche as the prophet of nihilistic humanism, and because Heidegger seems to condemn Nietzsche in these terms, it can be easy to overlook the extent to which Heidegger for his own part borrows Nietzsche's analysis of the earth- and body-denigrating (but exactly self-aggrandizing) consequences of the ascetic ideal practiced in religion, particularly the Judeo-Christian and Islamic tradition, with its doctrine of personal immortality.[47] Nietzsche summarizes this analysis in his *Anti-Christ* (a text that may instructively be alternately translated as *The Anti-Christian*), "the 'salvation of the soul'—in plain logic: 'the world revolves around me'."[48] The critique of humanism thus understood as a critique of the central role of humanity *tout court*, is thus the pendant to Heidegger's reading of Nietzsche's analysis of the same relative centrality of the human being as "the lord of the earth" which Heidegger articulates, using Nietzsche's words (although he was hardly the only one to do so), with reference to National Socialism and its broadest ambitions: "Man as a rational being in the age of Enlightenment is no less subject than is man who grasps himself as a nation, wills himself as a people, fosters himself as a race, and, finally empowers himself as lord of the earth" (H, 111//152). For Heidegger, regarded in terms of the "human" in the age of science, this nihilistic or nihilating impetus is not simply the matter of the human as the center of attention in a divine universe. Beyond even the wildest dreams of a medieval sense of the world, today "the human being becomes the relational center of that which is as such" (H, 88/128). Modern techno-science is a humanism.

From the perspective of human domination, this would be Nietzsche's (not Hegel's) lordship. The modern world is literally transformed into a picture, a transformation that supposes indeed that we have attained the fantastic position first proposed by Archimedes,[49] the ingenious Greek technician commemorated by Descartes at the start of his second Meditation.[50] Despite the fondness of theorists for discussions of time, this spatial dispositionality is the ecstatic locus for French philosophers of technology such as, on the hand, Bachelard, Simondon, and Lefebvre, and, on the other Ellul, Baudrillard, and Virilio. "Wherever we have the world picture, an essential decision takes place regarding what is in its entirety. The Being of whatever is, is sought and found in the representedness of the latter" (H, 89–90/130).

De novo, such a representational understanding of the material world (and of God) was Descartes' starting point when he began the process that shattered the scholastic conceptual relation of the human being to the world and to God. If the modern view of the world (as picture) differs from the medieval order of creation or chain of being, the modern scientific representation of the world is further still from the ancient Greek interpretation. For the Greek, as Heidegger reads Parmenides's coordination of thinking/ knowing and being, *To gar auto noein estin te kai einai*, the relationship to what is is almost perfectly reversed (i.e., at least as regarded from the modern perspective or point of view): "That which is, is that which arises and opens itself, which as what presences comes upon the one who himself opens himself to what presences in that he apprehends it" (H, 90/131). This more surrendered, dispositionally available orientation to that which is (for both Nietzsche and Heidegger, the key to the Greek world), is lost to our modern sensibilities: "in order to fulfill his essence, Greek human being must gather [*legein*] and save [*sōzein*], catch up and preserve, what opens itself in its openness, and he must remain exposed [*alētheuein*] to all its sundering confusions. The Greek human being is as the one who apprehends that which is, and this is why in the age of the Greeks the world cannot be a picture" (H, 91/131).

This same and vulnerable exposition to what is also constitutes the possibility of the modern perspective on the world as picture. The world is thus transformed into an objective correlate, set or posed as such: "ready" for a representing subject. Intriguingly, the same insurgence of the human as measure and reference point (objective representation and subjective relevance) is also the era of increasingly leveled uniformity. This humanism describes the trajectory of modernity for Heidegger: "In the planetary imperialism of technologically organized man, the subjectivism of man attains its acme, from which point it will descend to the level of organized uniformity and there firmly establish itself. This uniformity becomes the surest instrument of the total, i.e., technological, rule over the earth" (H, 111/152–53). Paradoxically, Heidegger points out, this technological dominion corresponds to the utter loss of liberty: the "modern freedom of subjectivity vanishes altogether in the objectivity commensurate with it" (H, 111/152–53).

The global delimitation of "reality," understood in a techno-practical fashion as the realm of what *works* and *can be worked upon*, belongs to the essence of modern science.[51] Scientific truth for Heidegger (as for the rest of us too, no matter where we stand, whether for or against Heidegger), proves itself via its efficacy, that is: "the efficiency of its own effects" (SD, 64/58). As the theory of what works, science is the theory of what is (or counts as) Real [*Wirkliche*] as "the working, the worked [*Wirkende, Gewirkte*]; that which brings hither and brings forth into presencing, and this which has been brought hither and brought into presencing. Reality [*Wirklichkeit*] means, then, when thought sufficiently broadly: that which, brought forth hither into presencing, lies before; it means the presencing, consummated in itself, of self-bringing forth" (VA, 45/160).

When we say "science is the theory of the real," we are already in the sphere of the transformation of the Greek meaning of theory into the Roman precinct of contemplation. "The Romans translate *theōrein* by *contemplari, theōria* by *contemplatio*. This translation, which issues from the spirit of the Roman language, that is, from Roman existence, makes that which is essential in what the Greek words say vanish at a stroke" (VA, 50/165). In the age of modern technological science, we have gone from an attentive regard for what presences to "observation [*Betrachtung*]" (VA, 51/167). We now have to do with an encroachment on the effective, technologically relevant domain of the Real, an ensnaring or a challenging forth, "specifically through aiming at its objectness" (VA, 52/167).

The result here is nothing less momentous for the very possibility of modern science than *causality* itself, that is: the very ideal of causal thinking that is of course the essence of calculation. Henceforth, in its modern aspect, for modern science, what is "real now appears in the light of the causality of the *causa efficiens*" (VA, 46/161). That is to say, "every new phenomenon emerging within an area of science is refined to such a point that it fits into the normative coherence of the theory" (VA, 53/169). Note here that, as such, this last sentence would hardly be out of place in any expression of the philosophy of science. But Heidegger reflects on the relevance of "objectness" implicit in the "essence of what is called 'end' or 'purpose.' When something is in itself determined by an end, then it is pure theory" (VA, 53/169). Thus Heidegger argues that the "assertion that modern atomic physics by no means invalidates the classical physics of Galileo and Newton but only narrows its realm of validity" (VA, 53–54 /169). This advance calculability is the essence of the "objectification of the real" (VA, 54/170). This transition is possible only because we are speaking objectively of the object that is now ineluctably (for whatever use or purpose we suppose we wish to make of it), disposed to use. To use the terms we recall from Heidegger's essay on technology, *standing reserve* encompasses either immediate or reserved use and this last represents nothing less than the ideal of *sustainable development* as we talk about globalization and the most efficient development of the resources or *reserves* of the earth today, East and West, and above all: North and South.[52]

So, far from the theoretical distance of Greek science, that is, the heart of reflective restraint or contemplation,[53] contemporary science "sets upon the real" (VA, 52/167). Science further orders reality "into place to the end that at any given time the real will exhibit itself as an interacting network, i.e., in surveyable series of related causes. The real thus becomes surveyable and capable of being followed out in its sequences. The real becomes secured in its objectness" (VA, 52/167–68). This utterly objective, technologically effective setting upon is the essence of modern science: The "pure" theory of science in every case will draw its applicable and working power from this same essence.[54] That is, once again, the ideal of a *causa efficiens*. Thus, Heidegger explains that "because modern science is theory . . . in all its observing [*Be-trachten*], therefore, the manner of its striving after [*Trachtens*], i.e., the manner of entrapping-securing procedure, i.e., [*the scientific*] *method*, has decisive security" (VA, 54/169, expansion and emphasis added).

It is in this securing-calculating context that Heidegger cites Max Planck's definition of "the real" as "that which can be measured" (VA, 54/169). The philosopher invokes the physicist's definition as a philosopher reflecting on science and thus in order to explain that "the decision about what may pass in science, in this case in physics, for assured knowledge rests with the measureability supplied in the objectness of nature and in keeping with that measurability, in the possibilities inherent in the measuring process" (VA, 54/169). It is important to emphasize that such calculation is of the essence of modern science as such, not only the putatively mathematical sciences but also the social sciences; hence the rage to appropriate quantificational rather than qualitative models is not merely the default of tradition.[55] Disciplinary specialization follows upon the nature of modern science as the theory of the real, delimited in each case in terms of object areas but no longer in terms of a focus on explanation (calculation) and understanding (interpretive or hermeneutic reflection).

Specialization is not then a result of the need to delimit and so to manage the sheer accumulated bulk of knowledge today, as we may tend to assume, in contrast say with the "amount" of knowledge to be had a century ago. There is not *less* to know in *superceded* knowledge schemes. One can, as we already noted, call these "folk" sciences but it is more honest if less incendiary to speak, as Foucault reminds us we must, of transcended, discarded or denigrated sciences, that is: "subjugated knowledges," those sciences we refuse to call sciences as opposed to those currently in favor. This Foucault clarified in his 1976 lectures to the Collège de France by defining "subjugated knowledges" to include "whole series of knowledges that have been disqualified as nonconceptual knowledges, as insufficiently elaborated knowledges: naive knowledges, hierarchically inferior knowledges, knowledges that are below the required level of erudition or scientificity."[56]

For Heidegger, rather than serving as a means of compartmentalizing and so dividing an otherwise unsurveyable mass of information, specialization is simply a "consequence of the coming topresence of modern science" (VA, 55/170). In this way, mathematics

> is not a reckoning in the sense of performing operations with numbers for the purpose of establishing quantitative results; but, on the contrary, mathematics is the reckoning that, everywhere by means of equations, has set up as the goal of its expectation the harmonizing of all relations of order, and that therefore 'reckons' in advance with one fundamental equation for all merely possible ordering. (VA, 54–55/170)

Whether we are speaking of classical physics or of quantum physics, whether of sociology or psychology, Heidegger can say, "nature has in advance to set itself in place for the entrapping securing that science, as theory, accomplishes" (VA, 57/172–73). This entrapping-securing is measurement or calculation, trading on a certain economically based resentment in the process. For Heidegger, everything including the subject–object relation, becomes "sucked up as standing-reserve" and an issue of "standing-reserve to be commanded and set in order" (VA, 57/173). Invoking his own essay on technology, Heidegger emphasizes that the constancy of standing reserve is "determined out of Enframing" (VA, 57/173). Thus the subject–object relation "attains to its most extreme dominance, which is predetermined out of Enframing" (VA 57/173).

THE QUESTION OF SCIENCE, THE QUESTION OF QUESTIONING, AND *GELASSENHEIT*

The watchword of Heidegger's philosophic reflection on modern science and modern technology is, so I have argued, nothing less than questioning, albeit and again, to be sure, questioning of a particular kind, expressed with an exigency that he expresses as "piety" or "devotion" as he concludes his essay on the nature of the question in the wake of technology. Heidegger had already suggested this "dedication" as safeguarding and would maintain this point throughout his life, "humanity will know, that is, carefully safeguard into its truth, that which is incalculable, only in creative questioning and in shaping man of the future into that 'between' in which he belongs to Being and yet remains a stranger amid that which is" (H, 96/136).

From the beginning as we have now seen, Heidegger contrasted authentic or genuine questioning to the answer-bound or problem directed inquiry: the kind of questioning corresponding to "curiosity" and investigative "research." By contrast with answer-bound inquiry, questioning as such, *authentic* or *creative* questioning, has always to overcome itself, that is, to open itself up into its own questioning after. In this way, questioning is always underway to questioning: Authentic questioning is inherently a matter of overcoming its own orientation and disposition.

Against logistics, and against positivism, and, above all, against technological scientism—and there is no other kind—Heidegger proposes the radical poverty, which he speaks of as the "impotence," of reflective thinking.[57] Heidegger (always stressing the modal dimension), suggests that we *might* yet find a way beyond calculative thought: "the poverty of reflection is the promise of a wealth whose treasures glow in the resplendence of that uselessness which can never be included in any reckoning" (VA, 66/181).

Heidegger's language in *The Question Concerning Technology*, of the "here and now," of the small (as opposed to the grand), which he invokes in terms of what "summons us to hope in the growing light of the saving power" (VA, 37/33) includes a restrained awareness that although one can reflect, "through this we are not yet saved," that is to say, as he says it again, to "foster the saving power in its increase," "here, now, and in little things" must "include holding always before our eyes the extreme danger" (VA, 37/33). Heidegger emphasizes that a resolution is exactly not a matter of human endeavor or will.

Heidegger asks, extending Kant as he does so, "What is the human being?" (GA 3, 207/141). This question in our modern age, conceived as Heidegger conceives it as the quintessential age of humanism, emphasizes the unremitting anthropocentrism of the current environmental crisis. This same concern with anthropocentrism is relevant for Heidegger in his debate with Cassirer and it is the reason Heidegger articulates *Being and Time* by means of an inquiry into that being that is always mine to be, always ours to be, *Da*-sein. The danger of modern technology is likewise the key to this fourth question, as Heidegger expresses this question beyond anthropology, beyond what he called humanism.

In the age of modern technology it is revealing as such that is threatened "with the possibility that all revealing will be consumed in ordering and that everything will present itself only in the unconcealedness of standing reserve" (VA, 38/33). In other words, "Enframing," that is, "the essence of technology, as the destining of revealing, is the danger" (VA, 32/28). As a "destining," Enframing, Heidegger had already noted, "banishes the human being into that kind of revealing which is an ordering." And we recall that wherever "this ordering prevails, it drives out every other possibility of revealing" (VA, 31/27). But this danger, both to Da-sein but also to what is, *cannot*, as we recall be countered by what we do. Hence in "The Turning," Heidegger reminds us, using Nietzschean, and indeed, a very Adornoesque trope to repeat this warning: "we do not yet hear, we whose hearing and seeing are perishing through radio and film under the rule of technology. The constellation of Being is the denial of world, in the form of the injurious neglect of the thing. Denial is not nothing; it is the highest mystery of Being within the rule of Enframing" (GA 79, 77/48–49). It is Being itself that refuses itself. The matter has everything to do with being, with "what is," (GA 79, 77/49) and this is the reason Heidegger (echoes of his reflection that "only a god can save us"), further reminds us that "whether the god lives or remains dead is not decided by the religiousity of men and even less by the theological aspirations of philosophy and natural science. Whether or not God is God comes disclosingly to pass from out of and within the constellation of being" (GA 79, 77/49).

DISCOURSE ON THINKING AND THE TASK OF GELASSENHEIT

If I hold Heidegger's reflections indispensable for a philosophy of science, I also think I do so for reasons shared by almost anyone who reflects on Heidegger and science. The contributions to the current volume bear witness to this collectivity and to its diversity. Heidegger offers us a set of reflections that the philosophy of science does not otherwise have at its disposal. If my views are stronger (or a bit more Nietzschean) than most, it is because I am persuaded that Heidegger's reflections take us to the very philosophical heart of the philosophy of science as such. As Patrick A. Heelan explains us in his reflections on Heidegger's method as hermeneutical, this hermeneutic method is "the search for the structures of meaning, explicit or latent, in the World horizons, particularly those created by science."[58] Hence, Heelan reflects that the "labors of the positive sciences precede the labors of philosophy and the labors of the philosophy of science is, as it were, a 'recapitulation' of the scientific effort."[59] In this sense, Heidegger offers us an *ontological* recapitulation of science and technology.

For Heidegger, we lack a philosophical comportment to the question of the nature of modern science and technology that inquires into the ground of the same. Thus, in *Discourse on Thinking* [*Gelassenheit*], Heidegger uses the enthusiasms of science (at the time his reference was to the proclamation of eighteen Nobel prize winners who proclaimed "modern natural science" the "pathway to a happier human life" [G, 17/50]),[60] as the occasion for a reflection on our lack of reflection on modern science (and modern technology). This reflection Heidegger articulates by attempting to pose the still timely (and still practical) question: "what is the foundation on the basis of which scientific techniques can discover and liberate new natural energies?" (G, 17/50, translation amended). Heidegger's frame of reference in his 1955 lecture was the atomic era, with all its promises and all its anxieties, just ten years after the U.S. bombing of Hiroshima and Nagasaki set a pair of unimaginable, still unsymbolizable and all-too-Real (in every Lacanian sense of the Real that endures to the present day) atomic explosions to outline the end of World War II (and this post-datedness should always be emphasized and it rarely is).[61] In the wake of just this legacy, Heidegger defines "nature" as "a gigantic gasoline station, an energy source for modern technology and industry."[62] If the enigmatic promise of developing nuclear power has lost none of its appeal (i.e., despite its ongoing unworkability, problems *then* problematic, i.e., storing or containing nuclear waste, remain largely unsolved today), the current day shows up the hype of 1950s and 1960s and 1970s futurology. Heidegger is guilty here less of *doubting* the technologists of his day than of taking the engineer's word for the deed.

Thus it is important to underline that Heidegger was as much a child of the nineteenth and twentieth centuries as anyone of his time, so that he could only suppose (as we ourselves still suppose), that "the decisive question of science and technology is no longer: where do we find sufficient quantities of fuel" (G 18/51) but the different problem of harnessing and above all of controlling "unimaginably vast amounts of atomic energies."[63] Additionally, it is essential to reflect on our own subscription to this same ideal as it is unquestioned on either side of the environmental debates on global warming and so-called sustainable development. Heidegger's point is that we have apparently rendered the world vulnerable to "the attacks of calculative thought, attacks" in our own imaginative vision and still more in our ongoing technological practice "nothing is believed any longer to resist" (G, 18/50). This confidence continues to inspire both our actions and our convictions today. To this day, indeed, we regard nature "as an energy source for modern technology and industry" (G, 18/50) and all our programs to solve our energy problems presuppose this same perspective.

But even with the consummation of our wildest technological dreams, there yet remains for Heidegger a key limitation: "in all areas of existence, man will be encircled even more tightly by the forces of technological arrays and automata" (G, 19/51). This is the substance of his lecture—and it is the reason the translators rendered the title of *Gelassenheit* as *Discourse on Thinking*. For Heidegger, like the different kinds of questioning with which we began this essay, there is a gap between what we think we know and what we genuinely (i.e., authentically and creatively) understand. It "is one thing" to take account of what we hear and read about, but "another thing to understand what we have heard and read" (G, 20/52). Such reflective regard is, of course, nothing less than hermeneutic thought. Where Heidegger cites the Nobel prize-winning biochemist Wendell Stanley who had predicted that "the hour is near when life will be placed in the hands of the chemist who will be able to synthesize, split and modify living substance at will" (G, 20/52), it is clear that merely to take account of this as a news report or a curiosity (or even as a phenomenon linking the promise of science then with current reality today) still fails the meditative reflection Heidegger asks of us.

For we remain despite all our enlightenment, the children of superstition and make believe, as Nietzsche always lamented. We remain, despite our scientific sophistication as confident as ever that where nothing is seen (or heard, or felt), there is nothing. Thus, we "believe in" science; better said, we believe in the expert testimony of industry. And we believe in the PR that the consequences of a "silent spring" will be nothing more egregious or more sorrowfully serious than the loss of the birds or the bees, the lamentable but somehow tolerable (or else we would not sanction as we do in our search for energy above all the destruction of habitat as we do—and what do we do when we name the living-world of beings other than ourselves a 'habitat'? For whom? How so?).

We then propose to leverage contemporary reflections and corporate- and government-sponsored research (it's all about the sponsorship isn't it?) on sustainable development into different reflections on sustainable habitation (thus we include ourselves with other animals whose collective lifeworld, including our own, is in jeopardy), through our own acts, perhaps offshore, perhaps as in the Jules Verne fantasies of a few decades ago, perhaps undersea, today, still fantastically, perhaps off-world (and when will this be and for which of us will this be?). All sustainability projects (call it development or call it habitation) presuppose that human society can only express itself (and historically this has been true for modern and premodern cultures alike) as so very many technologically sophisticated variants on slash and burn agrarian societies: cultures capable of no more than an unremittingly aggressive devastation of their own environment, followed by the urgent colonization of other territories. The frontier, the Wild West, Star Trek . . . Rather like a child faced with a milkshake and a straw.

We are not animals, we say to ourselves, at times thinking to quote Heidegger for confirmation. We are different (we are higher, whispered Nietzsche's birds). This is human hubris as humanism. The humanistic, even theistic meaning of modern technology and its science is proven by its achievements: *our* achievements. We have inherited the earth, we do indeed have dominion over it, over all the animals that crawl, fly, swim. And be it by hunting, poisoning, neutering, genetic modification, or ordinary sacrifice (all the many unsung, extinct equivalents of the snail darter, like the dodo, which we named for its stunning vulnerability to human rapacity, like the last living Galapagos tortoise of its kind, many carted off for food by sailors, many more, let us be clear on this: when they were not eaten, virtually *every single one* of the remaining "specimens" by competing scientists from different countries all intent on "research," a scientific collector's practice, let it be underscored that involves not the living the animal, but only the dead, the "specimen"), we have proven ourselves East and West, North and South to be consummate masters at emptying the world of as many species as we can. What upsets us is neither the enormity of our temerity nor the efficiency of our destruction of species or our pollution of land or water or our veritable alteration of the air. What upsets us, as it would upset a small child, is the thought that the things we do might not be doable for all time. That it cannot be sustained is what galls us, not what we do.

Sustaining means keeping on, maintaining an exhaustive depletion of the riches of the earth that we call a "harvest" when we do not call it development, as if it were somehow "incomplete" until being taken up or fulfilled for human purposes. And this humanism is also the reason modern science, so far from being the opponent of religion, is indeed, as Nietzsche argued, its latest and best expression. And historically at least Nietzsche's claim is patently true. Philosophers, historians, and public policy theorists have debated the accuracy of Lynn White's thesis, passionately so, since it was first published.[64] For me, it is significant that only such a very millenarian thesis makes sense of the conviction and the complacency of our thought on the environment.

Heidegger's perspective here reminds us that even as we attempt to reflect on the crises of our age, we fail "to get technology" in hand as we always mean to do, we fail to reflect on the question itself, in the wake of technology, to ask after to raise the question of technology. And to that extent, we fail to ask the same question that happened also to have been the very last public question Heidegger ever posed, addressed in a letter written to a group of North Americans, asking them for their part to revisit the question of the relationship between modern science and modern technology, a question revisiting the question of priority with which we began: "Is modern natural science—as is thought—the foundation for modern technology or is it already and for its own part the basic form of technological thinking, the determinative anticipation and constant intervention of technical representation in the fulfilled and institutionalized machination of modern technology?"[65] We remain in the grip of technological remedies or problem solving—but that only says, as he constantly repeats throughout *What is Called Thinking*, that we are still, all of us, and not only the scientists among us, not thinking. In "The Turning," Heidegger reminds us of what is involved in the project of reflective thinking, the project of catching ourselves in the act, an impossible move, even where the problem we mean to solve by means of such reflection is the problem of modern technological science itself

All mere chasing after the future so as to work out a picture of it through calculation in order to extend what is present and half though into what, now veiled, is yet to come, itself still moves within the prevailing attitude belonging to technological calculating representation. All attempts to reckon existing reality morphologically, psychologically, in terms of decline and loss, in terms of fate, catastrophe, and destruction, are merely technological behaviour. (GA 79, 76/48).

The challenge for thought is not that of technology (be it the challenge of advancing or mastering those same advances). What is entirely "uncanny" is not the increasing technicization of the world (whether mystified or demystifed) but our own lack of unpreparedness "for this transformation" of the world in the veritable image of technology as such. This is, again, "our inability to confront meditatively what is really daring in this age" (G, 20/52). This "daring" is what is really radical, really dangerous today.

It is enough to say here, but it would take much, much more to think it through, that when Heidegger turns his reflection on the challenges of modern scientific technology, on an occasion of commemorative recollection, turning to the groundedness of the works of man, Heidegger turns to reflect on the earth itself. For it is with reference to the earth that Heidegger suggests the strategy of *Gelassenheit*: to let and to let be. As a "comportment toward technology which expresses 'yes' and at the same time 'no,'" Heidegger's yes and no comportment, letting *and* letting be, means, as he tells himself "releasement to things" [*Die Gelassenheit zu den Dingen*] (G, 23/54).

With the paradox, using the *tactic* (if we can borrow a word from the sociological philosophies of de Certeau and Foucault, Bourdieu and Baudrillard) of letting and letting be, Heidegger proposes a way to step back, to step apart from, and thus, so we may surmise, a way of going with and beyond technology itself. The strength of the term and its promise he would have learned from his Asian students and interlocutors, but he had already heard this from Goethe (despite his critique of the latter's Romanticism), as he had it from Meister Eckhart, or and still more, from Angelus Silesius. But he would also have it, as he tells us too, from Nietzsche and from Hölderlin.

"When we look into the ambiguous essence of technology, we behold the constellation, the stellar course of the mystery" (VA, 37/33). Looking into the ambiguous essence, we are moved toward the paradox, the constellation of mystery. This insight is insufficient as it fails to take account of what Heidegger called the high sense of technology's ambiguity. It is when we are attuned to this "lofty ambiguity" that Heidegger's practice of *Gelassenheit* has its relevance, a practice that, like any practice, is always easier said than done. As a way of life, such practices require attention and mastery.

In this sense, *Gelassenheit* has an intriguing affinity with the Stoic disposition of equipollence or *ataraxia*. Yet there is nothing essential about this reference to the Stoics and I would urge that we should indeed think of the prior context for the Stoics themselves more than Socrates, the *physiologoi* as Aristotle spoke of them: the Pythagoreans and Empedocles but also including the Orphic tradition—and those with other tastes can invoke other cultural meditative practical traditions. What is important is that such traditions take us, as both Heidegger and Nietzsche urged us to go, beyond humanism.

Recalling mystery, reminding ourselves that there are unanswered questions, indeed, inherently unanswerable questions, for such questions are the essence of paradox, of mystery, of riddle, of enigma, we can "keep ourselves open to the meaning hidden in technology" (G, 24/55). And that very small effort, so easy to lose or to botch is precisely what there is for us: here and now, and in little things. Such an "openness to the mystery," (G, 24/55) the only option we have, is little enough, but for most of us, and it is worth thinking about this: It has been *more* than we can manage.

Perhaps, to end with a word of hope, until now.

NOTES

1. Motto to *Der Zeitbegriff in der Geschichtswissenschaft* (1915) in GA 1, 357.

2. WD, 4/8 et passim. For discussion, see Dmitri Ginev, *A Passage to the Idea for a Hermeneutic Philosophy of Science* (Amsterdam: Rodopi, 1997) as well as the contributions to Babich, ed., *Hermeneutic Philosophy of Science, Van Gogh's Eyes, and God: Essays in Honor of Patrick A. Heelan, S.J.* (Dordrecht: Kluwer, 2002) and Jean-Michel Salanskis, "Die Wissenschaft denkt nicht," *Tekhnema* 2 (1995): 60–85.

3. EM, 20/26, emphasis added. For a reading of Heidegger's discussion of philosophy and science in the 1930s, see Babich, "Heideggers 'Beiträge' zwischen politische Kritik und die Frage nach der Technik" in Stefan Sorgner et al., eds., *Eugenik und die Zukunft* (Freiburg: Karl Alber Verlag, 2006), pp. 43–69.

4. SZ, 10/31. See for a discussion of this constitution, from a variety of viewpoints, Patrick Heelan, as well as Theodore Kisiel, Patricia Glazebrook, Pierre Kerszberg, and Dmitri Ginev has contributed a useful monograph on the *Context of Constitution: Beyond the Edge of Epistemological Justification* (Dordrecht: Springer, 2006) in addition to his other work.

5. Martin Heidegger, "Wilhelm Dilthey's Research and the Struggle for a Historical Worldview," in Heidegger, *Supplements: From the Earliest Essays to Being and Time and Beyond* (Albany: State University of New York Press, 2002), p. 160.

6. Heidegger, "Wilhelm Dilthey's Research . . . ," 162.

7. See Thomas Seebohm, *Hermeneutics: Method and Methodology* (Dordrecht: Kluwer, 2004).

8. Recent philosophy of chemistry (and better said: the struggle to argue for the singular philosophy of science proper to chemistry) reminds us that there is considerable diversity and the reduction of chemistry to the new physics and its mathematics as Gaston Bachelard some half a century ago had argued, may in the future be more advanced by a nuanced reconsideration of the kind the little discussed

philosopher of chemistry Helene Metzger argued for in addition to the challenging conceptual work of Fritz Paneth. It is difficult to read an account of the history and philosophy of science that details such different sciences and the differences these differences make for the philosophy of science, although I undertake to do just this (despite considerable resistance) in a recent article that I also draw upon here. See Babich, "Continental Philosophy of Science: 1890–1920," in Alan Schrift and Keith Ansell-Pearson, eds., *The New Century Volume Three: History of Continental Philosophy* (Chesham, UK: Acumen Press, 2010; in press). Paneth, by contrast with Bachelard, and despite the enthusiasm for Paneth of even such an influential philosopher as Herbert Dingle, continues to be little read even in contemporary philosophy of chemistry. For Paneth's writings, see Dingle, and G. R. Martin, with Eva Paneth, eds., *Chemistry and Beyond: A Selection from the Writings of the Late Professor F. A. Paneth* (New York: Wiley-Interscience, 1964) as well as Eric Scerri's *The Periodic Table: Its Story and Its Significance* (Oxford: Oxford University Press, 2006).

9. See Pierre Duhem, *German Science: Some reflections on German Science and German Virtues*, J. Lyon, trans. (LaSalle, IN: Open Court, 1991). Duhem's discussion is rather more about French than German scientific styles and the salience of his discussion is patent in the recent efflorescence of contributions to French philosophy of science, such as represented, for one instance, by the contributions to Anastasios Brenner and Jean Gayon, eds., *French Studies in the Philosophy of Science: Contemporary Research in France* (Dordrecht: Springer, 2009).

10. Donald Worster contends, not unpersuasively, the ecology and environmental ethics overall presupposes a particularly *English* sensibility. See Worster, *Nature's Economy: A History of Ecological Ideas* (Cambridge: Cambridge University Press, 1985), see especially his first chapter on the Arcadian ideal of Nature in the writings of Gilbert White.

11. Heidegger, "Wilhelm Dilthey's Research . . . ," 163.

12. Ibid.

13. GA 29/30, 379–82/261–23. Note that Driesch had published his influential theory of organic development using the example of sea urchin development in 1894. Indeed, when I was a student of biology at Stony Brook in the late 1970s, Driesch's theoretical account was still given pride of place in then current textbooks of developmental biology and embryology.

14. There are political issues here. See for one account of the tension between mainstream and "alternative" medicine, Eveline Richards, "The Politics of Therapeutic Evaluation: The Vitamin C and Cancer Controversy," *Social Studies of Science*, 18/4 (Nov. 1998): 653–701 for the case of Linus Pauling and cancer in addition to Peter Duesberg et al., "The Chemical Bases of the Various AIDS Epidemics: Recreational Drugs, Anti-Viral Chemotherapy and Malnutrition." *Journal of Bioscience*. 28/4 (June, 2003): 383–412 regarding the political influences on accounts of aetiology of the AIDS virus. More broadly and more conventionally, see Timothy Lenoir, "A Magic Bullet: Research for Profit and the Growth of Knowledge in Germany Around 1900," in Lenoir, ed., *Instituting Science: The Cultural Production of Scientific Disciplines* (Stanford: Stanford University Press, 1997), pp. 179–202 and John Ziman's "No Man Is An Island: The Axiom of Subjectivity," *Journal of Consciousness Studies*, 13/

5 (2006): 17–42 as well as Sheldon Krimsky, *Science in the Private Interest* (Lanham, MD: Rowman and Littlefield, 2003). Feyerabend who argued for citizen direction of scientific research was, in retrospect, naïve about the extent of such financial interests—as most of us continue to be.

15. For a discussion of the epistemological implications of experimentation, see Shiv Visvanathan, "On the Annals of the Laboratory State," in Ashis Nandy, ed., *Science, Hegemony, and Violence: A Requiem for Modernity* (Oxford: Oxford University Press, 1988), 257–88.

16. I am asked, when I note that I began my studies as a student of biology, why I changed fields. It's a good question as the sciences continue to have more prestige—and more funding opportunities and higher pay—than the humanities My reason, in addition to indignation over biology's refusal to define *life*, as I heard even Bentley Glass at the time explain this exclusion, as operatively essential for the progress of biology as a science, a disinclination I took to be methodologically problematic for the *science of life*, had as much to do with the my own and very visceral refusal to guillotine hamsters for their ovaries.

17. See for example, Sandra Mackey, "Phenomenological Nursing Research: Methodological Insights derived from Heidegger's Interpretive Phenomenology," *Nursing Studies*, 42, 2 (2005): 179–86.

18. Heidegger, "Wilhelm Dilthey's Research . . . ," 163.

19. See Ansell-Pearson, *Viroid Life: Perspectives on Nietzsche and the Transhuman Condition* (London: Routledge, 1997) for an exploration of this theme with reference to Bergson and others.

20. Friedrich Nietzsche, *Sämtliche Werke. Kritische Studienausgabe*. Giorgio Colli and Mazzino Montinari, eds., (Berlin: de Gruyter, 1980), 15 volumes. Here Vol. 13, p. 374.

21. Nietzsche, *Sämtliche Werke*, 13/374.

22. SZ, 11/ 31. See Pierre Kerszberg, *Kant et la Nature* (Paris: Les Belles-Lettres, 1999) in addition to his reading of Heidegger and Kant in Kerszberg, *Critique and Totality*. (Albany: State University of New York Press, 1997).

23. SZ, 5/24; cf. SZ, 7/27. Nietzsche had deployed a similarly reflexive move in his self-critical preface to his first book, *The Birth of Tragedy*, when he characterized his original project as a "problem with horns," the "double dilemma" of "*the problem of science itself*, science considered for the first time as problematic, as questionable" (*Die Geburt der Tragödie*, "Versuch einer Selbst-Kritik," §ii *Kritische Studienausgabe*. Giorgio Colli and Mazzino Montinari, eds., [Berlin: Walter de Gruyter, 1980, Vol. 1]), hereafter cited as GT. Although Heidegger's approach differs, he concurs with Nietzsche's reflexive emphasis that "the problem of science cannot be recognized on the ground of science" (Nietzsche, GT, §ii).

24. VA, 9/3. This is presaged in *Being and Time* as Heidegger steps out of the logical circle that might be raised as an objection to the "question of the meaning of Being" into what will become the hermeneutic circle simply because "in answering this question, the issue is not one of grounding something by such a derivation; it is rather oen of laying bare the grounds for it and exhibiting them" (SZ, 8/28). Such a "laying bare the grounds" is a questioning that opens its own way.

25. WD, 128/212. This is Jean Salanski's point, which he makes in this case for the sake of a contrary or oppositional emphasis. Salanskis, "Die Wissenschaft denk nicht," *Revue de Métaphysique et de Morale*, II (1991): 207–31.

26. Martin Heidegger, "Die Neuzeit. 'Die' Wissenschaft. Wissenschaft und Denken." *Heidegger Studies*, 21 (2005): 9–14: 12.

27. For Heidegger, problem solving as a notion refers to Popper's definition of philosophy. Analytic philosophy has taken this definition over for its own part, a referential dependency that Steve Fuller's recent book on Kuhn and Popper presumes for its part. See Fuller, *Kuhn vs. Popper: The Struggle for the Soul of Science* (New York: Columbia University Press, 2004).

28. This is not the only type of such technology but it is the only type currently receiving contracts for installation but it is not only limited to that, it is also limited to the nature of electrical switching stations.

29. Paul Feyerabend, *Erkenntnis für freie Menschen. Veränderte Ausgabe* (Frankfurt am Main: Suhrkamp, 1980). Frontispiece, p. 2.

30. Both Slavoj Žižek and Stuart Elden suggest Alain Badiou as exception but we can also add Lyotard to the mix, including his enthusiasm for the liberating potential of modern information technology. See Babich, "Thus Spoke Zarathustra or Nietzsche and Hermeneutics in Gadamer, Lyotard, and Vattimo," in Jeff Malpas and Santiago Zabala, eds., *Consequences of Hermeneutics* (Evanston, IL: Northwestern University Press, 2009).

31. Žižek follows Lacan's praise of theory (qua talking cure) into the imaginary field of the capitalist meltdown qua economic schemata. Žižek, "Don't Just Do Something, Talk," *London Review of Books*, 10 October 2008.

32. This is also the reason that, to take a parallel example roughly contemporaneous with Heidegger, Bachelard's work, by contrast with Heidegger, appeals as much as it does to analytic philosophers of science like Gary Gutting and Mary Tiles, despite the overtly poetic and even indeed and emphatically or explicitly alchemical cast of Bachelard's work: Whatever Bachelard does, he challenges no part of the scientific project and all his subversion can be said to have been reserved for philosophy. Hence Bachelard has been part of the mainstream tradition of the philosophy of science, if indeed an uneasy part, given his flair and affection for poetry.

33. See, for example, Alan Woods and Ted Grant, *Reason in Revolt: Dialectical Philosophy and Modern Science* (New York: Algora Publishing, 2003). But see too Robert M. Young, *Darwin's Metaphor: Nature's Place in Victorian Culture* (Cambridge: Cambridge University Press, 1985) although one would be remiss not to note the resistance to Young's arguments among philosophers of evolutionary science. See too Richard Levins and Richard Lewontin, *The Dialectical Biologist* (Cambridge, MA: Harvard University Press, 1985) as well as, on Engels in particular, Paul Thomas, *Marxism & Scientific Socialism: From Engels to Althusser* (New York: Routledge, 2008). See too, including a chapter on J. D. Bernal, Helena Sheehan, *Marxism and the Philosophy of Science: A Critical History* (Amherst, NY: Humanities Press International, 1993). John Ziman would seem to be about as Marxist as one can be and still be mainstream.

34. Expressed in such terms, we are not speaking of the "science of the middle ages nor that of antiquity" (VA, 42/157).

35. Steve Fuller is the most informed source of this reading on the side of Kuhn; but from the side of Heidegger, see Patricia Glazebrook, *Heidegger's Philosophy of Science* (New York: Fordham University Press, 2000).

36. Thus Heidegger articulates many of the same distinctions made by the historian Peter Dear in Dear's disciplinarily historical reflections on Descartes and the difference between experiential recourse to the empirical and modern scientific experimental research procedure. See Peter Dear, *Discipline and Experience: The Mathematical Way in the Scientific Revolution* (Chicago: University of Chicago Press, 1995). Heidegger offers a useful philosopher's gloss on the subtle, historian's point Dear seeks to make.

37. Compare Heelan's phenomenological approach to the question of measurement with Nancy Cartwright. Heelan's forthcoming *The Observable* (Frankfurt: Springer, 2010) develops his original study of Heisenberg's complementarity in *Quantum Mechanics and Objectivity: The Physical Philosophy of Werner Heisenberg* (The Hague: Nijhoff, 1965).

38. Precise, scientific measurement is not "precise" observation because it reflects "a methodology essentially different in kind, related to the verification of law in the framework, and at the service of an exact plan of nature" (H, 82/122).

39. For a discussion of the German "*Wissenschaft*" in distinction to the English word "science," see Babich, "Nietzsche's Critique of Scientific Reason and Scientific Culture: On 'Science as a Problem' and 'Nature as Chaos'" in Gregory M. Moore and Thomas Brobjer, eds., *Nietzsche and Science* (Aldershot: Ashgate, 2004), 133–53.

40. It is to be regretted that Joseph Rouse misses an opportunity to make this broader connection in his essay, "Heidegger's Later Philosophy of Science" *The Southern Journal of Philosophy*. Vol. XXIII, No. 1 (1985): 75–92 and although twenty years later Rouse will again have occasion to note the connection to be made between Kuhn and Heidegger, he does not address the critical dimension in Heidegger's thought in his "Heidegger on Science and Naturalism," Gary Gutting, ed., *Continental Philosophy of Science* (Oxford: Blackwell, 2005), 123–41. It should be noted that in the two decades between these two articles, Rouse curiously introduces his readings as lacking precedent—in spite of the work done by the late Joseph Kockelmans, but also of Theodore Kisiel, and Patrick Heelan before 1985 and Patricia Glazebrook who, in the interval, wrote an entire monograph with the very plain title: *Heidegger's Philosophy of Science* directly addressing Rouse's theme (just to keep to Anglophone contributions).

41. I discuss Heidegger's emphasis on this apparently paradoxical character of scientific research in Babich, "Heidegger Against the Editors: Nietzsche, Science, and the *Beiträge* as *Will to Power*," *Philosophy Today*. 47 (Winter 2003): 327–59.

42. See Herbert Dingle, *Science at the Crossroads* (London: Martin, Brian and O'Keefe, 1972) and for a current discussion of the ideological concerns or "alternatives" at stake see the letters section of *Physics Today* December 2003, especially noting Dingle's "Modern Aristotelianism," *Nature* 139 (1937): 784. Dingle also notes, "Success in scientific theory is won, not by rigid adherence to the rules of logic, but by bold speculation which dares even to break those rules if by that means new regions of interest may be opened up." *Through Science to Philosophy*, Chapter XV, p. 346. See further, Rom Harré's "Science as the Work of a Community" in

Babich, ed., *Hermeneutic Philosophy of Science, Van Gogh's Eyes, and God* (Dordrecht: Kluwer, 2003), 219–29.

43. I refer here again to Steve Fuller's recent reprise of what he takes to be Popper's side in a battle that might be staged between Popper and Kuhn. See Fuller, *Kuhn vs. Popper*.

44. We hear this from Patrick Heelan but also from Don Ihde who has his own reasons for wanting to present his thought as more "evolved" than Heidegger's own (we recall that Heidegger regarded the question of the cutting edge as irrelevant to philosophy but central to the advertising gambits of the market place), and so on. To oppose this point of view *is not* to claim that Heidegger's knowledge of science was a knowledge of modern science in its latest most up-to-date phase. Indeed, if this is what we need, none but the scientists themselves (and not even they) may be counted as adequately au courant, *pace* Ihde.

45. See for one such analysis, Paul Virilio, *The Information Bomb* (London: Verso 2005).

46. This human centeredness is to be taken less in the Protagorean sense of measure, which for Heidegger (always influenced by Hölderlin in this regard), retains a fundamentally reticent character of balance [*Maß*] than this securing certification is to be heard in terms of Nietzsche's critique.

47. Indeed, as Heidegger read Nietzsche through and against the interpretations current at the time of his lectures on Nietzsche, it is important to distinguish between Heidegger's criticisms of what Nietzsche's interpreters argued in his name and Heidegger's own Nietzsche-indebted suggestions. See further, Babich, "Dichtung, Eros, und Denken in Nietzsche und Heidegger: Heideggers Nietzsche Interpretation und die heutigen Nietzsche-Lektüre." *Heidegger-Jahrbuch II* (Freiburg: Karl Alber Verlag 2005), 64.

48. Nietzsche, *Der AntiChrist*, § 43. See too the latter section of Babich, "Nietzsche's Philology and Nietzsche's Science: On The 'Problem of Science' and '*fröhliche Wissenschaft.*' " in: Pascale Hummel, ed., *Metaphilology: Histories and Languages of Philology*, (Paris: Philologicum, 2009), 155–201.

49. This Archimedean vantage point continues to intrigue philosophers and critics. It is the position Thomas Nagel innocently called the "view from nowhere," further identified by the biologist-turned-science-studies theorist, Donna Haraway, still more innocently but still more aptly as the *God-trick*.

50. For Descartes, this Archimedean vantage point need not correspond to an actual point beyond the limits of the earthly globe but could instead be "virtual": represented within one's own subjectivity as a thinking thing.

51. Thus science is the "theory of the real" (VA, 44/159).

52. See Glazebrook, "Global Technology and the Promise of Control" in David Tabachnick and Toivo Koivukoski, eds., *Globalization, Technology and Philosophy* (Albany: SUNY Press, 2004), 143–58 as well as Visvanathan, "Progress and Violence" in Lightman et al, ed., *Living with the Genie: Essays on Technology and the Quest for Human Mastery* (Washington: Island Press, 2003) and Aidan Davison, *Technology and the Contested Meaning of Sustainability* (Albany: SUNY Press, 2001).

53. Nietzsche, for example, reminds us to note that the Greeks had all the mathematical preconditions necessary for the development of modern science—and yet they did not.

54. Heidegger explains, "modern science as theory in the sense of an observing that strives after is a refining of the real that does encroach uncannily upon it. Precisely through this refining it corresponds to a fundamental characteristic of the real. The real is what presences as self-exhibiting" (VA, 52/167).

55. In an odd way, this means that the modern academic endeavor to which Kant seemingly gave voice as the effort to put whatever discipline (metaphysics or linguistics or sociology), on the secure path of a science, is far further along its way than we are usually inclined to think.

56 Michel Foucault, *Society Must Be Defended: Lectures at the Collège de France*, trans. David Macey (New York: Picador, 2003), 7.

57. See for discussion and further references, Babich, "The Essence of Questioning after Technology: *Techne* as Constraint and Saving Power." *British Journal of Phenomenology*. 30/1 (January 1999): 106–124.

58. Patrick A. Heelan, *Space Perception and the Philosophy of Science* (Berkeley: University of California Press, 1983), 17.

59. Heelan's citation of Heideggr's term *recapitulation* echoes the expansive sense of the biological apothegm: ontogeny recapitulates phylogeny. Heelan's footnote addressed to Heidegger and science is instructive for its generosity and the scope of his references. See Heelan, *Space-Perception and the Philosophy of Science*, p. 324. Indeed, and this is not true of most books, a general reading of the footnotes repays the reader's efforts.

60. See, however, chapter 4 of David S. Bertolloti *Culture and Technology* (Bowling Green: Ohio University Press, 1984) for an historical representation and discussion.

62. We remain far from the projected era when, as Heidegger supposed, "the unmediated procurement of new energies will soon no longer be tied to certain countries and continents, as is the localization of coal, oil, and the wood of the forests" (G, 18/51, emended).

63. G, 18–19/51. Heidegger then assumed, as we cannot, that the dream of the atom age had nothing to hinder it. Thus he imagined that "in the foreseeable future the world's demands for energy of any kind will be ensured forever." Technological triumphalism was common at that time and we seem to have learned slightly better to our pains, which is not to say that we are today committed to any lesser prospect.

64. See Lynn Townsend White, Jr, "The Historical Roots of Our Ecologic Crisis," *Science*, Vol. 155, Number 3767 (March 10, 1967): 1203–07 and his *Medieval Technology and Social Change* (Oxford: Oxford University Press, 1962).

65. GA 16, 747. Heidegger directed this question the participants of the tenth meeting of the Heidegger Conference (often called the Heidegger Circle) in Chicago, 14–16 May 1976. 25 years after Heidegger first wrote his letter, the current author invited participants, as the secretary-convener of the thirty-fifth annual meeting of the Heidegger conference, to begin to address the question he posed. Participants at the 2001 meeting who took up the challenge included Patricia Glazebrook, Ute Guzzoni, Patrick A. Heelan, Don Ihde, Theodore Kisiel, and others. It is noteworthy, on an ontic and empirical level, that a popular science or engineer's account has recently reprised Heidegger's thesis (without, to be sure, recognizing anything other

than Heidegger's critique of technology). See for an overview, John Markoff's review of Brian Arthur, *The Nature of Technology: What It Is and How It Evolves* (New York: Free Press, 2009): "Rethinking What Leads the Way: Science, or New Technology?," *New York Times*, October 20, 2009.

GELASSENHEIT

Beyond Techno-Scientific Thinking

Ute Guzzoni

As a motto for this section, I propose a reflection from Heidegger's Logos In Vorträge und Aufsätze: "nevertheless thinking changes the world. It changes it into the always darker fountain-depth of an enigma which, as being darker, is the promise of a stronger light" (VA, 221/78).

Here, I explore Heidegger's criticism of modern science not by considering its content, but by concentrating attention on selected aspects of what he contrasts against scientific-technological thinking. This latter formulation shows that I take science and technology as more or less coextensive. Heidegger stressed more than once that the essence of science is already technological.

Every criticism takes not only its justification but also its criteria and specific factual objections from a view toward something different, whether as something feared or as something desired. A critical view must always be guided by other experiences and by other attitudes. Thus, science and technology are examined and evaluated by Heidegger already from the viewpoint of a different thinking, that is, from the possibility of their counterpart. Heidegger himself would not speak here of "criticism." For he intends to show that techno-scientific thinking is not and should not be the only kind of thinking, and that the crucial thing is to contrast it against a different thinking that he calls *besinnlich*. As is well known, it is difficult to translate the German words *sinnlich* and *Sinnlichkeit*. *Sinnlich* means sensible, sensuous, sensitive, and sensual, but also concrete, physical, and corporeal. My use especially designates what is given to and perceived by the so-called "five senses." Thus, I render *sinnlich* with "sensual," but also sometimes with "sensitive." Furthermore, I translate the specific Heideggerian term *besinnlich* with "sensitive-reflective" in order to preserve its two different connotations, that is, its connoting of *sinnlich* and *Sinn* ("sensitive" and "sense"), and of *sinnen* ("reflect"), as in *Besinnung* ("reflection").

Accordingly, for Heidegger "there are two ways of thinking, both of which are justified and necessary: calculative thinking and sensitive-reflective thinking" (G, 13/46). What is at stake here is not, however, merely a side-by-side relation. Heidegger understands the scientific-technological attitude to the world as a metaphysical lack of a true human being-in-the-world, so he attempts to find a way out of this need by demonstrating the possibility of a more appropriate thinking and of a "poetical" being of humans.

These reflections first recall some of the basic traits of scientific-technological thinking that Heidegger uncovered. I clarify how they necessitate a different, reflective thinking. This alternative thinking is not a merely altered kind of science, but an attitude of thinking that is fundamentally different from science. Heidegger often engages this different thinking as a sensitive reflection on the main traits of the history of Being that generated and determined the scientific-technological world. But he also talks about a substantially different thinking that belongs to, as he calls it, poetical dwelling. Thus, sensitive-reflective thinking moves in two different directions that change its character. On one hand, it looks for the meaning at work in all things that are determined by technology; that is, it asks the question of Being by considering the oblivion of Being. But on the other hand, it turns its gaze toward a different region. As if looking from a different perspective, it views things that are in the world (e.g. the bridge one is crossing, the jug from which one is pouring, the house where one is dwelling). This section examines this latter mode of thinking in order to show in what ways Heidegger understands it in contrast to the scientifically determined thinking of the modern epoch.

In the letter with which Heidegger greeted the tenth Heidegger conference in Chicago, the last written text we have from his hand, he calls modern natural science "the determining fore-conception and constant intervention of technological thinking into the implementing and organizing *Machenschaft* of modern technology."[1] This description draws attention to the active, almost violent manner of science. Scientific knowledge is deeply rooted in technological thinking. In both, we meet the same grasping approach toward beings. This grasping character, which has its genesis in the essence of technology, culminates in scientific thinking, so scientific grasping and technological manipulation share a common root and purpose. Thus, technological thinking is not merely founded on science; rather, science is already a basic form of technology itself. Science is fundamentally technical, and is therefore violent or forcible in its technique. At other times, Heidegger calls techno-science a "challenging forth" (in German, Stellen), a "conquest" (H, 87/134), an attacking and mastering. Scientific determinations and categories hunt their object by dragging it into accessibility: They pursue it, and force it to submit to the a priori status of objects.

Heidegger's most frequent characterizations of scientific thinking are "calculating" and "measuring." These are modes of Stellen that force the object into a lasting presence so that it loses the possibility of revealing itself on its own terms. Every unpredictable and accidental occurrence, every particularity and peculiarity of a single fact is excluded, or at least rendered intelligible and manipulable at the very moment it comes into being. Only thus does knowledge gain certainty by connecting perceptions with explanations that are intersubjectively valid and provide every object with a guaranteed and secure place. The security of an all-embracing intelligibility and of the rational possibility for all and everything to be explained takes on an almost existential necessity in science.

Certainty is established by bringing the object to its ground, cause and principle and locating it in some structure of causality; that is, as Heidegger puts it in the Beiträge, through the "establishing of the correctness of a region of explanation" (GA 65, 149/103). The need to secure facts is accompanied by the need to have them at one's disposal, that is, to make them available. Here, as in other cases, Heidegger anticipated subsequent technological developments to an astonishing degree. For example, the Internet makes evident in a dizzying way the rendering at our disposal of everything for everybody, and for every application. Likewise, cell phones, in being handy, guarantee the accessibility of everybody for everybody.

Such boundless availability both necessitates and indicates an all-embracing leveling and equalizing. Heidegger refers to this homogenization with phrases like "distinctionlessness" [Unterschiedslosigkeit] and "organized uniformity" [Gleichförmigkeit] (VA, 92–3; H, 103/152 et passim).Through the "pattern of generally calculable explicability, everything shifts equally near to everything else" (GA 65, 132/92).Because all is submitted to a common measure, all is rendered calculable. The possibility of being measured has become a criterion of reality. The representing subject intends to get and keep things in his or her grip in order to be certain of them and dispose of them at will (cf. VA, 88–90).

The violence of this grasping is closely connected to a characteristic of seemingly minor importance that Heidegger often attributes to modern science. This is the "gigantic": "quantity as quality" (GA 65, 135/94). The gigantic is the measureless. Heidegger speaks of a "strange excess of the raging measuring and calculating" (VA, 197/228). The gigantic is a basic characteristic of contemporary times, and not peculiar to the world of science. "Nowadays the gigantic march of the calculating mode in technology, industry, economy and policy testifies to the power of a thinking obsessed by the *logos* of logic in a form that verges on madness. The full impact of calculative thinking culminates in the centuries of the modern epoch" (GA 79, 156).

This recapitulation of Heidegger's discussion of scientific-technological thinking suffices for the purposes of this essay with the addition of only one further essential aspect of his analysis. Heidegger repeatedly and urgently points out that this way of thinking is not a mere effect of human capabilities and activity. Rather, it corresponds to a destiny of Being that has the mysterious character of a withholding of and abandonment by Being. Heidegger calls this destiny *Machenschaft*, that is, "the domination of making and of what is made" (GA 65, 132/92). The "essence of modern metaphysics," determined by the destiny of Being, is at the same time the "essence of modern technology" (cf. H, 69/116), and as such demands the mathematical natural sciences.

The problem of the sciences, and more generally of human beings, is that this determination by Being under which the sciences are subsumed remains undetected, so human beings take themselves to be ruling masters, when in reality the sciences are only reliable components of an established order that has already absorbed them and within which they can only play an instrumental role. Heidegger, therefore, thinks it necessary to awaken a reflective sensitivity that considers the position of human beings in the world, and their relation to both Being and the world. In this sense, we read in *Der Satz vom Grund*: "Yet if we are to reach a path of reflection, we above all must find our way into a distinction that shows us the difference between merely calculating and sensitive-reflective thinking" (GA 10, 178). When Heidegger discusses the sciences and their way of thinking, he is not so much interested in the sciences themselves but in reflection on their "metaphysical ground," "which founds science as modern" (H, 70/117).

First of all, such sensitive-reflecting thinking points out the "fundamental features of technological-scientific world-civilization" in order to consider how metaphysical thinking and metaphysical reality have determined Western history. Sensitive-reflecting thinking goes back to the beginning of the intellectual history of the West to question the history of metaphysics and its different epochs, and to make visible the central features of its culmination in contemporary metaphysics. It answers to "the demand which immediately lets beings appear in terms of projectability and calculability, and challenges man into the manipulation of the beings that so appear" (GA 79, 124). Thus it opens the possibility of a future being that is evoked in a different way.

What I am interested in is, however, another kind of reflection. It is a sensitive-reflecting thinking that differs radically not only from calculating, pursuing and disposing, but differs also from the thinking that ponders these. At least implicitly, Heidegger assumed the simultaneity of two different ways of reflecting thinking, and he himself practiced this simultaneity. More importantly, in doing so, he also presumed the simultaneous reality of a different *matter of thinking*, a different world within which to dwell. Such thinking that questions world and things, nearness and releasement, that is, "on the way to language" and allows what Heidegger called "*Gelassenheit*," is not merely prophetically announced, and those things and regions are not just anticipatory visions. Rather, I am convinced that there is already today, and has always been, the possibility of experiencing and speaking in a different way. This possibility is evident not only because without it, no critical perspective on technological thinking would be possible, but also and especially because we ourselves in fact experience dwelling within the world and in exchange with things.

How should we conceive of this different kind of thinking that, although not itself scientific, is able to mark the borders of science? Why does Heidegger characterize it as a "simple saying," a "heartful" and "released" thinking? Why is it a dwelling and a wandering thinking, and, as I call it, a sensitive, metaphorical and pictorial thinking? I confine myself to two of these designations that both can be considered aspects of a counter-image to scientific technological thinking. First, I give a brief account of what Heidegger means by *Gelassenheit*—releasement. Second, I discuss what I call "pictorial thinking"—*bildhaftes Denken*—where "pictorial" does not mean representational thinking that renders its concepts by making use of merely figurative or even symbolic language, but rather means a thinking in, of and about visible and viewable things; that is, a *painting thinking*. Pictorial thinking is a departure from merely conceptual thinking.

The term *releasement* is first encountered in German mysticism. There, in Christian Hoburg for example, to be released means neither to know things, oneself, or even God, nor to want to have anything of one's own, but rather to release oneself from everything earthly in order to immerse oneself purely in God, who in this context becomes that which simply is not, or the nothing.[2] To be thus released, human beings must have nothing by which they are burdened. Willing such a complete unburdening is itself no longer even permitted to be a willing, and is grounded in the conviction that all sins arise from the harmful root of some particular thing.

For Heidegger, however, releasement is not a posture that "sinks into the nothing." With the phrase "releasement toward things," Heidegger describes the "posture of a simultaneous Yes and No to the technological world" (G, 23/54). In this posture, technological things are taken back into a transformed, open relation of human beings to the world, characterized as "meditative thinking" [besinnliche Denken] (G, 22/53) and "heartfelt thinking" [herzhaften Denken][3] (G, 25/56). For a heartfelt thinking, there can be no talk of a "harmful root of something." In this thinking, things are not harmful weeds that should be wiped out or eradicated. What is instead at issue concerns bringing oneself into a released relationship to the earthly, to those things that one encounters in the world and out from it. Releasement, as Heidegger once put it, is "something like calmness" [Ruhe] (G, 45/70; cf. 33/61; 40/67). This calm, however, does not oppose itself to the movement of things by positioning itself contemplatively across from them, but instead intends the most expansive openness to the enduring, moving play of the world, its things and occurrences. Like Adorno's "patience toward the matter," Heidegger's releasement toward things can be characterized by a long breath.[4] It is the long breath of a waiting that holds itself open for something still yet due, something still to come.

Heidegger writes, "In waiting we leave that for which we are waiting open. Why? Because the waiting releases itself into that openness, into the expansiveness of the distance in whose proximity it finds a while in which it remains" (G, 42/68). In contrast to the scientific, i.e. predominately active and grasping, tradition, waiting as a fundamental determination of thinking is an unusual conception because we are accustomed to see waiting as a merely receptive and passive posture. But waiting, as Heidegger would have it understood, leaves that for which it waits open, and releases itself into this openness, abandoning itself to it. Thus released waiting is not really passive. For as a releasing into, waiting orients itself toward what is coming, making itself ready for it. Such attentiveness and such placing properly into one's view surpasses or falls short of the alternatives of active and passive, spontaneous and receptive.

The thinking sought by Heidegger waits less to encounter a determinate object, some particular thing, but rather opens itself up to a realm, a horizon, a region. Thinking is released toward openness as such, that is, toward that out of which and from where something to be thought yields itself as a gift. Releasement, being released, means to release oneself not only toward, but also into the openness from out of which something can approach and encounter human beings. The openness, the open, is the still undetermined, free space into which something can happen, in which it can yield itself. It is the "openness of the open" both in the sense of what lies open to view, that is, what is visible, accessible, and thus not closed, and as what remains open in the sense of not determined, that is, what is still showing no determinate face, but is enigmatic and can still become this or that. This openness is itself nothing, but not nothing per se; rather, it is that which, in having the character of nothing, gives space for and access to something. If releasement as waiting releases itself into the open, it releases itself toward what in the openness of the open region has its place opened up by releasement, and what is therefore capable of addressing thinking from out of the open and saying something to it. That thinking itself should let something be said to it—this is exactly thinking for Heidegger.

As a space for and of things, the open is an *open world*, which Heidegger does not conceive as the totality of the categorically ordered relationship of beings as beings, but rather as the concretely experienced world of the "fourfold," that is, as the structurally enjoined regions of earth and heaven, divinities and mortals. In the world and as the world, the being with and toward one another of human beings and things, which comprise our Being or our life, plays itself out.

Heidegger does not say that the matter toward which releasement releases is transformed through this releasement, such that it is transformed by thinking. If the meditative gaze looks into essence, and if essence is nothing other than the event of the openness of the world, then such looking depends on more than just good will or striking out in the right direction. Rather, the look becomes genuine insight only when insight is *granted to it*, when something offers itself to the look: "The eyes catch sight of that which shines only insofar as they are already shone upon and looked at by that which shines."

Nevertheless, released thinking is not merely at the mercy of the occurring of the openness. It is no mere echo or mouthpiece of Being. This is clarified through Heidegger's speaking of something's "coming to arrive." When something does not reach someone or something, one cannot say that it comes, much less that it arrives. Thus, releasement is an admittance, a letting be, that is understood as an activity, indeed even as a "higher doing" (G 33/61). What comes to arrive in the openness and is present in it—or what does not come to arrive and remains absent—needs to be waited for. That which addresses needs somebody who listens and corresponds to it, just as that which is to be caught sight of needs the calm, waiting look. Moreover, openness itself and the open region cannot be what they are without a thinking that is released toward them, a thinking that Heidegger characterizes as a resolute openness and insistence. So it is not a thinking that can eliminate itself in the way that the mystic's *Gelassenheit* was a willing that does not will. What occurs and how it occurs are decided solely from what is in itself historically occurring. But only when thinking opens itself to the open region, when it is capable of what Heidegger calls "insistence [*Inständigkeit*] in the releasement toward the region" (G 60/82), is thinking capable of staying in the openness, and thus of dwelling in the world "in a totally different way" (G, 24/55.

When thinking gives up its metaphysically determined quest for principles, reasons, and causes that are everlasting and capable of guaranteeing certainty, when it abandons the forced and disciplinary attitude of the representational thinking of the subject–object relation, and finally leaves behind the ontological difference in favor of relations to the world and to things, then thinking will no longer rely upon and take as self-evident the language and method of scientific concepts, because these aim at what is universal and timeless, spaceless and worldless. Viewing pictures therefore can and must gain a particular and hitherto unknown importance for nonmetaphysical thinking.

Traditionally, it has been considered definitive of scientific and philosophical statements that they speak in abstract concepts. Art, on the contrary, expresses itself in and through concrete images. Kant said that poetry "is nothing other than a clothing of thoughts in pictures," and therefore observes that "Among all nations, the *Greeks* were the first to begin to philosophize. For they were the first to have tried not to cultivate the cognitions of reasons with the aid of pictures, but *in abstracto*; whereas the other nations always tried to make the concepts comprehensible to themselves *in concreto* exclusively through *pictures*."[5] Heidegger himself remarks, almost in resignation, the "saying of thinking is, in contrast to the word of poetry, without picture [*bildlos*]. And where there seems to be a picture, it is neither the work of poetry nor the visual moment of a 'sense,' but only the emergency anchor of an attempted but failed state of picturelessness" (GA 13, 33).

Yet Heidegger also says that the saying of thinking is not so much pictureless, as pictorial in its own way. His own later thinking and speaking, that often has been misunderstood as "pseudo-poetical," is itself pictorial to some extent. His frequent returning to the "wisdom of language" through etymology gives the impression of being a return to pictures. The concepts of traditional philosophical terminology or of everyday linguistic use are traced back to the pictures that are uttered in these words, that is, to the concrete connections and relations of the world that gain visibility in them. In these etymological returnings a trait appears that is typical of the later Heidegger: He does not try to provide concepts, but aims at helping the facts to speak in their concrete relevance to particular instances of questioning, so that his thinking images concrete connections of Being.

A sensitive-reflective and heartful thinking cannot exist without pictoriality. It is obvious that "picture" indicates seeing and the visibility of things. The picture is as well how a thing presents itself to seeing, feeling, and thinking, as what comes into being through a vision guided by sensitivity, understanding, and imagination. Thus the picture is the intersection of seeing and being seen. Pictures are not necessarily corporeally present. They also may appear, as Heidegger often remarks, before the so-called inner or mental eye. This picture may be evoked when we hear something special, or when we smell or touch something. Memory-pictures may have attached themselves inseparably to certain sense impressions. A well-known example of such a memory picture is the "Madeleines" of Marcel Proust. What is evoked in those moments is present and visible. The pictures are actually present for and to the capacities of imagination and sensitivity, whether something past, present or future appears in them. When we remember or anticipate something, it is present before us—whereas the concept, although this term has altered its meaning in the course of history, is essentially timeless, that is, outside of time. The most important traditional difference between the picture and the concept consists in the fact that pictures represent a particular thing, while concepts grasp the nontemporal universal.

The realm to which the picture belongs is the space of sensuality and sensitivity. As noted previously, *sinnlich* means sensible as well as concrete, physical, and corporeal. When I say, however, that pictures belong to the space of sensuality or sensitivity, it is not possible to differentiate this space in a precise way from the space of nonsensuality and insensitivity. This distinction, on which Heidegger says "all metaphysics are based," can no longer be seen as an exclusive and hierarchical opposition. There is no clear and readily traceable borderline between the sensitive or visible and the nonsensitive or nonvisible. There is even less of a dimensional difference between them that might serve to distinguish an earthly and visible world from a heavenly, invisible, nonsensual world. The visible and invisible interplay and permeate each other. There is no such thing as the invisible without the visible, and vice versa. We encounter the invisible within and *as* something visible. Thus visibility is the visibility of the invisible itself.

In ". . . dichterisch wohnet der Mensch . . . ," Heidegger says that the "nature of the image is to let something be seen" (VA, 194/226). He adds, the "poetic saying of images gathers the brightness and sound of the heavenly appearances into one with the darkness and silence of what is alien" (VA, 195/ 226). Here too what is visible in pictures cannot be separated from what is invisible in them, the sensual from the nonsensual, the bright from the dark, the sounding from the silent. According to the *Bemerkungen zu Kunst-Plastik-Raum*, philosophical thinking consists in such a letting-see that "brings to view the essential nature of things." The being pictorial of the letting-see and the bringing-to-view cannot be without that which remains secluded and invisible, without the dark and the secret that characterize the finitude of both human being and what it encounters. Thus we read in *Grundsätze des Denkens*: "Maybe this darkness takes part in every thinking. Human beings cannot eliminate it. They have rather to learn to accept the dark as unavoidable. . . . The dark keeps the bright with itself. The latter belongs to the former. . . . Mortal thinking has to climb down into the dark of the well-spring to get sight of the star in daytime" (GA 79, 93).

The sciences are able to perceive their borders, the finitude of human knowledge. But as sciences, they have the general intention of conceiving, making certain and making available, and so they are incapable of accepting the dark as something unavoidable, and preserving its pureness. The released distance that belongs to pictorial thinking and tries to preserve the visible's origin out of the invisible makes it wholly different from any representing, calculating, and measuring thinking the objects of which are themselves characterized by calculability and uniformity.

"I believe that you think much more soundly when thoughts arise from direct, tactile contact with things," says van Gogh to his brother.[6] Adorno speaks of a loving attitude toward things, an attitude that tenderly caresses the hair of what is encountered, and about the "long and non-violate gaze at the object."[7] Direct, tactile contact and caressing the hair nonetheless mysteriously preserve a distance that can be designated the distance of particularity and self-identity, or of the own proper being of the other. This sense of distance is consistent with what Heidegger intends, I suggest, when he speaks about the "coming near to the distant" (G, 42/68) and when he says that the brightness of the dark must be preserved for what is visible and bright to be accepted as coming out of darkness and invisibility. We are only able to have contact with things and catch sight of them without forcing and dominating them when we renounce getting close and seizing them, that is, if we leave them their own darkness, mystery, and distance.

It is also this preservation of distance, that makes pictorial thinking wholly different from conceptualizing, calculating, and measuring thinking. Distance, being alien, other or different, mystery—these are precluded wherever permanent availability and rendering for human disposal on one hand, and universal calculability and compatibility on the other, have to be guaranteed. Multiple meanings and openness belong essentially to pictorial thinking because it does not "drag its object into accessibility," but instead gives it the freedom to let what is invisible in it be seen in this or that way, or even not at all, when it hides in the picture. That the matter remains mysterious does not mean that there is a regrettable limit to the human capacities for understanding and technological reproduction. Rather, this mystery bears witness to something far more extensive, open, and multiple that happens between human being and the world and cannot be modeled in a linear and uniform way.

The uniformity, universality and—to borrow from Marcuse—one-dimensionality that belong to scientific technological knowledge directly negate the qualities that are specific to that which is encountered. The jug ceases to be a jug when it is scientifically identified as "a hollow within which a liquid spreads" (VA, 162/170). It is annihilated as a jug, as Heidegger says. Yet this negation and annihilation implies that the jug as such *is* something for itself. It is always possible to encounter it as a jug that contains and pours out whenever we let ourselves get involved in its being here and now in the concrete. Our days often may lack the leisure that would allow us to feel the jug really in our hands, and to empty it with a sensual awareness. That we are too hurried, distracted, or stressed does not mean, however, that the mystery of the jug is fundamentally hidden, or even has never reached us. We would not know the way a jug holds, nor even be able to use the word "jug" in a meaningful way, if a world within which we encounter jugs were fundamentally precluded.

Accordingly, releasement toward things is an attitude and behavior that is in no way a taking possession of and dominating them in order to set them in front of the subject (i.e., to seize and fix them in their alleged identity). Heidegger says, "I call the comportment which enables us to keep open to the meaning hidden in technology, 'openness to the mystery.' Releasement toward things and openness to the mystery belong together. They grant us the possibility of dwelling in the world in a totally different way" (G, 24/55). What is at issue here is a transformed being-in-the-world through a transformation in thinking; or more precisely, an attempt to find a way out of a thinking that has become technological and into a meditative thinking that is appropriate for the "calm dwelling of human beings between heaven and earth," or, as Heidegger says elsewhere, to the human stance between heaven and earth, birth and death, joy and pain, work and word. What is at issue is a different way of thinking, a "totally different" stance in the world, a transformed dwelling in the home that is the world.

Heidegger seeks truth content neither in the matter at hand, nor beyond it, nor in his own thinking, but rather in the occurrence of a coming to be unconcealed and a being sheltered in the openness of the regions of the world. Thinking does not stand in front of this occurring as some individual thing stands in front of another individual thing, or as a matter before another matter at hand; rather, thinking belongs as an essential moment to the play of openness. The "achievement" of thinking, if it is possible to speak of something like this, consists precisely in the capacity of thinking to give itself over to the event of the open and to its own being released into it. Thus, releasement is nothing that could have a will or be the object of will. Scientific-technological thinking as willing behavior has always already missed and neglected openness. In willing from and by itself, such thinking is not a released letting-be, but rather a demanding, contriving, and mastering behavior. Instead, released and sensitive-reflective pictorial thinking brings thinking into a realm of openness within which things are able to come forth and speak to humans.

NOTES

1. GA 16, 747. A bilingual edition is available at Martin Heidegger, "Neuzeitliche Naturwissenschaft und Moderne Technik," tr. John Sallis, *Research in Phenomenology*, 7 (1977), 1–4: 3.

2. *Gelassenheit* is used today in the sense of "composure," "calmness," and "unconcern." For Meister Eckhart and early German mystics it meant letting the world go and giving oneself to God. Cf. Anderson and Freund's note 4, page 54 of their translation. Heidegger uses the term in a way that draws on both sets of connotations (i.e., letting-go and calmness), but his usage differs from the mystics' in the ways Guzzoni clarifies below. [Ed.]

3. *Herzhaft* is translated by Anderson and Freund as "courageous." [Ed.]

4. cf. Theodor W. Adorno, "Anmerkungen zum philosophischen Denken," Ges. Schr. 10.2, Frankf/Main 1977, 602.

5. Kant, *Logik*, Kants Werke, Bd.IX, Berlin 1968, 27.

6. Vincent van Gogh, Briefe an seinen Bruder, Bd.1, Leipzig 1997, 184.

7. Adorno, *loc. cit.*

OPENING WAYS OF TRANSFORMATION

Gail Stenstad

This essay does some preliminary reflection on the deep relevance of Heidegger's thinking to our ways of living on the earth. This task is not as straightforward as it might seem, because this relevance does not unfold in the way that one might expect, but instead emerges within a tension that prevents facile interpretation while opening ways for radical transformation.[1] Consider this:

> eighteen Nobel prizewinners stated in a proclamation, "Science is a road to a happier human life." What is the sense of this statement? Does it spring from reflection?. . . . No! . . . The world now appears as an object open to the attacks of calculative thought, attacks that nothing is believed able any longer to resist. Nature becomes a gigantic gasoline station, an energy source for modern technology and industry (G, 17–18/49–50)

In this remark, made during an address given in his hometown, Heidegger alludes to a deep concern for the earth and nature, and casts doubt on claims that modern science and technology will solve our problems. In fact, later in the same speech, Heidegger warns that we may also, in unreflective acceptance of scientific and technological progress, lose our own humanity. In "The Question Concerning Technology," this issue is couched in even stronger terms, when Heidegger says that in our attempt to make everything calculable and controllable we, too, may become no more than interchangeable and expendable units in a gigantic technical framework (VA, 21/18).

Heidegger's deep concern and critical questioning are quite clear. Yet we know also that he does not intend any of this to have ethical import or to be taken as expressing value judgments. Again and again, Heidegger refused any request to construct or validate an ethics. Although Heidegger explicitly rejects any attempt at theoretical normativity, there is from early to late an emphasis on the *transformative power* of thinking. For example, there is the "saving power" that may avert our becoming no more than controlled controllers of natural resources; there is the "soft power" of the country path (and path of thinking), in its own way stronger even than the huge force of atomic power (VA, 32/28). Too numerous to cite are the explicit references to transformation—of thinking, of language, of us, of our relations to the things of earth and world—throughout Heidegger's work. The fruitless hunt for a Heideggerian ontological basis for any traditionally constructed normativity misses what *is* there: a rich and complex account of *ways* of opening up to transformative possibility, ways that neither depend on nor construct ethics, ontology, or any kind of "-ism."

But what then are we to make of Heidegger's critical references to the ways we are abusing nature and diminishing ourselves? And what is the import and relevance to this of Heidegger's lifelong emphasis on the question of being? It is in the centrally important *Contributions to Philosophy (From Enowning)* that we find the way to bring these issues together.

> the question of the *truth of be-ing*—is and remains *my* question, and is my one and only question. . . . [T]he task remains: *to restore beings from within the truth of be-ing* [*Seyn*]. The question of the "meaning of being [*Sein*]" is the question of all questions. (GA 65, 10–11/8)[2]

On the surface, it may sound as though this is an ontological quest. Not so. The questioning concerns ontology, but does not yield an ontology. There, is however, yet another allusion here to a transformed relation to beings. Everything Heidegger says about transformation emerges from his attempt to come to grips with the question of the meaning of being (and truth of be-ing). The most extended discussion of this matter is in *Contributions to Philosophy*, and unfolds there in ways that explicitly link the question of being with possibilities of transformation. *Contributions* speaks of mindfully confronting the historical origins of Western philosophy as metaphysics, enacting an opening toward, as Heidegger puts it, "a completely transformed relation to beings and to be-ing" (GA 65, 184/129).

The apparent tension I have sketched, between fairly blunt critical description of the way we live and rejection of the kind of normative theorizing that would ordinarily address such issues, finds its context in the deeper tensions and complexities that emerge and unfold in *Contributions to Philosophy*. The question of being and the thinking that opens toward the manifold transformations just mentioned are not two different matters, but instead emerge in us as a "between" wherein "be-ing and beings in their simultaneity" are transformed (GA 65, 14/11). There is more than one way, however, to enter this "between" and transformatively engage the matter for thinking. On one path, engaged with the question of being, thinking comes to grips with abandonment of and by being, and the absence (and refusal) of ground that emerges for thought in this abandonment. This is more than a mere flatly conceived lack of ground, but is the dynamic arena of transformation. "Ab-ground is thus the in-itself temporalizing-spatializing counter-resonating site for the moment of the "between" as which Da-sein must be grounded" (GA 65, 387/271).

Heidegger is quite clear, however, that the approach by way of the question of being is not the only way into this transformative thinking.

> The other way is most securely to be taken in such a manner as to interpret and make manifest the spatiality and temporality of the thing, of the tool, of the work, of machination, and of all beings—all as sheltering of truth. Throwing this interpretation open is implicitly determined by the knowing-awareness of time-space as ab-ground. But proceeding from the thing, the interpretation itself must awaken new experiences. . . . The way that begins here [moving into ab-ground, from the question of being,] and the way that begins with a being have to come together. (GA 65, 388/271)

Hence, the issues to be taken up here are as follow:

1. Thinking the historical unfolding of being. (The way from the question of being.)

2. Technology, technicity, machination. (One way "from a being," from our way of living with beings.)

3. Bringing the two ways together, as openings to transformation.

ONE WAY: COMING TO GRIPS WITH THE FIRST AND OTHER BEGINNING

Here, we become aware of how the possibility of a transformative "other beginning" for thinking opens up in an encounter with the first beginning of what we usually call Western thinking, an encounter that for the first time genuinely retrieves the movement of thinking in that early beginning. Consider a brief sketch of this retrieval. It begins with the recollection that we inherit our idea of "being" from the Greeks, and with Heidegger's having noted, already as early as *Being and Time*, that we have a longstanding tradition of simply taking this heritage and "being itself" unquestioningly for granted. Hence we have the early unfolding of the *question* of being in Heidegger's thinking.

The question of being, however, calls for deeper inquiry into the *origin* of the idea of being. The Greeks, Anaximander through Plato and Aristotle, found themselves in the midst of beings, without knowing *what* these beings *are*, without a knowing awareness of the "is" that these beings "are." This not-knowing astonished them, and moved them to deep wonder at the being of beings. Attuned by this astonished wonder (which is, says Heidegger, the grounding attuning of the first beginning), they pondered this guiding question: What is a being? What is this beingness of beings? *What is it that is common to all beings as beings?* (Notice this quest for what is common to all, rather than what is unique in each; this becomes rather important later in the discussion.) The thrust of this questioning is to conceive being from out of some aspect of an understanding of beings. Thus, the response to the Greeks' question emerges in the determination of being as presence, as the being that, constantly present as what is common to all beings, grounds beings in their being, their presence.

Being is here first differentiated from beings, as their ground, but the differentiating move itself is not explicitly thought or questioned, either by the Greeks or in the subsequent history of metaphysics, the centuries-long history that multiplies names and interpretations of being. Moreover, that philosophical differentiating of being and beings (the creation of what in hindsight can be called the "ontological difference") is forgotten. Subsequently, the grounding function of being thought as a being, as *the* being, is simply assumed. Its meaning does not become a matter for further questioning. And the notion of an *origin* of (the idea of) being remains unthought, and—within metaphysical parameters—unthinkable (GA 45, 152–80, 205–6; GA 65, 75–7/52–54; 232–33/164; 423–24/298–99).

Following Heidegger this far, we can see the guiding question of the Greeks *as* guiding question, in such a way that their "answer" also comes into question. As this questioning-thinking unfolds, being (its meaning, its origin, its function as ground) can no longer so simply be taken for granted. This guiding question of the early Greek thinkers, when explicitly thought as such, evokes what Heidegger calls the grounding question of an other beginning: What is the meaning or truth (originary disclosure) or *Wesen* (the enduring emerging-as-such and holding-sway) of being? This is said to be the *grounding* question because it wants to inquire into the ground of being. But what could *that* "be"? How can there be a ground of what has for more than two millennia been taken as the highest and ultimate ground? What begins as a rather straightforward sketch of the historical origin of our received notion of being takes here a rather startling turn. To think that far already suggests that ground and grounding may not "be" what we have assumed them to be. And that being, if not ground, is . . . *what*? But if we must ask those questions about the purported ground of beings, we must also ask: What about *beings*? We hardly even give them, as such, a thought.

This thoughtlessness about beings is no accident. It has deep roots in our philosophical heritage. Even though the guiding question of the first beginning was about beings, once their beingness is determined as their rising into presence and into view, each unique being becomes less and less significant.

> [T]he more questioning the question becomes and the more it brings itself before beings *as such* and thus inquires into beingness and is consolidated into the formula *ti to on*, the more *technê* is in force as what determines the direction. . . .
>
> In order for Plato to be able to interpret beingness of beings as *idea*, not only is the experience of the *on* as *physis* necessary, but also the unfolding of the question under the guiding thread of the counterhold of *technê* . . . (GA 65, 190–91/133–34)

In astonished wonder, the Greeks came up before *on ê on*, beings as beings, rising up (*physis*), revealing themselves (*alêtheia*), coming forth into view (*ousia* as *eidos*). But the dynamic rising and disclosing gets consolidated in the question *ti to on*, "what [is] a being?" and *technê* takes over.

Technê (originally, relating to beings so as to understand and preserve their being) shapes the first beginning in such a way that astonished wonder at the beingness of beings gives way to a manifold change in both understanding and action. The Greek understanding aims at being—at the constant presence common to all beings. After Aristotle, this beingness is grasped as the union of *morphê* and *hylê*, form and material, adding another layer of meaning to what can be thought as "in common" to beings, later emerging as *essentia*, essence.

All later forms of metaphysics play out, each in its own way, these distinctions. Along the way, ideas (instead of *physis*) and representability (instead of *alêtheia*) become the measures of knowledge and truth. Beings become objects of representation; truth becomes correct statements about beings; humans begin to think of themselves as rational animals; astonishment, wonder, and questioning give way (in philosophy and eventually in modern science) to a drive for calculable knowing and even certainty; *technê* (preserving-making) becomes technique, machination (GA 65, 133/93, 158/109, 191/134). This begins to sound rather familiar, as well it should. It also becomes, for the first time, truly questionable. We asked, what is being's origin, what is its ground? In one sense, its origin is the creative thinking that took place with the Greeks. But then is being "itself" historically contingent and *without ground*? Yes, if ground is understood as it usually is, metaphysically.

> *That* beingness was grasped as constant presence *from long long ago* counts already as grounding to most people. . . . But the inceptual and early character of this interpretation of beings does not immediately mean a grounding. . . . [T]his interpretation is not grounded and is ungroundable—and rightly so, if by grounding we understand an explanation that goes back to another being (!). (GA 65, 195/136–37)

What then? Was the conceiving of the ontological difference sheer invention?

Let's take the questioning deeper. Our taking being and beings for granted rests on an earlier forgetting, *Seinsvergessenheit*, the forgetting of the originary move whereby being was first differentiated from beings. As soon as we mindfully consider the first beginning, however, that which was forgotten begins to emerge, but not in the way one might expect. The original conceiving of being is thought, but "being itself" cannot be found, shifting us into an awareness of what Heidegger calls *Seinsverlassenheit*—abandonment of and by being.

But if being serves as ground, then where is the ground now? Nowhere, apparently. No-thing as well. That is, "being" is not a being, in fact being *is* not, at all. But if that is the case, if we can see that the *ontological difference no longer has actual ontological import*, then grounding can no longer be taken for granted. The security and certainty of ground and grounding—the traditional function of being—is refused or denied, not by us, but apparently (so it seems at first) by "being itself." But how can *that* be so? What "is" *being*, that it can refuse to manifest as ground? Did I not just say that being is not a being? This is no longer the guiding question of the first beginning, but an emergent question of grounding that opens up within and toward an other beginning for thinking. It calls on us to confront abandonment of being and think what was *unthought* in the first beginning and subsequent history of metaphysics. It calls us to an encounter with grounding that is un-grounding (or ab-ground, *Ab-grund*, absence of ground in a strong and dynamic sense). And here, the thinking begins to converge with issues also raised in "The Question Concerning Technology."

> What is this abandonment? It is itself arisen from what is precisely not ownmost to be-ing, out of machination. . . . The abandonment

of being [is] brought nearer by being mindful of the darkening of the world and the destruction of the earth in the sense of *acceleration, calculation* and the *claim of massiveness*. (GA 65, 107/76; 119/83)

TECHNOLOGY, TECHNICITY, MACHINATION

The dominance of *technê* in shaping the first beginning effaces the uniqueness of beings in favor of what, held in common, can be re-presented in ideas. This dominance and the ensuing consequences in the history of metaphysics are no mere philosophical abstractions. If we had lived in medieval Europe, we would have assumed that beings are God's creations. God is being, *the* being, grounding the being of all other beings, as their maker, their creator. If we were well-educated enlightenment-era Frenchwomen, heavily influenced by Cartesian rationalism, beings would be extended substances, to be known through clear and distinct ideas, grounded on the (presumably) undeniable existence of nonextended substances (mind, and perhaps also God). All extended substances are in principle measurable and calculable, which means also controllable (GA 65, 111/77, 131–32/92). Science (shaped not only by Descartes, but also by Francis Bacon) is understood as the means by which to firmly secure the god-given human domination and control of nature.[3] And so it goes, through the various permutations of metaphysics, with beings grounded on some idea of being, while that idea of being, just as it was in the first beginning, is determinable from the understanding of beings that is in play. So we ask, what are beings *now*, in our technology-driven era? Heidegger opens up this field of questioning most clearly in "The Question Concerning Technology."

The question here is: What is the *Wesen* of technology? That is: What is ownmost to technology? What holds sway in the emerging and enduring of technology as such? We are used to thinking of technology as our possession, as available means to our ends. Reflecting on this instrumental means-and-ends definition of technology pulls thinking into a consideration of causality, specifically of the four modes of causality outlined by Aristotle. What holds material, formal, final, and efficient cause together? They are ways of being responsible for bringing something forth into appearance. Bringing-forth (*poiêsis*) may manifest as *physis* (arising of itself) or *technê* (making); both of these are ways of revealing (*alêtheia*) (VA 9–16/3–12). Here, a key insight emerges.

Technology, even modern technology, is a *way of revealing*, not merely a means subject to our control and mastery. There is, however, an important difference in how modern technology reveals things. It is no longer *technê* as *poiêsis*, but technicity as a setting-upon nature and a challenging

(*Herausfordern*) of all things to be constantly on hand for predetermined use. Things are represented, ordered, and calculated in advance. They are interchangeable (for any given use) and disposable (disposing as ordering, and disposing as discarding the expendable). At the beginning of "The Question Concerning Technology," Heidegger urges us to pay particular attention to how the language moves and reveals the matter for thinking.

At this stage of the discussion, he calls our attention to some of the ways that we describe the things around us. Fifty years later, what he says is startling in its accuracy. We can all too easily elaborate on it, from our own experience and reading. Earth and soil: mineral deposits, land for development; Farming: agribusiness; Food: nutriceuticals (On mentioning to my veterinarian that green beans were helping my dogs to lose weight, she replied, "Yes, the nutriceuticals are becoming more important."); Students: full-time equivalents (VA 18–19/14–15).

> The hydroelectric plant is set into the current of the Rhine. It sets the Rhine to supplying its hydraulic pressure . . . sets the machines in motion whose thrust sets going the electric current for which the long distance power station and its network of cables are set up to dispatch electricity. In the context of the interlocking processes pertaining to the orderly disposition of electrical energy, even the Rhine itself appears to be something at our command (VA 19–20/16)

And even the river "itself" is only something to be viewed and photographed by a tour group organized and herded there by the travel industry. What emerges in this language is not something random or accidental. Note that the river, the soil, the farm and the food, are subject to a particular way of being-revealed. Caught in this web of interlocking processes that order them to stand by for disposition, beings are now revealed as *standing reserve*, which says "the way in which everything presences that is wrought upon by the revealing that challenges. Whatever stands by in the sense of standing-reserve no longer stands over against us as object" (VA 20/17).

This web of interlocking processes is now becoming globalized. The World Trade Organization and North American Free Trade Agreement (NAFTA) aim to impose standardized regulations on the global economy, overruling even national and local governments' ability to protect workers, the environment, and the health of their citizens. All is, as Heidegger says, ordered toward "maximum yield at minimum expense" (VA 19/15). Chapter 11 of NAFTA allows corporations to demand compensation from governments if governments have in some way interfered with their ability to maximize profits. Already there have been cases where the government of Canada, and the state of San Luis Potosí (Mexico) have had to pay compensation after respectively banning a cancer-causing gasoline additive and refusing to let a metals factory dump toxic wastes near a village water supply. Still pending is Sunbelt Water's claim of $10.5 billion against Canada, which revoked its license to export supertankers of water from British Columbia to California (the Canadians had begun to have second thoughts about the environmental impact of such a project).[4]

All things emerging from earth can, it seems, now be represented (linguistically and mathematically) as less even than objects, as mere material for use and disposal. And this is just where the question of what holds sway in technology converges with the historical unfolding of the question of being (thinking of the first and other beginnings).

> The planning-calculating makes a being always more re-presentable, accessible in every possible explanatory respect, to such an extent that for their part these controllabilities come together and . . . broaden a being into what is seemingly boundless, but only seemingly. In truth what is accomplished . . . is a relocating of the gigantic into the planning itself, by what is subordinated to the planning. And in the moment when planning and calculation have become gigantic, a being in the whole begins to shrink. The "world" becomes smaller and smaller, not only in the quantitative, but also in the metaphysical sense: a being as being, i.e., as an object, is in the end so dissolved into controllability that the being-character of a being disappears, as it were, and the abandonment of beings is completed. (GA 65, 494–95/348)

So: the thought of abandonment of and by being, arrived at earlier by way of coming to grips with the historical unfolding of being, also emerges here in "thinking from a being," from our ways of understanding, naming, and interacting with beings (GA 65, 388/271). The way of revealing through modern technology, revealing beings as standing-reserve, is the culmination of the history of being.

In that long history, being, the constant presence that grounds (our understanding of) beings, has had many names, from *eidos* and *ousia* to God, substance, mind, noumenon, and absolute spirit. All of these say *ways of revealing* beings, whether as shadows of the forms, the unity of form and matter, creatures of God, clear and distinct ideas, extended substance, phenomena, and so on. What about now, when beings have been reduced to less than objects, in standing-reserve? What now is *being*? Heidegger gives us the word *Ge-stell*, which says a gathered setting-upon and setting-in-place, or *enframing*. It says the ways in which we are gathered into a way of being that challenges us to calculate, manipulate and order all things into the interlocking webs of the standing-reserve. Beings have their standing as enframed in this way (VA 23–24/19–21). The word Heidegger uses to say this in *Contributions to Philosophy* is *machination* (*Machenschaft*); this usage makes it a bit easier to see the link to the history of being. When *technê* shaped the Greeks' answer to the "what is a being?" question, it laid the ground for the ensuing history of metaphysics to culminate in *technê*-at-an-extreme, shorn of any sense of things arising in themselves (*physis*). Instead, beings are now revealed through machination that enframes everything as representable, calculable, orderable, and disposable, with no conceivable limit on the degree of quantification and control that can be expected. Science itself becomes an adjunct to machination, subordinated to the claims of ordering all things for production; it provides the specialized and ever-increasing refinement of rigorous accuracy that is called for. With no thought of conceivable limits, all questions are merely problems to be solved. Here, we have circled back to what Heidegger said in the "Memorial Address," calling into question the notion that science can solve all our problems. Such science ("modern science") has become reductionist science, caught in the claim of machination (enframing and disposing, accelerating and enlarging the scope and scale of technical application)[5] (GA 65, 108–9/76; 135/94; 145/101; 155/107; VA 23–26/19–23; G, 12/45–46; 18–19/51).

At first, it seems that we, who presumably benefit from all this planning and producing, are in charge of it. Having seen and dealt with the "human resources" (formerly "personnel") offices at our workplaces, we know that this is not necessarily the case. We, too, are subordinated to the web work of planning and production. Heidegger alludes to this: "the forester . . . is made subordinate to the orderability of cellulose, which for its part is challenged forth by the need for paper" for the newspapers and magazines that mold public opinion in precalculated directions (VA 22/18). *Downsizing* has recently entered the English language, referring to the disposing of expendable human resources. We are indeed very close to ourselves being little more than units in the standing-reserve. And Heidegger refers to this as dangerous. But this danger unfolds from within a more fundamental danger, which is that the sway of enframing machination enchants and overwhelms us to such a degree that it could drive out any other possibility of revealing, and any way of thinking that is not calculative (VA 29–31/25–27; G, 25/56).

It is at this point in "The Question Concerning Technology" that Heidegger mentions the "saving power" that, from within enframing, could emerge and reopen other, incalculable possibilities. This possibility emerges from within a fundamental ambiguity in being's holding sway now as enframing machination. This ambiguity has already been hinted at in that machination is a way of revealing beings, a way of revealing that, however, follows upon *Seinsvergessenheit* (forgetting of being) in such a way that it manifests abandonment of being in our thoughts and other acts.

What is established in Plato, especially as priority of beingness is laid in terms of *technê*, is now sharpened to such an extent—and raised to the level of exclusivity—that the fundamental condition is created for a human epoch in which "technicity"—the priority of the machinational, of the rules for measuring and of procedure . . . necessarily dominates. The self-evident character of being and truth as certainty is now without limits. Thus the *forgettability* of be-ing becomes the principle, and the forgetfulness of being that commences in the beginning spreads out and overshadows all human comportment. (GA 65, 336–37/236)

As the forgetting and abandonment of being becomes thinkable, a deeper forgetting also emerges, a forgetting of something so long hidden as to be not just forgotten, but rather unthought.

TRANSFORMATIONS: THE UNTHOUGHT EMERGES INTO THOUGHT

Lurking behind the forgetting of the positing of the difference between being and beings, is the *forgettability of be-ing*. In their wonder at the beingness of beings, the Greeks did not inquire any further into the coming-to-be of beings, of be-ing, of the enowning that holds sway in all appearing. In hindsight, we can almost imagine that *physis* and *alêtheia* could, perhaps, have opened this matter to questioning, but that was not the path followed by the thinking of the Greeks, or that taken by the metaphysical unfolding subsequent to them.

So now it is we who must ask: What about the wherefrom and wherein of coming-to-be? As the grounding question of an other beginning emerges from the attempt to come to grips with the first beginning, an attempt to bring this unthought wherefrom-and-wherein to language first becomes possible. This is challenging in that, as Heidegger affirms again and again, the truth of be-ing cannot be said directly in metaphysical (reifying, conceptual, systematic) language, and yet some merely invented or artificial language would also not *say* (show) the emerging thought. Instead, what is called for is a transformation of language and thinking that first opens up in hearing the emerging thought of be-ing (GA 65, 78/54). But what does *this* say: be-ing?

In section 50 of *Contributions to Philosophy*, on the same page where Heidegger tells us that our experience of abandonment of and by being arises from machination's hold, he also says that this abandonment carries within itself an echoing hint of be-ing (GA 65, 107/75). This opens up a distinction between the *being* of metaphysics and its wherefrom-and-wherein, *be-ing*. Be-ing, here, is not thought—as was being—from beings; that is, be-ing is not a being or a property of beings. Be-ing, he says, is "nothing at all, but holds sway" (GA 65, 255/180; see also GA 65, 13/10; 235–36/166–67). Be-ing's holding-sway is, in German, *Wesung*, which says emerging-as-such, holding-sway ("itself"), enduring coming-to-pass. Coming-to-pass also hints at *Ereignis*, enowning of what arises or comes to pass, which again, "is" not, but *does*: "Enowning *enowns*," says Heidegger, reducible to no being or event (GA 65, 349/244). This proliferation of names for be-ing, for the unthought, says, each in its own way, the same, moving away from constant presence as ground, into ab-ground, moving thinking to engage with the retreat and refusal of ground. Here we encounter something elusive, something ungraspable in conceptual terms, no-thing that is not just nothing, something that calls for thinking that can say (that is, show) enowning while surrendering any claim to immediate comprehensibility (GA 65, 4, 14, 56, 64–65).

We must be careful not to mislead ourselves while we try to hear what these words say. Although not thought from out of some characteristic of beings in the manner of the first beginning, neither do these words name something extra, something *beyond* beings. Be-ing is not some sort of dynamic hyper-being. Our forms of language, especially philosophical language, come to us from 2,500 years of metaphysics. We must be mindful of the necessity and difficulty of avoiding reification, which would nullify the movement suggested in these *tentative* words. They evoke something that retreats as it comes forward, something that eludes our thinking grasp as it opens and makes a way for that thinking. They are spoken from a reservedness that accords with that elusive disclosure.

Just as the first beginning accords with a grounding attuning—astonished wonder—so too the other beginning is attuned in its own way. However, this attuning cannot be so simply named, as it is attuned by and to "something" that refuses to be represented in a name, as if it were a being. There are many evocative names, but no grasping concept or representation, either for "what attunes" or for the attuning. Some of these names are: startled dismay, awe, reservedness. But "there is no word for the onefold of these attunings" (GA 65, 14/11). "Startled dismay" marks the thinker's being-shifted from the received assumptions about being into the refusal of ground mentioned earlier, the thought of abandonment by being, which hints at or echoes a deeper abandonment.

> What is abandoned by what? Beings [are abandoned] by be-ing, which belongs to them and them alone. . . . (GA 65, 115/80)

Be-ing, Wesung, and Ereignis all say the unthought of the first beginning. They say the same, in a nuanced, nonidentical way.[6] This saying-the-same emerged from thinking the *difference* of be-ing from being and beings. But then what are we to make of *this* statement? Apparently we can and must also think being, beings and be-ing *together*, as also saying the same. Being emerges within and is enowned by be-ing; be-ing holds sway as being. They emerge *in* their differing in the thinking of the first and other beginnings, to name what was first brought to thought (being) and what remained unthought (be-ing) in the first beginning. Both of them can only be thought *as such* from a position already under way within an other beginning.

Be-ing, Wesung, Ereignis, being, in their very differing, say the same, when thought carefully. But in this *same*, they are also *not* the same, in a way that goes deeper than mere nonidentity. This thought draws us back into ab-ground, the refusal and staying-away of ground.

> . . . when being abandons beings, be-ing *hides itself* in the manifestness of beings. And be-ing itself is essentially determined as this self-withdrawing hiding.
>
> Be-ing already abandons beings in that *alêtheia* becomes the basic self-withholding character of beings and thus prepares for the determination of beingness as *idea*. (GA 65, 112/78)

When *alêtheia* functions in the first beginning as revealing, as unconcealing, it opens up a way of access to beings in their revealing themselves to perception and thought. Beings arise and emerge into the open to be examined regarding their whatness. However, this unconcealing is also fundamentally concealing in that it opens up access to beings by closing off access to (i.e., the possibility of thinking) disclosing itself. And this is left unthought altogether: the *originary concealing* that is always in play *in* the unconcealing, the self-withdrawing sheltering *of* the clearing and opening for the revealing of beings. This "unthought" of the first beginning harbors what now begins to be sayable, emerging into language, as be-ing, Wesung, Ereignis. These names hint at and evoke the thought of "what" they name, but they resist any conceptual grasping that attempts to define and systematize them. Perhaps it is helpful here to remind ourselves again that the ontological difference has no actual ontological import in the thinking of be-ing as enowning. It only helps us think the first beginning (GA 65, 250/176–77; 465–67/327–28). The resistance just mentioned echoes and hints at the self-withdrawing that is in tension with and at the same time is necessary in clearing a way for the emerging and revealing of beings. This clearing is *opening*: for beings/things and for thinking. To reify this opening in any kind of conceptual grid would break the resonance between word and thought and that which the word is trying to say (show). Opening would once again retreat into philosophical closure.

If, however, we respect this resistance, and heed the resonance in the movement of thinking, we are shifted into this opening. From where we are now, we can think abandonment of and by being (and be-ing) from under way on two converging paths: (1) the thinking of the history of being, and (2) the mindful consideration of our epoch of enframing machination that is the culmination of that history. This abandonment reveals itself in the self-certainty that rejects ambiguity, denies all distress, refuses any limit or indeed any "no" or "not" in what is encountered, and covers machination with a veneer of "values." As abandonment of being reveals itself, it also conceals itself in our enchantment with the apparent scientific and technological progress yielded by calculation, and the rapidly accelerating movement from problem to solution to "the next thing," cutting off any questioning or doubt or hesitation (GA 65, 59/41; 117–23/82–86). This stupefying enchantment by technicity effaces beings in their unrepeatable uniqueness, with "the utmost squandering of be-ing occur[ring] in the most ordinary publicness of beings that have become all the same" (GA 65, 238/168). They are all the same, hence the continuous and desperate hunt for new experiences (fodder for the entertainment industry). We, too, are all the same in our orderability and expendability. Mystery and wonder seem to be gone, banished from consideration. But it is the nascent awareness of this banishing that moves us, even compels us, to engage more deeply with this mystery, with the unthought that is only now emerging into the possibility of an other grounding, in ab-ground.

Ab-ground names the movement of thinking into opening, as it encounters what Heidegger calls "hesitant refusal of ground." *Hesitant* refusal does not deny any thought of or inquiry into grounding. In fact, its movement is just the contrary of such closure. "Ab-ground is the in-itself temporalizing-spatializing-counterswinging site for the moment of the 'between,' as which Da-sein must be founded." (GA 65, 387/271; cf. 298/210–11; 311/218–19; 321–22/226; 342–43/240). This denial of ground, says Heidegger, opens up the possibility of our being-shifted from Da-sein to *Da*-sein, that is, from the *being* that is there in the midst of beings, to being t/here. T/here, where? T/here in the *opening* of be-ing, making way for beings/things in their saying/showing, which also makes way for language and thinking. To move into this manifold opening, this t/here, this "between," is to be undergoing transformation already. We cannot calculate or plan the transformations that may unfold, but can only let ourselves be attuned as we thoughtfully attend to the paths that open up.

Heidegger says that this attuning is marked by deep awe and reservedness in the face of the not-knowing that confronts us in the refusal of ground. I would suggest that we also put this into play with the grounding attuning of the first beginning, astonished wonder. In this "between," where the not is not nothing, where the saying names are the same and not the same, where what always withdraws in self-sheltering nevertheless hints at and echoes its ownmost working, surely deep and astonished wonder is an appropriate response. Such wonder can move us to deep questioning that does not seek an answer, but instead lets itself be attuned by reservedness, seeking a thinking-experience of the holding-sway of be-ing as it overtakes and emerges into beings (ourselves included).[7] Attuned in wonder, thinking attempts to attend to be-ing "itself," which "belongs to" beings in their uniqueness, not their generality or commonality (GA 65, 4/3–4; 10/7–7; 13/10; 66/46; 86/59–60; 115–16/80–81; 241–42/171; 249/175–76; 429/302; 484/341).

It is April in upper East Tennessee. The peach trees are finishing blooming. Half the blossoms, as usual, were damaged by frost. Last year, the fruit set was fairly good, then the half-grown green fruit turned brown, then black and shriveled. The experts say a copper soap spray might help, if done just at petal-fall. But that may well kill the honeybees. The pears are beginning to bloom. Last year, fire blight killed all the pears on the only tree to set fruit. Spray with streptomycin? Again, not while the bees are working. Speaking of bees, what is the latest scientific research on the varroa mite problem? Trees and bees: It is difficult to move away from problem-solving thinking that calculates according to representable commonalities.

Walking up the hill, emerging into the higher meadow, I stand wordless. Passing the grape arbor, I look up past a grove of dark cedars, and there against a startlingly blue sky is a mass of white blooms. The world's biggest bowl of popcorn? No. This huge old cherry tree does not "look like" anything else. Walking closer, I hear the bees working the blooms. To let the cherry tree stand there in its unrepeatable uniqueness is to already be shifted away from complete enclosure by enframing machination. This shift is not just a matter of a "changed attitude" to the tree, or to trees. It is also a hint of the shift from Da-*sein* to *Da*-sein mentioned above. Attuning by reservedness and wonder allows stillness; no word is needed, but words may come and actually say, that is, show, something of this tree, of this relation of me and the tree, of me and earth (GA 65, 34–35/24–25)

Two things need to be seen in their nearness at this point.

1. Be-ing cannot be reduced to beings, or be thought in terms of some characteristic of beings, and yet is inseparable from beings.

2. This thought of be-ing (by whatever name we call it) works in us as opening-to a transformative opening-for beings (or better said, perhaps: things, earth, world).

This opening is, as Heidegger puts it, attuned to and by something that simply cannot be squeezed into the mold of any form of theoretical thinking. Since theoretical thinking is the thinking of enframing machination, to think in another way is already necessarily transformative (GA 65, 13–14/10–11; 65/46). And it is this transformative thinking, attuned in wonder and reticence, that shapes so much of what Heidegger says:

1. Dwelling thought as the "sparing and preserving" of things in "Building Dwelling Thinking." (VA 139–56/145–161)

2. The reversal of the priority over beings of calculable time and space, to timing-spacing that co-emerges with the thinging of things, in that same essay. (See also section 242 of GA 65, as well as the entire text of *Time and Being*.)

3. The emphasis on openness to mystery (G, 24/55) and heeding that which withdraws from our attempts to think it (VA 142/350, throughout GA 65, and many other places).

4. The work on language—nearly all of it!—wherein Heidegger unfolds the notions of claim and response (*Anspruch* and *Entsprechen*), of the importance of silence, of a hearing (*hören*) that arises from within our belonging (*gehören*) to the cleavage of earth and world, of our being the "between," the midpoint for beings/things to reveal themselves.

Where do we go from here? We can think much more carefully what Heidegger says of the issues concerning technology and science, the "strife of world and earth," and dwelling, putting these matters into play with that shift from Da-*sein* to *Da*-sein. No polemics, no ethics, no theory. What then? Do the thinking, letting go of calculative control over it, and allow oneself to be surprised and transformed by what emerges (GA 65, 247–8/174–5). This also means abandoning the compare-and-contrast, calculative assessment of Heidegger's thinking in relation to theoretical approaches to environmental thinking. Instead, we would do well to take up the most provocative of those theories (deep ecology, the land ethic, Holmes Rolston's attempt to broaden the scope of "value," and ecofeminism) and address to them these questions: What remains *unthought* in their thinking? How do they open thinking to the wonder of be-ing? Even as they may fail to do so, in what ways do they open transformative possibilities? How do these thoughts change as we attempt to *think* with them rather than calculatively assess them? How does this change *us*? We can also be mindful of "hearing" and thinking earth and the things of earth: what they say/show, and what is not shown. In what Heidegger refers to as the "strife of world and earth," disclosure and revealing make what we call world, while the arising within earth is also always a sheltering-concealing (GA 65, 71/49). We can bring

Heidegger's work on language to bear on this mindful attentiveness to earth, opening to the possibility of letting earth speak through us.

Why does earth keep silent in this destruction? Because the earth is not granted the strife with world, because the earth is not granted the truth of be-ing. Why not? Because, the more gigantic that giant-thing called man becomes, the smaller he also becomes?

> Must nature be surrendered and given over to machination? Are we still capable of seeking earth anew? (GA 65, 277–78/195)

No, and yes. There is nothing inevitable about the complete overtaking of world by calculation and machination. Neither is averting this repellent eventuality a given. It is, as Heidegger says, a matter of *seeking*. It is a matter of entering any of the transformative ways of thinking. It is a matter that calls on us to release our urge to nail down every thought and thing in a word, for use, and to rekindle a spirit of open and questioning wonder.[8]

NOTES

1. Don Ihde and Andrew Feenberg yield examples of what I mean by "facile interpretation." In a short three pages, Ihde dismisses Heidegger's thinking on technology as "romantic," "romanticism," and "romantic reductionism," all the while conflating technology, technicity, and *Ge-stell* (enframing), as well as merging the discussions of artworks and technology as if they addressed precisely the same matters. Feenberg is a bit more careful in his reading. Nevertheless, he emphasizes Heidegger's "fatalistic resignation to technology," completely ignoring the transformative intent of Heidegger's thinking. See Don Ihde, *Expanding Hermeneutics* (Evanston: Northwestern University Press, 1998), 53–67, and Andrew Feenberg, *Alternative Modernity: The Technical Turn in Philosophy* (Berkeley: University of California Press, 1995), 23–26.

2. I will generally follow Parvis Emad and Kenneth Maly's translation of GA 65. Given the difficulty of translating the text, and also the possibility (necessity!) of multiple meanings of some of the words that carry the deepest meaning, I also occasionally alter the translation.

3. Lynn White's well-known 1967 essay, "The Historical Roots of Our Ecologic Crisis" (*Science*, Vol. 155, No. 3767, *physis*. 1203–207), gives a clear and brief account of the coming together of science (theoretical, "aristocratic") and technology (empirical, "common") in the sixteenth through nineteenth centuries. He also indicates how this shift in thinking yielded the technological superiority that enabled Europe to dominate the rest of the world politically and economically, setting the stage for the eventual worldwide domination of "Western thinking." For a more extended and detailed discussion of these developments, see Carolyn Merchant, *The Death of Nature: Women, Ecology, and the Scientific Revolution* (San Francisco: Harper and Row, 1980). These historical discussions are helpful in clarifying the sense in

which science becomes subservient to technology in the era of what Heidegger refers to as machination or enframing.

4. William Greider, *The Nation*, April 30, 2001, 5–6.

5. This is not a claim that all science being done now is necessarily reductionist. A detailed discussion of the relationship of science and technology (as distinct, and in interplay) is beyond the scope of this essay. However, a few points of clarification, with some suggestions regarding directions in which such a discussion might go, can be briefly made here. (1) *Technik* can mean technique, technology, or technical/applied science. When I say "technology" I am referring to the apparatus, the machinery, and techniques; when I say "technicity" I am referring to the whole array of machination, enframing, standing reserve, and disposability discussed in the technology essay and GA 65, which includes the understanding of "modern science" as discussed. (2) In these parts of Heidegger's writing, "science" means modern mathematical science, which is, as I said, reductionist (and also unfolds within subjectivity, whereas enframing and machination, as the extreme emergence of *being*, unfold also beyond our subjectivity). (3) These are not always *precise terms* (i.e., there is in Heidegger some fluidity in usage). The emphasis is not so much on the terminology as on the questioning-questing movement of thinking.

6. I am using "same" in the sense that Heidegger opens up in *Identity and Difference*. That is, "same" does not mean identity. It means a belonging-together that does not suggest conceptual unity, but emerges from and echoes the movement of thinking as it responds to what shows itself (says itself).

7. The sense of "experience" here is *Erfahrung*, not *Erlebnis*. In several places in *On the Way to Language*, Heidegger uses *Erfahrung* for the thinking-experience as a journey on paths of transformation. In *Contributions to Philosophy*, *Erlebnis* names "lived experience" as something that blunts our capacity to become aware of abandonment of being; they are "new and exciting" experiences that are produced and consumed within the scope of enframing machination.

8. I am deeply grateful to Kenneth Maly for his careful reading and commenting on a draft of this essay.

V

REVISITING *BEING AND TIME*

HEIDEGGER AND THE EMPIRICAL TURN IN CONTINENTAL PHILOSOPHY OF SCIENCE

Robert P. Crease

The philosophy of technology began with a venerable first generation of thinkers, whose members included Heidegger, Jaspers, Jonas, Ellul, and Marcuse, who carried out the valuable task of pointing out the philosophical suppositions and historical conditions from which modern technology sprang.[1] This, it turned out, was only a first step to the development of a full-fledged philosophy of technology, for the work of this first generation had several shortcomings. One was that it treated technology—rather gloomily—as monolithic: Technology with a capital 'T'. If the thinkers in this first stage mentioned specific technologies, these were generally invoked in passing as examples. Another shortcoming was that the accounts proved too abstract, sweeping and naïve; different technologies have quite different effects. For these and other reasons, a second generation of philosophers of technology found it necessary to make an *empirical turn*, and look, not at Technology with a capital 'T,' but at the nuances of specific technologies—technologies with a lower-case 't'—to chart their concrete development and impact. These thinkers adopted and adapted the tools of the first generation, with the tools changing in the process, giving rise to a more mature philosophy of technology. This process of adapting and transforming what has been historically and culturally transmitted to us in striving to look at the things themselves is how scholarship works.

It is time for something similar to happen to Continentally inspired philosophy of science. The first generation of Continental philosophers who attempted to address science in a comprehensive way—Husserl, Heidegger, and Merleau-Ponty among them—tended to take science as something monolithic, and focused on how science, its attitude and practices, were rooted in the lifeworld. These thinkers often displayed a rather paternalistic and superior attitude toward science, deeming it to be an impoverished form of revealing. This approach had shortcomings. For one thing, it treated science as something monolithic—Science with a capital 'S'—tending to lump all sciences together. Furthermore, its accounts were often abstract, sweeping, and naïve, lacking contact with how science is actually done. It is time to carry out an empirical turn in Continental philosophy of science, and apply and adapt the tools that have been given us, taking a more nuanced look at specific sciences—with a lower-case 's'—and specific practices of these sciences, leaving ourselves open to revising the accounts of the first generation in the process. If we do this, it will give rise to a much more nuanced picture of science than the one Continental thinkers tend to work with at present. It will also allow us to speak about scientific practice in a way that scientists themselves are more likely to recognize and appreciate.

To put it crudely, if we want to do genuine philosophy of science, we cannot keep returning to the words and ideas of the Old Masters. Why should we? They had other interests—and they, too, sometimes nodded.[2] To keep thinking freshly, we cannot keep articulating what we already have, but have to keep responding to alterity, transforming what we have in the process.

Heidegger, for instance, famously had other interests: Being, the open, the clearing, the manifesting; rather than beings, what was opened up in the clearing, the manifested. He took science seriously, particularly in his earlier writings such as Being and Time, but understood it chiefly as a matter of calculating, predicting, and controlling what has been manifested. "A scientific investigation constitutes itself in the objectification of what has somehow already been unveiled," he writes in *Basic Problems* (GA 24, 456/320), which it does by way of mathematical projection. Or, in the *Origin of the Work of Art*, he writes that science is "not an original happening of truth but always the cultivation of a domain of truth that has already been opened." Science works by "the apprehension and confirmation" of what shows itself. If in doing so it comes to affect the open, to that extent "it is philosophy." What interested Heidegger about science during this period was not how specific concepts emerged out of the background of the lifeworld, but how the scientific-theoretical approach *tout court* emerged out of the background of the lifeworld.

Nevertheless, many of the concepts Heidegger developed en route to the *Seinsfrage* can be adapted for understanding scientific practice in such an empirical turn. This essay will point to one, *formal indication*. Although Dreyfus has advanced this concept in playing a role in his so-called "robust realism,"[3] my point is that the value of this concept would become much greater still with an empirical turn.

THE EMPIRICAL TURN IN CONTINENTAL PHILOSOPHY OF SCIENCE

Taking the empirical turn in Continental philosophy of science requires a few orienting steps. Step 1 is to note that scientific practice is an ongoing inquiry within a changing situation. Each scientist is historically situated within a particular historical, concept-laden discourse and within a particular historical, technological environment, interpreted according to available historical resources. Moreover, it is not a question of a solitary cognitive subject confronting a specific object or problem, or collection of objects and problems, but of a historical community of living, practically engaged researchers confronting a historical, holistic situation. *This* generation of researchers in a given field is not only a different community but faces a different situation from the previous generation, and the next.

Simple as this is to state, and as obvious as it may be to the hermeneutically sensitive thinker, it's all too easy to miss when thinking about science history. For then we are reconstructing, looking back at a set of past products and trying to use them to understand a present, self-creating process. It is hard to do so in a way that helps us understand the concrete present, what's happening now, what is driving science forward at this moment. The impulse of that moment is too easily lost, and when it is, we lose the present. When it does, time stops moving, and with it, science. Real science is a force, a pressure, that extends outward. Seeing the vestiges of that force in the past does not necessarily help us to understand it in the present. We can call that force curiosity, ambition, desire to help people, but that's a static conception of science. We edit out this force, and see only the remainders. Science is a process by which new forms, new concepts are created; growth and change, a movement of differentiation, is integral to it. A true philosophy of science has to not just *allow* for there to be fresh creations, new forms, continuous enrichment of the lifeworld, but to show the drive, pressure by which these are produced as a natural movement in inquiry.

Step 2 is to note that inquiry is spurred by dissatisfaction within that lived situation. Such dissatisfaction takes the form, not merely of discontent, nor of the experience that some knowledge is outstanding—of a lack of understanding. Rather, the dissatisfaction of inquiry is the experience of a collision between our expectations and what we encounter, of the sense that we should understand something that we do not. Dissatisfaction in science arises in many ways. In each case it involves the experience of something outstanding or obscure that, we sense, can and must be brought into the open. Inquiry is a response to this experience, in which we apply what has been historically and culturally transmitted to us in our attempts to transform the dissatisfying situation. We transform the situation in making it our own, so we are more "at home" in it. In Theodore Kisiel's clever phrase, this is the "eigen-function" of research. Yet every clarification does not finalize appearances but brings new dissatisfactions and continued inquiry.

Oliver Heaviside's remark, when he took Maxwell's frustrating and torturous equations and revamped and simplified them, resulting in a huge practical and theoretical advance in the understanding of electromagnetism, that he had inherited something from Maxwell and was simply trying to take what his predecessor had done and "see it clearly"—these words could be said by any genuine researcher, including Maxwell, who took what he had inherited about electricity and magnetism and made it clear for himself.[4] We see such examples multiplied throughout the history of science. Richard Feynman created his famous diagrams for particle interactions, he said in numerous interviews, because he was confused about what he was hearing and wanted to put it in a form *he* could understand; now these diagrams has become the inherited way *we* understand.

The third step involves characterizing such a process of inquiry as hermeneutical. The concepts involved undergo a process of evolution as they are reinterpreted again and again in different research contexts. To understand such a continuously unfolding process, we have to pay attention to this practical process of inquiry, rather than to the beliefs or theories that are its epiphenomena or products. The world is never fully transparent, we always meet it with historically and culturally transmitted assumptions that reveal and conceal. But what appears in the acts of inquiry serves to unearth the presuppositions, expose the assumptions, that are standing in the way of the thing itself, retuning the connections between our expectations and what we encounter, and deepening our engagement.[5]

Step 4 involves identifying the role of technologically mediated experimental performances in this practical process. To pursue this kind of inquiry, it is not sufficient to consult what we already have. It is not enough to read more books or talk to more people—this will not change the situation! To further practical inquiries of science, we have to stage events that show us, that give us back, more than what we put into them. What then appears in such experimental acts forces us to reinterpret our situation, recasts our resources, and reshapes our understanding. The "how" of this process includes the designing, enacting, and witnessing of events in laboratory situations.[6] These events are not fully transparent, and do not deliver us objects with Cartesian clarity, permanently segmented from each other, and independent of the performances that stages them.[7]

Moreover, we grasp the meaning of such events in relation to everything else that we know; these events are implicated in our understanding of the world. That is why what appears in these experimental acts is not merely a set of idle facts that we are free to ignore; they place claims on us. We are already related to these experimental events, even if they are enacted in special laboratory contexts. Thus, the fact that these events may only take place in special laboratory contexts does not mean that they are abstract and unworldly. It's the other way around: the special laboratory contexts are what make these events part of the world, and therefore are responsible for them being of pressing concern to us. A laboratory is like a garden where special cultivation and conditions allow things to grow that may not grow "in the wild," so to speak, yet the mere existence of the things grown under such conditions makes a claim on our understanding of the "wild."

The fifth step involves appreciating the temporality of the scientific process; how it relates to its past and future. The history of science often seems like beautiful ruins: Much of it appears in the form of structures that were obviously useful and important in their day, and that seem once to have been coordinated and mutually interdependent although in an inefficient and sometimes even incomprehensible way. In any case, these structures are no longer fully useful to us, and not coordinated and integrated with the reality we face. The real is what we encounter in the present—what we can confidently and even unavoidably reach, what appears in our horizon inevitably, and what we cannot turn away from. Yet even our present seems a little unclear, somewhat discordant, not fully grasped, with hints of another, deeper order just over the horizon. This discordance is why we inquire, and what makes newly achieved discoveries seem, strangely, to be both discovered and invented. When these discoveries arise, the greater unity thus achieved promises to turn our present into ruins. Each generation of researchers comes to terms with the unique historical world that it has inherited, and hands over a different one to the next generation.

These steps form the basis of the Continental philosophy of science, whose framework recently has been elaborated and critiqued by Dimitri Ginev.[8] One of its earlier innovative proponents is Patrick Heelan, who identifies both Heideggerian and Husserlian elements in scientific practice.[9] The Heideggerian element is the moment prior to object-constitution, the context or horizon or world or open space in which something appears. The Husserlian elements are the intentionality structure of object constitution, the presence of invariants through which we grasp what appears as an object in a horizon, and in the correlation of noetic and noematic poles. "The noetic aspect is an open field of connected scientific questions addressed to empirical experience; the noematic aspect is the response obtained by the scientific experiment from experience. The totality of actual and possible answers constitutes a horizon of actual and possible objects of human knowledge and this we call a World."[10] The world then becomes the source of meaning of the word "real," which can be defined as what can appear as an object in the world. The ever-changing and always historical laboratory environment with all its ever-to-be-updated instrumentation and technologies belongs to the noetic pole; it is what makes the objects of science real by bringing them into the world in the act of measurement. As we develop and improve empirical practices (the instrumentation and techniques for handling electrons) and the background horizon (electromagnetic theory), data and object will appear differently.[11]

This twofold picture of scientific practice appears in other Continental philosophers of science as well. Another example is found in Hubert Dreyfus, where this twofold structure supplies the basis for what he calls "robust realism." For Dreyfus, *both* moments are Heideggerian, if not fully Heidegger's own. The horizon or world or open space is one moment, involving what Dreyfus calls Heidegger's "practical holis," (or the "claim that meaning depends ultimately on the inseparability of practices, things, and mental contents"), and is captured in the idea "that human beings are essentially being-in-the-world." This is the moment, according to Dreyfus, that "repudiates both metaphysical realism and transcendental idealism."

But for Dreyfus there's another moment as well, one in which Heidegger acknowledges that experiences such as breakdowns show us that "entities *are* independently of the experience by which they are disclosed, the acquaintance in which they are discovered, and the grasping in which their nature is ascertained" (52, 183/228). This can happen when, for instance, we deworld entities and recontextualize them within theories that do not refer to our everyday practices. Then it is quite understandable how we can refer to entities as having nothing to do with these practices. Dreyfus finds that Heidegger fails to provide a satisfactory account of how this might happen, which is why Dreyfus finds Heidegger to be only a wannabe robust realist. Heidegger does not yet have, according to Dreyfus, "a practical form of non-commital reference that could refer to entities in a way that both allowed that they could have essential properties and that no property that *we* used in referring to them need, in fact, be essential."

Dreyfus then supplies his own account, involving the notion of formal indication as a methodological principle. This, he says, allows one to "designate something by its contingent properties and then be bound by that designation to research its essential properties." It allows us to "make sense of the strange as possibly having some necessary unity underlying the contingent everyday properties by which it is identified."

Rather than follow the arguments and counter arguments to Dreyfus's defense of robust realism, I would like to suggest that this is precisely the place where the empirical turn might be of immense value. For right here—what happens when inquiry encounters and explores the strange—is a problem in traditional, non-Continental philosophy of science. The exposition of this issue in mainstream philosophy of science, which treats scientific entities as rigid objects, is accompanied by numerous well-known problems involving incommensurability and theory change (for a discussion see Solar, Sankey, and Hoyningen-Huene 2008). When theories change, we seem to be in the awkward predicament of having to say that our previous concepts did not actually refer, and of having to prepare to say the same of our own concepts in the light of future theories. The implication, in Hilary Putnam's words, is that "just as no term used in the science of more than 50 (or whatever) years ago referred, so it will turn out that no term used now (except maybe observation terms, if there are such) refers."[12] Despite attempts to fudge the problem by principles of charity, such an implication is bound to haunt any view of science that envisions researchers as ahistorical, cognitive subjects who theorize about and represent objects in the world from nowhere—objects that have Cartesian independence and clarity—so that every discovery is just another instantiation of what we already know and represent, and scientific progress involves representation, failure, and re-representation.

Empirical investigation would not treat this issue as a logical puzzle to be solved by logical means, but would consult the experience of scientific researchers. What actually happens when scientists confront the strange and "bring it home?" This is only rarely, and uninterestingly, a case of correcting technical error, like instruments misbehaving—and equally uninteresting if it involves incomplete information or data of inadequate resolution. It's also uninteresting if it springs from biases towards information one already has. In a recent discussion of misdiagnosis among doctors, for instance, Jerome Groopman distinguishes between *anchoring*, or the tendency to overvalue the first data you encounter, *availability*, or having your judgment colored by recent or dramatic cases; and *attribution*, or having stereotypes prejudice thinking.[13] These biases are all ascriptions of strangeness and change to the subject, not to a transformation of the noetic-noematic encounter; they wouldn't apply to the strangeness of the encounter with a new disease.

But it is not the experience of theoretical replacement, either. For scientists, the temporality of science is not a staccato-like series of discontinuous representations, nor a series of jumps from error to truth, but a modulation of the real (in a way I have called *covariant realism*[14]), a continuous flow of adaptation and alteration in which something often grasped dimly at first is clarified. Moreover, this often happens in a way that does not involve a single object observed with greater magnification by steadily improving technologies, but a transformation of how a phenomenon is viewed as relating to clusters of other phenomena. Phenomena that seem to come from different parts of the scientific landscape (electricity and magnetism, mass and energy) come to be linked, while seemingly unified phenomena are seen to be composed of several different ones. Finally, what we grasp through our concepts at one time provides the grounds for more complicated concepts to develop in a more enriched grasp.

One tool in such an empirical turn is Heidegger's notion of formal indication. He meant for it to apply only to philosophical research. But might not some sort of ontic analogue also characterize scientific research? Might not scientists, *in practice*, hear the language of their discourse not as ground-plans of nature, not as "cognitive essences" that are operationalized, but as indications allowing them to pick out phenomena that become bodily present in instrumentally mediated ways with possibilities of appearing otherwise given other mediations? If so, might it not go a long way toward understanding the temporality of scientific understanding?

FORMAL INDICATION

Heidegger developed his notion of *formal indication* as part of his "hermeneutics of facticity."[15] His motivation was the "problem of access," or how to characterize a phenomenon in a way that does not prejudge it, but leaves open the possibility that it may appear concretely in different ways in different contexts. It's a provisional discourse, one that calls to the attention of others phenomena that they can also experience. The discourse is "formal" for it describes phenomena in specific enough ways that it can be recognized, activated, and experienced by others—the phenomenon is not grasped "as it is." Yet the discourse is only indicative because it does not make something present in a definitive way but only provisionally. In short, it's part of a process of unveiling, not of ordering something already unveiled—an unveiling of something with which we are already involved and have already interpreted, but that is incompletely present.

Let me sketch out four aspects of formal indication that seem relevant to the scientific context.[16] First, formal indication involves the primacy of the phenomenon indicated, not the primacy of the concept. Within traditional philosophy of science, concepts are viewed as tightly constrained in a theory, with their reference changing when the theory does. Such concepts, too, are fulfilled or not, with the real corresponding to what fulfills the concepts. If these concepts do not grasp the entire phenomenon, it is because they grasp only what is essential, and leave out the contingent. But formally indicative concepts can point to the real without complete fulfillment, and are revisable. Formally indicative language, Heidegger writes, does not make something present as "the thematic object of a straightforward and exhaustive account."[17] The object is not something calculated and worked out and characterized fully in advance, but always already interpreted and anticipated, an "anticipatory apprehension," an "anticipatory forehaving" of a unity "which prepares a path of research in advance." Formal indication is akin to the markings on a map; we use it to get around, to be revised in the light of what we find, rather than to dictate to what we find.

Second, in formal indication there is always a surplus of what's sayable about a phenomenon over what's said. Each step of the inquiry raises new questions, and implies that there is more to be found. This is Heidegger's notion of *Wiederholen*, that a major step of the inquiry can demand that the inquiry needs to be restarted. As Heidegger says in §62 of Being and Time, the hermeneutical situation is one in which you have to repeat the inquiry. The map is not only able to be revised; we always find more with which we can revise it.

Third, the formal indication is actively questioned by what is indicated, even and especially when fulfilled. The phenomenon itself suggests the deficiency of the indication. For it is not that the indication is done by an objective and independent observer; the inquirer is transformed by the inquiry, and needs to repeat it after every advance. What we find with the map changes the "we" who use it, in ways that call for new maps.

Finally, formal indication is a call, not to a biological entity or theorizing subject, but to a historically situated human being, and in a way that puts its being at stake. Responding to such a call is more than curiosity or scholarly propriety; it involves an affective dimension. It also makes the researcher answerable to a general, and even infinite community—submitted to inexhaustible critique.

These four aspects of formal indication are readily apparent in history of science, once we know to look for them. Episodes in history of science suggest that a particular kind of interaction takes place between two elements at work in the evolution of a scientific entity—a formal (conceptual) part and an experiential part—in a way which exhibits the aspects just mentioned.

First, laboratory experience, like ordinary experience, does not have a givenness that is fully conceptual; the weight is instead on the phenomenon experienced. The revisability of scientific concepts is the case even in "revolutionary" transformations. Historian Mara Beller, for instance, has described how the development of quantum mechanics was not at all the product of a simple switch over from one "paradigm" to another. Rather, she shows, it was a complex process in which many of its founders grasped the phenomena they were examining using different confused and often incompatible perspectives, often within the same talks, correspondence, and published papers, and in a way which was slowly revised over time in an ongoing dialogue. "Science is rooted in conversations," as Beller likes to quote Heisenberg.[18] The quantum realm is so different in its fundamental characteristics that it is difficult to imagine how this transition might have occurred otherwise.

Much of science historian Max Jammer's work has consisted of describing how fundamental concepts of physics (including space, force, mass, and simultaneity) have changed in a way that resembles less a string of revolutions descending from nowhere than a continuous rethinking—sometimes radical rethinking—based on previous concepts and on what scientists have encountered.[19] These descriptions recall Heidegger's remark in §3 of *Being and Time* to the effect that the level of maturity of a science "is determined by how far it is *capable* of a crisis in its basic concepts," which can lead to "more or less radical revision" of its domain (52, 9/27).

Second, the experiential part of history of science is not simple observation, verification, or fulfillment, but the experience of an event in a laboratory context in which one is aware that what one has witnessed could be staged in different ways, with different kinds of instruments. Phenomena do not usually appear at first with Cartesian clarity, but as incomplete, able to show themselves differently in different performances, and as embedded in a background. Phenomena show themselves as being able to be looked at with better instrumentation, and in different ways. The formal part thus does not fully specify the phenomenon, but rather points to the potential presence of phenomena that are able, even in need of, being interpreted—phenomena that show themselves as always already interpreted, in the sense of being realized or made present in this particular way at this particular time, with the possibility, even necessity, of being realized or made present in other ways. When Robert Hooke introduced the *cell* concept into biology, it was to characterize the honeycomb-like empty structures, reminding him of monk's quarters, that he saw through his microscope in cork, and that he thought explained what made it buoyant. He hardly thought he had made a final description, something that subsequent investigation would not elaborate or change. "[S]ome diligent Observer, if help'd with better Microscopes, may in time detect" much more in them than he had, Hooke wrote in *Micrographia* (1665). He had simply described what he saw, anticipated that more would show up with different instrumentation, and was under no illusion that he was describing an "object" that would be apprehended exactly as he had by inquirers with other instrumentation. A similar discretion appears in later biological researchers. They, too, knew that they were interpreting and anticipating; they, too, were aware that they were only taking initial steps in an ongoing process that would eventually overtake and transform their own work. This is not like the appearing of different profiles of a fully grasped invariant—the hidden profiles of a cup—but of something *more* about the phenomenon itself.

Third, the phenomena that show themselves in laboratory events may even cause us to alter the assumptions or theories that were used to program the equipment. Even and especially in the case of revolutions, advances can provoke more questions than they answer.

Finally, the appearance of a new phenomenon is a call, affects way of being. A new and unexpected result—parity violation, CP violation (change and parity violation), stars not in right place, electrons not in right place, narrow resonances—all these have a power, a power felt if you are member of scientific community. Exert more like power of face than a piece of dead matter that doesn't fit into the existing bins. There's more going on, that we have not captured in our theories. What is this affective power? These results speak, and that speaking has something to do with how science works, why we engage in it, why it is compelling, why people take it up even though not a lot of money, it has a kind of demand, I am called, addressed, we feel humiliated, and so on. And this allows us to identify *befindlichkeit*, not from simply a general regard for "Nature" with a capital 'N,' but to the satisfaction and pleasure and awe in wonder of particular cases, particular developments, even particular stages.

CONCLUSION

Continental philosophy of science provides resources for exploring issues that often are perplexing within the framework of traditional philosophy of science. Heidegger's notion of formal indication is an example of such a tool, with the potential to clarify the nature of scientific entities as the way we conceive them changes over time. The value of the notion seems to be that it avoids what Heidegger calls "premature formalism," a danger that philosophers of science should learn to heed. The value of something like an ontic analogue of Heidegger's notion suggests deeper possibilities for continentally inspired approaches to understanding science practice than have hitherto been explored.

NOTES

1. "The classical philosophers of technology occupied themselves more with the historical and transcendental conditions that made modern technology possible than with the real changes accompanying the development of a technological culture." Hans Achterhuis, *American Philosophy of Technology: The Empirical Turn*, tr. by R. Crease (Bloomington: Indiana University Press, 2001), 3. See also *What Things Do: Philosophical Reflections on Technology, Agency, and Design*, by Peter-Paul Verbeek, tr. by R. Crease (University Park: Pennsylvania State University Press, 2005).

2. Recall the point in the *Phenomenology of Perception* where Merleau-Ponty speaks of dancing as if it were a matter of "discovering the formula of the movement in question." This is unforgiveably obtuse from the philosopher of the lived body; it would be as if one spoke of painting as if it were a matter of envisioning ahead of time an array of colors, and then putting them down one by one—something

Merleau-Ponty eloquently showed us was not so. Maurice Mearleu-Ponty, *Phenomenology of Perception*, 126.

3. Hubert L. Dreyfus, "How Heidegger Defends the Possibility of a Correspondence Theory of Truth with respect to the Entities of Natural Science."

4. Oliver Heaviside, *Electromagnetic Theory*, vol. 1 (New York: Chelsea, 1971), vii.

5. *Hermeneutics and the Natural Sciences*, ed. R. Crease (Boston: Kluwer, 1997).

6. R. Crease, "Inquiry and Performance: Analogies and Identities Between the Arts and the Sciences," *Interdisciplinary Science Reviews* 28:4 (2003), 266–72; *The Play of Nature: Experimentation as Performance*. Indiana University Press, 1993.

7. R. Crease "What is an Artifact?" *Philosophy Today*, SPEP Supplement 1998, 160–68.

8. Dimitri Ginev, A Passage to a Hermeneutic Philosophy of Science, Amsterdam, Rodopi, 1997; The Context of Constitution. Beyond the Edge of Justification, Dordrecht: Springer, 2006; The Tenets of Cognitive Existentialism (forthcoming).

9. Patrick A. Heelan, *Space-Perception and the Philosophy of Science*, Berkeley, University of California Press, 1983; for an exposition, from which some of this is adapted, see R. Crease, "Experimental Life: Heelan on Quantum Mechanics," in B. Babich, ed., *Hermeneutic Philosophy of Science, Van Gogh's Eyes and God: Essays in Honor of Patrick A. Heelan, S.J.*, Kluwer, 2002.

10. Patrick A. Heelan, *Quantum Mechanics and Objectivity*, Nijhoff, 1965, x, also 3–4.

11. But doesn't measurement observe a symbol of the real (i.e., data) rather than the real itself? No, Heelan (1983) says, for "we take the observable symbol to be the criterion of reality for something whose nature is known only as part of a complex relational totality expressed symbolically in linguistic or mathematical terms"; the object is the invariance underlying all theoretically possible data presentations. This sounds complex, Heelan continues, but it is a process we carry out easily and everyday, as when we speak of the city of Dublin or New York or Chicago as a real worldly entity, but cannot comprehend it except as what is intended in a series of connected but partial views. In the case of quantum mechanics, Heelan writes, "deterministic and statistical elements are organically and inseparably united." Deterministic elements are involved in the wave function, "which is an idealized formula from which the results of individual and concrete acts of measurement can be computed and statistically correlated." Thus the difference between quantum and classical physics does not lie in the intervention of the observer's subjectivity but in the nature of the quantum object. For while in classical physics deviations of variables from their ideal norms are treated independently in a statistically based theory of errors, the variations—statistical distribution—of quantum measurements are systematically linked in one formalism. Heelan's discussion of quantum mechanics illustrates the kind of inquiry that might take place, extensively and systematically, in an empirical turn in Continental philosophy of science.

12. See Lena Soler, Howard Sankey, and Paul Hoyningen-Heune, *Rethinking Scientific Change and Theory Comparison: Stabilities, Ruptures, Incommensurabilities*. New York: Springer, 2008.

13. H. Putnam, "Meaning and the Moral Sciences (London: Routledge and Kegan Paul, 198), 184.

14. Jerome Groopman, "Diagnosis: What Doctors Are Missing," *New York Review of Books*, November 5, 2009, 26.

15. R. Crease, "Covariant Realism," *Human Affairs* 2 (2009) 223–32.

16. The genesis of this idea has been traced by Daniel O. Dahlstrom, Heidegger's Method: Philosophical Concepts as Formal Indicators, *The Review of Metaphysics*, 47 (4), 775–95. Theodore Kisiel, *The Genesis of Heidegger's Being and Time*, Berkeley, CA: University of California Press, 1993.

17. Thanks to Joydeep.

18. Martin Heidegger, *Ontology-The hermeneutics of facticity*, tr. John can Bu7ren. Bloomington: University of Indiana Press, 1999, p. 13.

19. M. Beller, *Quantum Dialogue: The Making of a Revolution*, University of Chicago Press, 1999.

20. M. Jammer, Concepts of Space: The History of Theories of Space in Physics, Harvard University Press, 1954; Concepts of Force: A Study in the Foundations of Dynamics, Harvard University Press, 1957; Concepts of Mass in Classical and Modern Physics, Harvard University Press, 1961; Concepts of Simultaneity: From Antiquity to Einstein and Beyond; Johns Hopkins U.P., 2006.

A SUPRATHEORETICAL PRESCIENTIFIC HERMENEUTICS OF SCIENTIFIC DISCOVERY

Theodore Kisiel

I had occasion to review Joseph Kockelmans' *Heidegger and Science* shortly after it appeared in 1985.[1] Kockelmans came to write his book by way of his long-standing interest in developing a phenomenological ontology of science comprehensive enough to cover the entire field of the sciences, not only of the natural sciences but also of the human sciences (*Geisteswissenschaften*) in Dilthey's distinction. It is the distinction that the neo-Kantian Windelband tellingly (and quasi-ontologically) distinguished into the nomothetic and the idiographic sciences, and that Kockelmans, in his book extends to a detailed analysis of the intermediate domain of the behavioral and empirical social sciences. But Kockelmans's book assumes a diffuse quality, not because of the breadth of the applications in the philosophy of science that it musters, but because it now and then only mentions in passing, and thus tantalizes us with, a central question that should have been made the unifying guiding thread of his book: namely, why the theme "Heidegger and Science" is not a tangential issue for this self-proclaimed "thinker of Being," but strikes at the very nerve of his lifelong thought. For at least one oddity of that lifelong thought of our thinker of Being is the frequency and intensity with which he addresses himself to the phenomenon of "science" without ever really intending to do "philosophy of science" (except for a final student "trial lecture" on July 27, 1915, on the concept of time in the idiographic-historical and the nomothetic-natural sciences). Thus, at first glance, his various meditations-on-the-sense (*Be-sinnungen*) of science appear to be mere "spinoffs" of his more overtly ontological intentions. And yet these apparent byproducts were fundamental and penetrating enough to anticipate insights into the historicity of science that will recur decades later (beginning in the radical 1960s) in the revolutionary "new philosophy of science" of Polanyi, Toulmin, Kuhn, Hanson, Feyerabend, and the like.

Trish Glazebrook's *Heidegger's Philosophy of Science*, which appeared fifteen years later, does not make the mistake of omitting to examine this unifying guiding thread that is the source of Heidegger's recurring meditations on the "essence" of the sciences.[2] But she tends to displace and so shortchange this central thesis, which bears on the essential historicity of the sciences, in order to accommodate a second aim of her book, not only of "demonstrating the significance of science to Heidegger's thought and the contribution of that thought to philosophy of science,"[3] but also of "locating his thinking in the analytic discourse,"[4] of showing "that issues crucial to Heidegger's analysis are central in the analytic tradition of philosophy of science,"[5] in short, of "bridging"[6] Heidegger's work to, of putting it "into dialogue with,"[7] the analytic tradition of philosophy of science, which is mainly an Anglo-American tradition currently dominated by figures like T. S. Kuhn, Paul Feyerabend, Ian Hacking, Karl Popper, and Imre Lakatos. Because the mathematical natural sciences are paradigmatic for analytic philosophy of science, Glazebrook is especially taken by Heidegger's thesis that "science is the mathematical projection of nature,"[8] which she finds in incipient form already in the 1915 trial lecture on "The Concept of Time in Historical Science" (GA 1). This thesis then recurs ever more insistently and so persists in several permutations through six decades of thought in Heidegger's later accounts of scientific theory, scientific representation, and scientific experimentation to the point of the essential convergence of modern science with technology, all of which constitute central topics of current analytic philosophy of science. Recoiling back into Heidegger proper, Glazebrook finally makes the claim that these central topics in close proximity to Heidegger's other noteworthy contributions to philosophy like

> his overcoming of metaphysics, his rereading of the ancients, . . . his vision and revision of language, truth and thinking—have at their core an inquiry into science that drove his thinking for sixty years. I am not arguing for a new reading of a few texts, or for adjustments and refinements of existing readings of Heidegger. Rather, I am bringing to light a new basis on which to interpret his work as a whole. . . . Heidegger may be right that "Every thinker thinks only one thought," but . . . [there are] . . . multiple possibilities for envisioning such a thought. I read Heidegger's [one] thought as a philosophy of science.[9]

Heidegger's re-newed, re-viewed question of be-ing *is* the question of science? This hyperbolic claim clearly begs a qualifier, a qualifier that is even more comprehensive and universal in its recurrence in the Heideggerian opus than the theme of science, and that better serves to account for Heidegger's reading of the history of metaphysics, "his re-reading of the ancients . . . his re-vision of language, truth and thinking" as well as his re-vision of our sense of science. Following Heidegger, I identify it as the *hermeneutic* qualifier, and take this traditional term in the full ontological depth that Heidegger found in the later Dilthey, who identified hermeneutics with human life itself, and saw it as the way that life spontaneously develops itself in its very being and meaning—*Das Leben selbst legt sich aus*: "Life itself lays itself out, explicates itself, interprets itself." It is this "missing link" of the hermeneutical that I wish to bring out in the following discussion.

PHENOMENOLOGY AS PRETHEORETICAL PROTOSCIENCE

To regain our footing on what has become a slippery slope into the Americanization of Heidegger's thought, let us return to the very first and more continental of two central theses that can be read as pivotal in Heidegger's repeated struggles with the essence of science, namely, the thesis that philosophy itself is a science. Heidegger remains close to the theme of science because he first took up Husserl's program of establishing "philosophy as a strict science" by getting back "to the matters themselves" or "to what matters itself" (i.e., be-ing). If Heidegger is to be called a philosopher of science at all, it can only be of the "protoscience" or "primal science" (*Urwissenschaft*) of philosophy itself, and this only during his overtly phenomenological decade of 1919–1929. Science for Heidegger is first the *logos* of phenomeno-logy, which for him is *ipso facto* the "logic" of onto-logy, the "science of being." Heidegger's phenomenology of being is to be a fundamental ontology whose single matter of concern is variously called "factic life(-experience)," the "historical-I," the "situation-I," and finally "Da-sein" (my/our unique "being here").

Heidegger's phenomenological decade is framed on the hither side by his course in 1919 on "The Idea of Philosophy and the Problem of Worldviews" (GA 56/57) in which he first laid down the ideal of phenomenological philosophy as a supratheoretical primal science of an original experience which cannot be objectified. As a pretheoretical science (is this phrase not a "square circle"?), it is like none of the particular positive sciences and therefore problematizes the theoretical, epistemological, and objective nature of these sciences by tracing their eidetic genesis from their initially pretheoretical and nonobjectifiable matters. On the yonder side of the phenomenological decade is the lecture course of 1928–1929, "Intro-duction to Philosophy" (GA 27) in which the very radicality of the matter of philosophy, now called the "historical happening of transcendence," dictates the abandonment of the ideal of a strict, original, and primal science, misleading in part because it makes us think of particular positive sciences like mathematical physics and theology, which radically contradict the essence of philosophy.

Thus, after a decade of vacillation over this strange pretheoretical primal science so unlike any other science, Heidegger definitively abandons the project of developing philosophy into a strict science. "What science on its part is, resides in philosophy in an original sense. Philosophy is indeed the *origin* [*Ur-sprung*, the "primal leap"] of science, but precisely for that reason it is *not* science, not even the original science" (GA 27, 18). He observes that it is not a science not out of lack but rather out of excess, since it springs from the ever superabundant and ebullient "historical happening of Dasein's transcendence" itself, the fundamentally dynamic but chiaroscuro "evidence" of life. Superlatively a science from its abiding intimate friendship (*philia*, GA 27, 22) with this comprehensive evidence, "scientific philosophy," much like the formula "round circle," becomes a misleading and even dangerous redundancy, deceiving us into pursuing the wrong tasks in both philosophy and philosophy of science. Philosophy should be regarded in its finite tentative (and so inventive) character as ever "under way" in its transcending movement, as ever philosophi*zing* in response to its ever unique situation with its ever unique, fundamentally chiaroscuro evidence. Philosophizing becomes explicit transcending by letting transcendence happen, repeatedly enacting the transition from the preconceptual understanding-of-being to a precursory conceiving of being. In this way, it repeatedly actualizes the ontological difference between be-ing and beings without objectifying be-ing itself (as in GA 24, the course of 1927, upon the "horizon of time").[10] Philosophy in this frenetic transcending nevertheless continues to function as the foundation (now, however, as a *fundamentum concussum*) that makes sciences and their regional ontologies possible, and moreover in its epochal time and history also accounts for the periodic revolutions in their fundamental concepts (GA 27, 16–19; 219 ff).

These shifting relationships between philosophy and science evolve into the later contrast between meditative thinking and calculative "thinking," and the statement of provocation that "science itself does not think." Scientists themselves, however, on occasion do think more or less basically, particularly in the moments of crisis in the fundamental conceptuality of their disciplines . . . when they, in being forced to deliberate on the fundamental sense (*Besinnung*) of their concepts and so of their science, in fact become philosophers! The early Heidegger's litany of scientific revolutions in progress in his time is also worthy of note, since some of his readers have been captivated by the germinal Kuhnian insights embodied in this litany.[11] Significantly, this recognition of the pervasiveness of scientific revolutions in the "hermeneutic situation" of his time first emerges in the wake of a confrontation of Dilthey's philosophy of science and is repeated throughout the period of the drafting of *Sein und Zeit*.¹² Heidegger therefore does not restrict himself to the crisis of revision of the fundamental concepts in the mathematical natural sciences in the wake of Einstein's theory of relativity and the strife between formalism and intuitionism that had precipitated a crisis of the foundations of mathematics. He also points to the crisis over the fundamental concept of life in biology and the attempt made by Dilthey himself to revive the lived sense of the actuality of history and tradition that is basic to the historical human sciences, and the search in the science of theology "for a more original interpretation of the human being's be-ing toward God" (SZ, 10/30). All of this groundlaying work that re-views and re-vises the fundamental concepts of sciences in crisis calls for a "productive logic" of concept formation that "leaps ahead into a particular realm of being in order to first disclose the constitution of this being and to make such structures available to the positive sciences as transparent frameworks that orient their inquiry" (SZ 10/30–31). But these different frameworks of regional ontologies that guide the different inquiries of the ontic sciences—Kuhn will call them paradigms—require a master framework, which Heidegger calls his fundamental ontology. A hermeneutics of Dasein, of the unique being that already understands (and so spontaneously interprets) be-ing—is to serve as a "guideline" for the overall "ontological task of constructing a non-deductive genealogy of the different possible ways of being" (SZ 11/31) into which the various sciences inquire. For the scientific revolution of major concern to Heidegger is in fact the ongoing crisis of philosophy itself as a science, as "a revolution in the very way in which philosophical questions are to be posed and formulated"¹³ as a result of the confluence of Husserl's phenomenology with Dilthey's hermeneutics of historical (factic) life, along with other philosophies of life (Bergson, Nietzsche) and of existence (Jaspers) in postwar Germany in the Weimar 1920s.

The neo-Kantian forms of scientific philosophy then in vogue were notorious for beginning, in the spirit of the Enlightenment, with the accepted *fact of science*—and then analyzing science as an already finished product. The most notorious fruit of this would be the logical positivist image of science as a nomothetic system of laws and logical structure of theoretical proofs that map and order its sense data. But by way of the genetic phenomenological reduction, science is no longer accepted as a given fact but is made into a problem that is to be resolved by tracing the eidetic genesis of the theoretical from its pretheoretical protopractical roots (SZ, 357/408), a genealogical problem that the early Heidegger made central to his hermeneutics of facticity. The sciences are to be viewed, not after the fact as a fixed structure of finalized results, but before the fact as an ongoing research process in a concrete "problem situation" that is to be interpreted and resolved against the historical background of inherited presuppositions that impart direction and sense to all the aspects of the science. In the same radically historical vein, the teacher Heidegger advises his beginning students on opening day of the winter semester of 1923–1924 (GA 17) to approach their chosen *Wissenschaft*, be it physics, historiography, or theology, not as a finished theoretical structure attempting to correspond to reality, but more "phronetically" as a praxis, as a historically unique practical situation in which they find themselves caught up and already under way, where each of them is called on to become "native" in an ongoing science *in via* by making its presuppositions their very own and by developing a passion for the questions that it generates, thereby "resolutely" appropriating the relevant opportunities and choosing the problems that their particular discipline is transmitting to them.[14]

Heidegger thus antedates by decades the sense of science-in-process-and-revolution and science as hermeneutical praxis to which Anglo-American philosophical historians of science like Kuhn, Toulmin, and Shapere would later acclimate us. In the language of current philosophy of science, such a hermeneutic approach already at the initial interrogative level of choosing the vital problems at the cutting edge of one's science in its current "problem situation" would put the priority on the "context of discovery" over the "context of verification" that logical positivism made paradigmatic in its formal understanding of science. Making discovery the paradigmatic problem itself at once recalls the shift in the primacy of "truth" from propositional correspondence to a more historically precedented fallibilistic uncovering so dominant in Heidegger's thought. Science-in-genesis of discoveries is hermeneutical precisely insofar as it necessarily understands, interprets and painstakingly resolves its problem situations against the background and in terms of the presuppositional contexts of meaning that give rise to its problems and at once provide the resources for their resolution. Although it is not amenable to the analytic rationality of formal logic, scientific discovery is not ipso facto irrational, as the positivists would have it, accountable only in reductively psychologistic terms like intuition, inspiration, the "flash of genius" the "Eureka!" experience, the workings of the unconscious, and even the "serendipity" of "chance," as the evolutionary epistemologists are wont to put it. Scientific discovery is a rational practice that develops in accord with a situational logic of interrogative demand arising from its particular "hermeneutic situation" and soliciting an appropriate human response to that situational demand. This demand-response logic of the hermeneutic situation is governed by a contextual rationality ("The context decides") and a practical rationality, a hermeneutic rationality that calls for the explication of the appropriate sense lying fallow and latent in the situational context of discovery.[15]

Heidegger's "hermeneutic essence of science" caught the attention of early commentators on Heidegger and science like Caputo and Heelan, but more recently, Glazebrook's analysis has muddied the waters by misconstruing it in an unfortunately sparse and highly selective gloss of §69(b) of *Sein und Zeit* on "The Temporal Sense of the Modification of Circumspective Concern into the Theoretical Discovery of Objectively Present Things Within the World."[16] As announced in its section title, §69(b) first traces an "existential genealogy" of theoretical and objective science as a latter-day phenomenon emerging from protopractical and preobjective contexts already articulated by factic life experience, thus already "interpreted" by the spontaneous "hermeneutics" indigenous to human facticity. The "hermeneutic essence" of theoretical and objective science is thus derived from its rootedness in this pretheoretical and preobjective "hermeneutic situation" of unique historical Da-sein, from which it always arises and to which it repeatedly returns. The unique historical occurrence of this hermeneutic situation of human existence is the sole topic of Heidegger's hermeneutic phenomenology, which is not to be conflated with the more traditional, theoretically oriented phenomenology of intuition propounded by Husserl. Glazebrook's misconstrual of §69(b) stems in large part from her imperfect grasp of the distinction between two phenomenologies, one traditionally oriented toward the "self-evident eternal" truths of intuition and the other toward the more comprehensively temporal and hermeneutically chiaroscuro truths of unconcealment.[17] Her neglect of *hermeneutic* phenomenology extends to the point of complete omission of the term *hermeneutic* from an otherwise amazingly thoroughgoing Index bringing up the rear of this work. What follows is a brief corrective gloss of §69(b), a key section of *Being and Time* that is pivotal not just for Heidegger's "philosophy of science" but for the development of the entire project of *Being and Time* itself.

It is in §69(b) that Heidegger first clearly makes the above-discussed distinction between the logical conception of science fixated on the ex-post-facto analysis of its nomo*thetic* and hypo*thetic* results, as opposed to the "existential" (i.e., hermeneutical) conception of science concerned with its more genetic moments of becoming a science by cultivating the "theoretical attitude" out of (and therefore still within, Heidegger stresses in SZ 358/409) the practical natural attitude. But he also postpones the task of a

> completely adequate existential interpretation of science until the meaning of be-ing and the 'connection' between be-ing and truth have been clarified in terms of the temporality of existence. The following deliberations [on the existential conception of science and the ontological genesis of the theoretical attitude] are a

propaedeutic to understanding this central problematic [of joining be-ing and truth in a temporal continuity, i.e., on the horizon of Temporality]. Within this central problematic, *moreover*, the idea of phenomenology, *as opposed to* its preliminary conception already indicated in our Introduction (in §7[c]), will also first be developed. (SZ 357/408, emphasis added)

It may be recalled that Heidegger in §7(c) distinguishes a "formal" definition of phenomenology based on its etymology of "letting what shows itself be seen," which doubly accentuates the truth of intuition central to Husserl's phenomenology, and a "deformalized phenomenological" (SZ 35/59) sense of phenomenology whose task is to hermeneutically expose, "wrest" (SZ 36/61) out of concealment, "that which first and foremost does *not* show itself . . . but at the same time belongs to what first and foremost does show itself so essentially that it in fact constitutes its sense and ground" (SZ 35/59). It is this hermeneutic sense of be-ing and of its correlative phenomeno-logy that the Third Division of *Sein und Zeit* (the entirety of §69 is projecting its several tasks) is to elaborate in the full glory of its ekstatic-horizonal temporality (this Division was never published). In a footnote shortly after spelling out this task of "developing" the idea of phenomenology as a science of concealing-unconcealing be-ing, in fact as an aside on the thematizing process that objectifies and "makes present" the mathematized objects of natural science (SZ 363/414), Heidegger observes that Husserl, in fulfilling a long Western tradition of a metaphysics of constant presence, takes its central epistemological thesis that all knowledge aims at "intuition" to have the temporal meaning that all knowing is a making present. Heidegger then suggests that not every science, in particular "philosophical knowledge," has the intuitive aspirations of making present. The exception he has in mind is clearly a hermeneutic ontology along phenomenological lines, concerned with the truth of unconcealment, its concealments and absences, in order to "let them be" in and through the explication of their tacit ways of constitution. At one point along the way at least, Heidegger observes, contrary to Western tradition, that the most basic form of all knowing is not intuition but rather expository interpretation (*Auslegung*)[18] of what is already understood simply through living, and be-ing.

In §69(b), Heidegger *contrasts* his preliminary conception of phenomenology, which he had *already developed* in §7(c), from the more hermeneutical "idea of phenomenology" left undeveloped in §7(c), which he is now in a position to develop in its full temporality, once the ekstatic-horizonal structure of the full tensorality (*Temporalität*) of be-ing has been worked out (schematized in §69[c]). Phenomenology is for Heidegger the "method" or way of philosophy, which in the context of Kant and German idealism is still understood as a tensoral transcendental science. But in 1928–1929, the very idea of a scientific philosophy becomes for Heidegger a redundantly misleading "round circle" still associated with metaphysics and onto-theology, both of which are to be overcome by way of meditative thinking, which science itself cannot do. In this context, I would suggest that it is not the question of science that in fact dominates the entirety of Heidegger's *Denkweg*, but rather the way of doing philosophy, and so the metaquestion of the nature of philosophy. This dominant thread is captured very nicely in the title of William Richardson's book as amended by Heidegger's letter prefacing the book: not "From" but "*Through* Phenomenology to Thought" (emphasis added) suggesting that phenomenology is never really left behind.

Accordingly, the question that dominates the entirety of Heidegger's career of thought is the "question of phenomenology as a possibility," first, of philosophy as strict science, and later, of philosophy's transformation into the "poverty of thought" faced with an ineffable matter which in its very withdrawal (*Entzug*) draws us to thought, "calls us to think." It is therefore important to note the lingering phenomenological vestiges in the various traits (*Züge*) or vectorial "pulls" that draw the thinking that outstrips science, first of all in its persistent pursuit of sense, in its meditation-on-the-meaning (*Be-Sinnung*) of the historical situation in which each of us as individuals and as communities (of Germans or North Americans, of scientific researchers or philosophical freelancers, etc.) happens to find ourselves here and now, in the epoch of globalization. Vestiges of the free variation of eidetic reduction toward Husserl's *Wesensschau* are surely to be felt in the repeated ventures of thought to establish the dynamic historical identity of such individuals, communities, and human endeavors, not in terms of generic universals but by way of temporally distributive (*jeweilige*) universals ultimately to be defined according to the individuating historical context, "*je nach dem*," in order to determine, for example, how modern science in the age of globalization "perdures and prevails, rules and administers itself, comes to be and passes away—in short, how it 'essences.'"[19] By way of such historical variations, one arrives at an aspect of science that science by itself cannot reach, its hard-core "uncircumventable, that which cannot be gotten around,"[20] the *Sache* of each respective science and how it is to be approached, nature for mathematical physics, the psyche for descriptive psychology, and so on. Instead of a re-duction, eidetic or transcendental, Heidegger now speaks of a "leap" of thinking, respectively to "essential thinking" or to "originary thinking."

Natorp's word play on the "primal leap" to and from the "origin" (*Ur-Sprung*) is, however, very much in evidence already in GA 56/57 in 1919, when Heidegger first defines phenomenology as a "pretheoretical primal science of original experience," which already puts its status as a science in question. In its reductive regress back to the matter of be-ing *itself*, in its unendingly circular deepening of the historically unique, hermeneutically "intentional" *relationship* of the understanding-of-being, in its exposition of *limits* at which something begins to be what it is in its *possibilities*, where it accordingly "essences," fundamental thinking still bears the traces of the phenomenological approach from which it receives its initial start. While discarding the inappropriate concern for science and research in favor of a more fundamental logos, fundamental thinking retains not only many of the means acquired from its phenomenological discipline, but especially the common end of the pursuit-of-meaning (*Be-Sinnung*). The movement "*through* phenomenology to thought" dominates Heidegger's path from start to finish.

HERMENEUTICS OF THE POLITICS OF SCIENCE IN THE GERMAN UNIVERSITY

This section presents the concept of science in the hermeneutic context of the German university in the Nazi 1930s.[21] Situating the sciences in the context of the university naturally places them in a community of interpretation, so that one can in the same context appeal to the intellectual function of the university as the site for preserving and transmitting a tradition across generations. But Heidegger, as university rector, not only invoked the tradition of thought transmitted from the Greeks to the Germans as a "nation of poets and thinkers." In the first wave of enthusiasm for the political revolution of 1933, he also modeled the university after the worker-state projected by the National Socialist German Workers Party, projecting it as a kind of Platonic Idea but modifying its ideality to accommodate a uniquely German folk ethos, its traditional "work ethic." The new German university student, as a future leader of the nation, is to engage in work service and military service as well as in the main service of the university, the service of knowledge and science, which as the "work of the brain" does not differ in kind from, and so is no higher than, the two levels of the "work of the hand and fist." All work is intellectual or "spiritual," a knowledge-laden deed and action that incorporates a craft know-how and an ordered understanding of its place in the world. But work in particular involves a "capacity of resoluteness and perseverance in carrying out the undertaken task to its conclusion, in short, *freedom*, which means: *spirit*"[22]; to which we might also add the prized German trait of *Gründlichkeit*, thoroughness, and even a related word then current in Nazi jargon, "hardness." What is "science" in this context?

> Science is but the *more rigorous* and thus *more responsible* form of that knowledge which the entire German people must seek and demand

for its own historical Dasein as a [worker] state, provided that this people still wills to secure its continuance and its greatness and to preserve these in the future. The knowledge of genuine science is *in essence in no way* different from the knowledge of farmers, foresters, miners or gravediggers, and handworkers. . . . For knowledge means: *to know our way around* the world in which we are placed as individuals and in community.

Knowledge means: in decisiveness and initiative *to be equal to* the task to which each of us is consigned, be it the task of plowing the field, felling the tree, digging the grave, interrogating nature in its laws, or expositing history in the power of its destiny.

Knowledge means: *to be master* of the situation in which we are placed. (GA 16, 234ff)

One can agree that Heidegger "treats the university as the institutionalized expression of the human desire to know,"[23] but in his speeches as Rector beginning with "The Self-Assertion of the German University," which are primarily intended for domestic consumption in a time of domestic national crisis and its revolutionary resolution, this desire to know assumes a uniquely German accent, more specifically, the folk accents of German idealism concretized through Heidegger's *protopractical* and multivalent sense of Dasein as *care*. The will to know, learn, question, discover on the level of the university takes on the form of *Ent-schlossenheit (phronêsis)*, resolute openness, at first actively strenuous in its volitional rigor in responding to the demands exacted by a time of national crisis and, on its other face, receptive in its openness in order to let the "hermeneutic situation" of post-Weimar Germany be. This resolute openness is the very *spirit* of the German university in its willful self-assertion, where its will to science is the will to *question* the various sciences in "their boundless and aimless dispersal into particular fields and niches" in order to expose them once again to the full comprehensiveness of overwhelming "world-shaping powers of the human-historical Dasein" of a people "*in the midst of the uncertainty of beings as a whole*," that is, in the interwoven contexts of nature, history, language, and the state that appear to be most suited to it, as well as the law, custom, economy, technology that it is to develop. Such a will to science "will create for our people its world of most intimate and extreme danger, which is its truly *spiritual* world." This *spiritual world* of the people exposed by the German university is not a superstructure of high culture or a depository of useful information and values; rather, "it is the power that most deeply preserves the people's earth-and-blood-bound energies and, as such, it is the power that most deeply moves and most profoundly shakes the Dasein of a people."[24] It is this resolute power of indigenous spirit that guarantees each particular people its possibility of greatness, for it to choose in resoluteness or to allow to lapse and fall into decline.

Heidegger from his university podium continues to pose this fateful choice as late as 1935 to the German people, challenging it to recover its autochthonous spirit and so reclaim the spiritual world indigenous to it. Germany, this nation of poets and thinkers caught in the land-locked vice-like grip of central Europe, now lies in the great pincers between the metaphysical twins of America and Russia, both of which are caught up in "the same hopeless frenzy of unchained technology."[25] It is thus metaphysically threatened on its Western front by the international "spirit of capitalism" (Max Weber's phrase) and on its Eastern front by the international "specter of communism" (opening line of *The Communist Manifesto*) then "haunting" Europe and the entire planet. Germany, the most metaphysical of peoples, is by the same fact best equipped spiritually to reverse the drift of the disempowerment of the spirit through scientism, positivism, materialism, utilitarianism, and other identifiable versions of nihilism incurred by the industrial revolution, and so to arrest "the decline of the West." For the "inner truth and greatness" of the indigenous German *movement* called "national socialism," born of the "spirit of the front" (*Frontgeist*) in World War I, resides in its promised autochthonous resolution, through its *völkischen* worker-state, of "the encounter between global technology and modern man" (EM, 152/199)

It would take Hitler's announcement of the "Four Year Plan" in September 1936 and the impact that this "total mobilization" of the German military-industrial complex, tacitly in preparation for a total war in four years, would have on the universities before we find the first clear evidence of wholesale, albeit (as usual) discreet, resistance to state policy and planning on the part of Heidegger. This resistance is sustained by a full-scale meditation (*Besinnung*) on the essence of modern science (thus of the increasingly specialized and fragmented "research university") as technological, that is, on the institution of scientific research as a "business" activity, in which method takes precedence over the domain of objects that is itself being projected by the particular scientific specialty. It is within such a meditation that the essence of modern metaphysics is first seen to culminate in the essence of modern technology.[26]

The courses from the summer semester of 1937 on display increasing concern over the technical organization of the sciences toward useful results for the "benefit of the people and nation,"[27] and over the "superpower" of technology and the total technical mobilization of the entire planet (GA 6.1, 404/186). This takes place against the background of Germany's Four Year Plan impacting on the university, whereby even the human sciences are being made into pedagogical tools to inculcate a political worldview, and the "technology of vast libraries and archives" are placed at the disposal of news media and information services, or used primarily to avoid unnecessary duplication of costly lab experiments (GA 6.1, 236/16). "Today the major branches of industry and our military Chiefs of Staff have a great deal more 'savvy' over 'scientific' exigencies than do the 'universities'; they also have at their disposal the larger share of ways and means, the better resources, because they are indeed closer to the actual'" (GA 6.1, 237/16).

Consternation over the Four Year Plan, especially among the younger faculty at Freiburg, led to a series of "working meetings" among them, independent of the party sanctioned discussions of the matter, beginning in the autumn of 1936. Heidegger took over such a working session meant especially for instructors in the natural sciences, mathematics, and medicine in the winter semester of 1937–1938, launching it on November 26, 1937, with a keynote lecture on "The Threat to Science."[28] Long before the external threat of the Four Year Plan, there had been an internal threat to science that comes from itself in its modern development, in its giving of primacy to method over its matter, which has led to the progressive technization and specialization of the sciences, to the detriment of their relation to their respective domains of being. The resulting threat of groundlessness is only amplified by the "unusual emergency of our people" (Heidegger, 1936: 8), which confirms the political need for further technical organization. Industry now takes over science, after the American way of scientific pursuit and in competition with it. Scientific organization is becoming a world process, such that in the future, it is no longer the countries with the richest natural resources, but "the countries and peoples with the greatest and most impressive inventions that will seize world leadership" (9). As industry takes over science, science in its fulfilled form of abundantly equipped facilities no longer finds any deep and inner roots within the university. Yet it is not just industry that takes science away from the university, but also the absence of the long overdue "self-assertion of the German university." For Heidegger, there is a deeper sense to the self-assertion of the German university, over against its coordination into the ever expanding military-industrial complex of the Third Reich, than a mere clash of the basic institutions of the army, industry, the university, and the state. Self-assertion of the German university does not mean clinging to a past academic tradition [of the Humboldtian university], nor, to begin with, the political organization of the university, but rather "the will to put itself into question and thereby, and only thereby, to win back its proper task for itself, and in a higher form: thus to be the site in which *science itself, on the strength of an original knowledge of itself*, secures and continually renews and augments itself. This knowledge of itself can only grow out of the communal meditation on the meaning of the different albeit interrelated domains of science and groups, out of the will to a historically spiritual ground" (Heidegger, 1936: 10). Without such a self-assertion, the escalating threat to science at the university is further intensified by the necessity of political education and of the creation of a new generation of Leaders in the party. The recent announcement of plans for a new kind of supreme technical school, "with the will of the Führer behind it" (Heidegger, 1936: 11) does not necessarily make the university superfluous, but it will certainly dilute its initiative. With the multiplication of new departments in the university like military science, racial science, ethnology, Germanic prehistory, and space research (i.e., the geopolitical problem of *Lebensraum*), and the establishment of new chairs in them, "the great will toward a meditation on what is essential will become more and more impossible" (Heidegger, 1936: 11). But ultimately, the essential threat to science comes not from political measures against it nor from the new utilitarian goals set for it, but solely from itself, its inability and unwillingness *to renew and transform itself from within.*

Heidegger's notes for and from these "working meetings" turn again and again on the political constellations that relate science to the National Socialist "worldview" (and not to the "movement," as in 1935!). One choice example:

> There is not even a transformative will for this new organization [of science as a spiritual power]. The farcical 550th jubilee celebration at Heidelberg University: forced and inflated without ground and background. And the Führer? Stays away! Instead, on August 16 [1936] he closes the Olympic games in Berlin; on the same day, he organizes the preparations for the Tokyo games! . . . The Olympic games are better suited for foreign propaganda. The sports greats from all lands are courted for their approval—one is more among one's own kind! 'University people' of the old style also know too much. (Heidegger, 1936: 22)

One of his greatest mistakes in the rectorate: "that I did not know that the ministry cannot be approached with creative projects and large goals" (Heidegger, 1936: 24). Now that the "coarse and nonsensical and naive outburst of a 'new *völkischen* science' has totally gone awry," the pendulum has swung the other way.

In demanding undisturbed quiet for supratemporal science, one finds a new common ground for compromise. From the side of science, one concedes that there is no such thing as pure theory, that there is room for a worldview. From the side of the *völkischen* representatives, one concedes that one must concentrate work on the "matters themselves," but also that the demand for a worldview is indispensable. Both sides are now saying the same thing, but the compromise thereby diffuses all the forces of questioning that would bring us to "the moment of true inception and a real change"

(Heidegger, 1936: 24). What to do in this stalemate? Running away solves nothing. Best to remain and exploit the possibility of meeting like-minded individuals while willing one's own individuality. "This not to prepare the university—now hopeless—but *to preserve the tradition*, to provide role-models, to inspire new demands in one or another individual—somewhere, sometime, for someone. This is neither 'escape' nor 'resignation' but the necessity that comes from the essential philosophical task of the second inception" (Heidegger, 1936: 24 ff). In this situation, "we must put on the mask of the 'positivists' and be confused with them. We thereby enter into the 'circle of the *Lanthanontes* [secret ones].' But these Lanthanontes can only be those who know that and why they must be secretive. Not the game of the misunderstood or of those passed by or of the 'long-suffering.' Resignation? No. Blindly agreeing to everything? No. Accommodation? No. Solely to build for the future" (Heidegger, 1936: 25). The university is at the end and so is science, "but this is precisely *because philosophy has its second essential inception before itself*. That what we have called *science is running its course and technologizing itself, perhaps for a whole century*, proves nothing to the contrary!" (Heidegger, 1936: 26). In view of its uselessness, philosophy's positions and chairs are being reduced or cancelled. "But with the abolition of philosophy, the Germans—and this with the intention of gaining their *völkischen* being!—are committing suicide in world history" (Heidegger, 1936: 27).

With this "total" entry into the industrial arms race in preparation for war, National Socialism, purportedly in search of geopolitical "living space" and scarce natural resources, has unequivocally placed itself on the same plane as capitalism and communism. The "movement" in search of its uniquely German roots and common unity has become, like them, a technological worldview. At this point, Heidegger gives up his fading hope in a difference in the decisions made by narrow-minded party functionaries and by Hitler himself, the statesman (*phronimos*) whose originative deeds create a new state and a higher order of be-ing. After he develops a more refined sense of the essence of technology as completed metaphysics, Heidegger will characterize Hitler as the supreme technician of a System as much being imposed upon him as manipulated by him, by way of a shrewd calculative thinking totally devoid of any vestige of the meditative thinking (*phronêsis*) required of the statesman.[29]

In his first approximation of the metaphysical essence of technology in "The Age of the World Picture," which revisits the pincers passage of 1935 three years later in order to characterize the situation as a "struggle of worldviews" (GA 5, 87/134), Heidegger identifies the "national socialist philosophies . . . the laborious fabrications of such contradictory products" (GA 5, 92/140) as among them (but as usual, discreetly, in an appendix that was not read!). He singles out the phenomenon of the "gigantic" (*das Riesige*: also the "titanic, colossal, mammoth . . . monstrous!") that appears in various guises and disguises in the course of the technological "conquest of the world as picture" (GA 5, 87/134)—referring not just to the oversize machines like particle accelerators merely to "smash" the miniscule atom, but also to the gigantic numbers of atomic physics as well as astrophysics, the annihilation of mammoth distances by the airplane and radio, and so on—and observes that this manifold phenomenon of giganticism cannot be explained by the catchword "Americanism" and its presumed worship of bigness (GA 5, 87/135). For "Americanism itself is something European" (GA 5, 103/153) and the modern worldviews that come from Europe develop their own gigantic displays "when the tallies of millions at mass meetings are a triumph!" (EM, 29/38). Giganticism is but one of the results of the "global" thrust of modern technology, already manifesting its totalizing consequences in the early twentieth century in global phenomena like the world war and the worldwide economic depression.

GE-STELL

That giganticism is not just American but in the end "something European" will return with a vengeance for Heidegger as the 1940s progress and he first develops his sense of the essence of modern science and technology as *Ge-Stell*, the syn-thetic com-positioning of man and nature alike into the technological grid. One of the first examples that he gives of the forced and challenging extremes to which technological exploitation is carried, along with examples like the extraction and processing of gigantic amounts of uranium ore for the fabrication of a few atomic bombs and the conversion of agriculture into a motorized food industry, is the "fabrication of corpses in gas chambers and annihilation camps. (GA 79, 26). How to comprehend in an existential way the "megadeaths" involved in such systematic killings? "Hundreds of thousands die in masses. Do they die? They are done in, annihilated, 'wasted,' killed. Do they die? They become component parts of a stockpile com-posed [and decomposed] from the fabrication of corpses. Do they die? They are surreptitiously liquidated in annihilation camps. And even without camps—millions are now reduced to famine in China and end by starving to death" (GA 79, 56). So much for the folkish worker-state which, like the truth of the New Testament, "shall make you free" (*Arbeit macht frei*).

Perhaps due to his early long-standing association of "Americanism" with mammoth feats of technology like the Hoover dam and skyscrapers, the old Heidegger, undeterred by the later Russian achievement of Sputnik, developed a penchant between 1966 to 1976 of repeatedly posing the question of the technological *Ge-Stell*, mercifully without ever using this difficult to translate artificial Teutonism, to a series of American conferences that he was told would be devoted to his "thinking." The formulation chosen as the anniversary question of this past year's meeting (in the millennial year of 2001) of the Heidegger Conference of North America is perhaps the clearest and most direct version of this "question of technology":

> Is modern natural science the foundation of modern technology—as is commonly assumed—or is it itself already the basic form of technological thinking, the determining preconception and continual incursion of technological representation repeatedly being geared into the operative and adaptive machinations of modern technique? The latter's ever escalating "efficiency" impels the oblivion of be-ing to extremes and so makes the question of be-ing appear trivial and superfluous. (GA 16, 747)

Glazebrook regards *Ge-Stell* as the culminating projection of science within a series of three projections that historically constitute the very essence of science.[30] The first is the *ontological projection* that predetermines the basic concepts of the particular region of beings that serve to guide the investigations of the particular regional science, understood to be situated within the more comprehensive context of a unified field of be-ing that is left to the "science" of philosophy to investigate. The second is the *mathematical projection* of nature that constitutes *modern* science, making mathematical physics into a paradigm for all the positive sciences. It is therefore the first of the reductive projections in its extension of Galilean-Newtonian mechanics, mathematized by the CGS system of calculation, not just to quantum mechanics but to all the "positive" sciences. Both "mechanics" and "calculus" already suggest that the mathematical projection is but a "transitional phase" destined to ripen into the *technological projection* that is currently "globalizing" science as well as humanity.

The *Ge-Stell* (best translated etymologically as syn-thetic composit[ion]ing) can also be espied in filigree in the grand metaphysical systems of modern philosophy that aspire to a *mathesis universalis* and take being to be absolute position (Leibniz, Kant). Even the passage through absolute negation in the grand dialectical systems of late modernity only temporarily postpones the inevitable impression of a giganticism and an ultimate standoff of absolute object standing over against absolute subject, who has in advance "set up the object before" itself, re-presented (*vor-gestellt*) it. The standing object is the hypo-thesized component of the nomo-thetic composite of science, while the stock item standing in reserve (*Bestand*) is the ever uniform and disposable (*be-stellbare*) component of the technological com-posite. Its syn-thetic stock of "natural" resources extends even to the uniform units of reserve "manpower." It is easy enough to espy the shifting patterns of such syn-thetic com-posites in the year 2001, at the axis of the millennium, of which a few have already been named: the military-industrial complex, corporate globalization, the totally mobilized "war machine" at once coordinating Atlantic and Pacific fronts by way of space satellites, the Missile Defense System, "Houston Control," satellite tracking systems, the air traffic control network, the all-pervasive and multi-utilitarian Global Positioning Systems, the comprehensively coordinated weather forecasting model, the regionalized electric power grids controlled by global corporations cascading into blackout or bankruptcy, the donor "organ bank" system, the internetted WorldWideWeb and the 24/7 global network CNN, the Human Genome Project, and so on. It is sometimes difficult to look beyond the global com-posite and find the right words to formulate the "question of be-ing" inherent in this millennial "hermeneutic situation." "For there still looms like a specter over all this uproar the question: what for?—where to?—what then?" (EM 29/38).

NOTES

1. Joseph J. Kockelmans, *Heidegger and Science* (Washington, DC: Center for Advanced Research in Phenomenology & University Press of America, 1985). My review of it appears in *International Studies in Philosophy* 21, no. 1 (1989), 96 ff.
2. Trish Glazebrook, *Heidegger's Philosophy of Science* (New York: Fordham University Press, 2000).
3. Glazebrook (2000), 5.
4. Ibid., 4.
5. Ibid., 5.
6. Ibid., 3.
7. Ibid., 253.
8. Ibid., 1.
9. Ibid., 13.
10. Space does not permit the detailed treatment of this idiosyncratic attempt, so untypical of Heidegger as a whole, "to objectify be-ing itself on the horizon of time." In taking note of this idiosyncratic formulation, Glazebrook (2000: 29) makes the interesting observation that such an "objectification" does not posit be-ing, since the *temporal* and *transcendental* science of phenomenological ontology is *not* a *positive* science, but rather seeks to elucidate "why the ontological determinations of be-ing have the character of apriority" (GA 24, 462/325). For a more detailed treatment, see my "Ontology and Fundamental Ontology," *The Encyclopedia of Philosophy Supplement*, ed. Donald M. Borchert (New York: Simon & Schuster Macmillan, 1996), 384–86; "Fundamental Ontology," *Encyclopedia of Phenomenology*, ed. Lester Embree et al. (Dordrecht: Kluwer, 1997), 253–57.

11. See, for example, Glazebrook (2000), 15, 87, 109, and 217.

12. The then widespread phenomenon of productive revolutions prompting revision of the basic concepts of the various sciences, as well as of philosophy, is first discussed in the opening paragraph of the Kassel lectures of April 1925 and reiterated in the lecture courses GA 20, §1; GA 21, §3; GA 25, §2b; and especially in SZ, §3.

13. From the opening paragraph of the first of the Kassel lectures of April 1925, which bear the general title, "Wilhelm Diltheys Forschungsarbeit und der gegenwärtige Kampf um eine historische Weltanschauung," transcript of Walter Bröcker, ed. Frithjof Rodi, *Dilthey-Jahrbuch* 8 (1992–1993), 143–80, esp. 144.

14. GA 17, 2 ff. This phronetic point and telling hortatory appeal to the students is much clearer in the highly detailed student transcripts recorded from the *vox viva* of the actually delivered course than in the incompletely collated German edition of this, the first of the Marburg lecture courses, which, based on the principles of an *Ausgabe letzter Hand*, relied more heavily on the initial version written in Heidegger's own hand, a highly truncated first draft of the course composed in great haste since he was in the rushed process of making the move from one university to another, from Freiburg to Marburg. See the Editor's Postscript for the faulty rationale of this edition of GA 17.

15. On a hermeneutics of the natural sciences based on Heidegger's "existential" and historical conception of science, with special attention to the logic of scientific discovery, see Theodore Kisiel, "Scientific Discovery: Logical, Psychological, or Hermeneutical?" *Explorations in Phenomenology*, ed. David Carr and Edward Casey (The Hague: Nijhoff, 1973), 217–34; "Hermeneutic Models for Natural Science," *Phänomenologische Forschungen* 2 (1976), 180–91; "Heidegger and the New Images of Science," *Research in Phenomenology* 6 (1977), 162–81, reprinted in *Radical Phenomenology: Essays in Honor of Martin Heidegger*, ed. John Sallis (Atlantic Highlands, NJ: Humanities, 1978), 162–81; "The Rationality of Scientific Discovery," *Rationality Today/Rationalité Aujourd'hui*, ed. Theodore F. Geraets (University of Ottawa Press, 1979), 401–11; "Ars Inveniendi: A Classical Source for Contemporary Philosophy of Science," *Revue internationale de philosophie* 131–32 (1980), 130–54; "Paradigms," *Contemporary Philosophy: A New Survey*, under the auspices of UNESCO and the Institut International de Philosophie (The Hague: Nijhoff, 1982), Vol. 2, 87–110; "Scientific Discovery: The Larger Problem Situation," *New Ideas in Psychology* 1 (1983), 99–109; "A Hermeneutics of the Natural Sciences?: The Debate Updated," *Man and World* 30 (1997), 329–41. On the dissolution of any logic of scientific discovery in a deconstructive "hermeneutics," see my review of John D. Caputo's *Radical Hermeneutics* in *The Modern Schoolman* 67 (1990), 223–28.

16. Glazebrook (2000), 96–100; cf. 210.

17. Just a few examples of the results of this confounding, taken from a single page of Glazebrook (2000) suggest that their intent is to justify the claim that it is the "question of science" as such which is in fact central to Heidegger's thinking throughout his career of thought rather than the question of a scientific philosophy: ". . . it is in his discussion of the theoretical attitude that he himself sees his conception of phenomenology being developed. . . . Phenomenology is to be developed in the context of natural science [understood as 'the mathematical projection of nature itself': SZ, 362/413–14], not in the context of philosophy as a science . . . , because he holds that the sciences are phenomenological. . . . Heidegger

chooses natural science as the place to develop a phenomenology of 'letting beings be' . . . evidently natural science is the home of such a phenomenology. . . . Yet that Heidegger in 1927 understands his account of science to be the place where his conception of phenomenology will be developed for the first time is a clear indication that the question of science is central rather than peripheral to his thinking" (Glazebrook, 2000: 97).

18. GA 20, 359/260. For more on this revolutionary contrast, see Theodore Kisiel, "From Intuition to Understanding: On Heidegger's Transposition of Husserlian Phenomenology," *Études Phénoménologiques* 22 (1995), 31–50. Reprinted in Theodore Kisiel, *Heidegger's Way of Thought: Critical and Interpretative Signposts* (London/New York: Continuum, 2002), 174–86.

19. Glazebrook (2000), 212, citing the essay on technology.

20. VA, 59–61/174–77; cf. Glazebrook (2000), 217 and 238.

21. Glazebrook (2000), 119–62 addresses this topic but gives no real indication in what sense the institution of science is in fact "a politically and historically situated hermeneutic project."

22. GA 16, 239. Cited here are speeches given by Rektor Heidegger in January 1934 to the citizens of Freiburg, "from gown to town," as it were, where Heidegger tended to stress that there is only one class in the new Germany, the worker class.

23. Glazebrook (2000), 132.

24. Martin Heidegger, *Die Selbstbehauptung der deutschen Universität*, ed. Hermann Heidegger (Frankfurt: Klostermann, 1983), 13–14; "The Self-Assertion of the University," tr. Lisa Harries in *Martin Heidegger and National Socialism: Questions and Answers*, eds. G. Neske & E. Kettering (New York: Paragon, 1990), 5–13, esp. 9.

25. EM, 28/37. On the political dimensions of this lecture course of the summer semester of 1935, including the hermeneutical politics of science, see Theodore Kisiel, "Heidegger's Philosophical Geopolitics in the Third Reich," *A Companion to Heidegger's* Introduction to Metaphysics, eds. Richard Polt and Gregory Fried (New Haven: Yale UP, 2001), 226–49.

26. This is a quick summary of "The Age of the World Picture," the talk first delivered in Freiburg on June 9, 1938, which constitutes Heidegger's public philosophical response to the Four Year Plan and which some regard as his first unequivocal critique of National Socialism.

27. GA 6.1, 323/103. The following collation of scattered "technical" passages in the summer semester of 1937 is drawn from GA 6.1. Remarks on the technologizing of science and scholarship continue into the winter semester of 1937–1938 in the course entitled *Basic Questions of Philosophy* (GA 45, 3/5, 53–55/49, 110–13/96-99, 141–43/123, 179/154).

28. Martin Heidegger, "Die Bedrohung der Wissenschaft: Arbeitskreis von Dozenten der naturwissenschaftlichen und medizinischen Fakultät (November 1937)— (Auszüge)," *Zur philosophischen Aktualität Heideggers, Vol. 1: Philosophie und Politik*, eds. D. Papenfuss and O. Pöggeler (Frankfurt: Klostermann, 1991), 5–27, hereafter cited in text as Heidegger (1936). Page references here are first to the delivered talk itself (5–11) and then to an accompanying set of loose "notes on the working circle" that was held *privatissime* since the fall of 1936, which the editor, Hartmut Tietjen,

has entitled "Philosophie, Wissenschaft und Weltanschauung" (14–27). A middle section entitled "Besinnung auf die Wissenschaft" (11–14) is not cited in the above.

29. VA, 71–99, esp. 94 and 96. English translation by Joan Stambaugh in Martin Heidegger, *The End of Philosophy* (New York: Harper & Row, 1973), 84–110, esp. pp. 105 and 107; reproduced in the Wolin edition, *The Heidegger Controversy*, "Overcoming Metaphysics (1936-1946)," 67–90, esp. 85 and 87; references are to no. XXVI of this collection of notes, a note that was written no earlier than late 1942. Heidegger's last reference to Hitler as *bona fide* statesman, tinged with a mild critique, occurs in the Schelling course of the summer semester of 1936: "It is in fact evident that the two men who have initiated countermovements in Europe for the political formation of their nation as well as their people, that both Mussolini and Hitler are essentially determined by Nietzsche, again in different ways, and this without the authentic metaphysical domain of Nietzschean thought having an immediate impact in the process" (GA 42, 40–41).

30. Glazebrook (2000), 5.

HEIDEGGER'S PHILOSOPHY OF SCIENCE

The Two Essences of Science[1]

John D. Caputo

The later Heidegger criticized the kind of thinking that, bent only on control and exploitation, threatens to overrun the contemporary technological world. He directed this critique, not precisely against technology, but against what he called the "essence" (*Wesen*) of technology, where *Wesen* had the old verbal sense of coming to be, coming to pass, coming about.[2] He was concerned, not with technological instruments themselves, but with what is coming to pass in and through a world filled with these instruments. What mattered for Heidegger is our understanding of ourselves and of the world in an age governed by the paradigm of technological control. He offered an important critique of the totalitarian tendencies of contemporary culture, in virtue of which science and technology are swept up in a repressive, totalizing movement that threatens to run out of control.

It often goes unnoticed, however, in the light of the later Heidegger's ringing pronouncements against the "essence of technology," that there is in fact "another essence" of science and technology in Heidegger's writings, an important and affirmative conception of science to be found in *Being and Time* (1972), which bears suggestive similarities with the work of Kuhn, Feyerabend, Polanyi, Hanson, Hesse, and other post-positivist theories of science. Indeed, I think, the later Heidegger himself neglected his own earlier reflections, and drifted, ironically, into a positivistic conception of science.[3] But this move, I argue, only weakened his incisive critique of contemporary culture. His own later warnings against the "essence" of technology would have been more properly served had Heidegger continued to argue, as he had in *Being and Time*, against a misunderstanding of what science is in the first place, a misunderstanding of its essence in the more straightforward sense—of what scientists do in their workshops.

Instead, driven no doubt by the imminence of the threat, Heidegger shifted his attention away from his own suggestive "hermeneutics" of science in *Being and Time*, toward the danger of a totalitarianism by which science and civilization are being subverted. And it is because of the shrillness of these later warnings that we—and that includes Heidegger himself—have tended to forget the insightful theory of science defended in *Being and Time*.

Thus, I think there are in fact "two essences" of science in Heidegger, the first of which was suppressed by the second. The first, which is to be found in *Being and Time* and which we will call the *hermeneutic* essence of science, consists in a suggestive "existential genealogy" of science, its genesis in the historical life of the scientific investigator, which can not be separated from an allegedly "pure logic" of science. The second essence of science and technology is its *Wesen* in the later, more radical sense—let us call it here its *deconstructive* sense—which signifies an entire understanding on man and world, of being and truth, which Heidegger wants to delimit, to critique, to disrupt.

My aim in this essay, which turns on these two essences, is twofold. First, I want to sketch out the main lines of the hermeneutic essence of science proposed in *Being and Time*. In so doing, I hope to restore to its proper place a Heideggerian analysis that Heidegger himself tended to forget, and thereby correct a misunderstanding about Heidegger's alleged hostility to science, which Heidegger himself has in part fostered. Second, I want to provide a preliminary introduction to the second essence, elaborated in his later essays on technology, and thereby put in context a critique of totalizing and totalitarian projects—which is in fact what I take the later work to be—which I regard to be of the utmost importance. And I want to show that the first sense supports the second; the hermeneutic account serves the deconstructive critique. Supported by the hermeneutic interpretation of science defended in *Being and Time*, his later warnings about a dangerous historical momentum are not humanistic jeremiads about science and technology, but a thoughtful protest against a growing momentum toward totalization that misunderstands science itself. In my view, the later Heidegger's protest against the "essence of technology" converges with the emancipator and liberationist impulse that has emerged in French philosophy in the last two decades. His critique of metaphysics has a sociopolitical cutting edge that he himself failed to exploit, even as a deconstructive purpose is served by his earlier hermeneutic work, which he tended to forget.

I argue here that we need both the early Heidegger and the late, both hermeneutics and the more radical deconstructive critique of his metaphysics, both Gadamer (whose point of departure is Being and Time) and Derrida (whose point of departure is the later critique of the history of metaphysics) and recent French philosophy as a whole. I hope to show that there is a fruitful interplay, and not an antagonism, between the two essences of science. And if I cannot in fact show this in the limits of this essay, I can at least make some pointers in that direction.

AN "EXISTENTIAL CONCEPTION OF SCIENCE"

In section 3 of *Being and Time* Heidegger speaks of a "productive logic" that leaps ahead of the concrete work that the positive sciences do and breaks new ground, as opposed to the standard "logic" "which limps along after, investigating the status of some sciences as it chances to find it, I order to discover its 'method'" (SZ 10/30). This productive logic differs in principle from a merely reproductive one, which can do nothing better than describe the work of science in its creative past, vainly attempting to formulate that past into an epistemology by which scientific practice should presumably be guided in the future (SZ, 10/30). Such logics and methodologies are always written at dusk, to borrow Hegel's image, just when they have become obsolete. Methodologists are constantly being disrupted by the unorthodox turn of events awaiting them around the next historical bend.

We learn more about this new logic in section 69(b), where Heidegger discusses what he calls an "existential conception of science:"

> This must be distinguished from the "logical" conception which understands science with regard to its results and defines it as "something established on an interconnection of true propositions— that is, propositions counted as valid." The existential conception understands science as a way of existence and thus as a mode of Being-in-the-world, which discovers or discloses either entities or Being. (SZ, 357/408)

There is thus a difference between the standard logics of science, which treat science as a constituted result, attempting only to display the chain of connections among its propositions, and the existential genealogy of the origin of science from prescientific life.[4] It is the latter that interests Heidegger, and it is this shift in perspective that gives rise to a fundamentally different conception of science in *Being and Time*. The attempt to treat science as if it were a pure logic, as if it dropped from the sky, is an illusion, indeed a transcendental illusion, which vainly tries to endow science with a pure transcendental status. Science derives from the concrete historical life of the scientific investigator.

It is this new logic of science that Heidegger describes as "hermeneutic." But Heidegger's hermeneutic conception of science differs considerably from Dilthey's. In the first place, Heidegger rejects Dilthey's "objectivism," which regards science as somehow or other able to seize upon a thing in itself, whether that be nature in itself, as in the natural sciences, or indeed some cultural object in itself, as in the *Geisteswissenschaften*. Second, Heidegger rejected Dilthey's division of the sciences into two qualitatively different sorts, natural and human, which confines hermeneutics to but one side of the distinction. Heidegger insists that the sciences constitute a unity, albeit a hermeneutic and not a positivistic one.[5] In *Being and Time*, hermeneutics is a universal, ontological structure that included all the sciences in its sweep. "Hermeneutics" is not the name of a method of the human sciences, but of the ontology of the understanding itself. Heidegger takes hermeneutics out of the domain of the purely methodological or epistemological and shows that it ultimately has an ontological sense. We understand as we do because we are as we are. Understanding follows Being; *intelligere sequitur esse*. There is no field of pure epistemology for Heidegger but only of the ontology of knowing or understanding. Every science is a way the historical investigator has of "casting" things, of framing them out, within a certain conceptual framework. Far from attempting to leave one's own interpretive framework behind, as in Dilthey's naïve objectivism, scientific understanding is not possible without it. Hence, instead of differentiating natural science from hermeneutics, Heidegger is interested in showing how all science, natural and human, is made possible by an anticipatory, hermeneutic fore-structure.

The fundamental feature of the understanding, and hence of the hermeneutic theory of science defended in *Being and Time*, is its *projective* character (*entwerfendes Verstehen*). For Heidegger, to understand is to contextualize, to situate a thing within the contextual arrangement in which it belongs. And that is what Heidegger means by "projecting" a thing "upon its horizon," or "projecting a being in its Being:" to set it forth in or on the horizonal backdrop that it requires in order to be manifest as the thing that it is, to "cast" in the appropriate terms. Indeed the English word "cast" captures a good deal of what Heidegger means by "projection." It operates within the rule of the same metaphor (*werfen* means to hurl or cast): to cast a thing in a certain light, to thrust it into a certain framework. To understand a hammer is to situate it within the chain of equipment, the equipmental context, to which it belongs, just as in the foreground–background analysis of perception in Gestalt psychology.[6] Accordingly, scientific or theoretical thinking consists in projecting an adequate conceptual horizon, one that allows, not tools or everyday things, but scientific objects to appear.

Hermeneutic understanding proceeds from a network of presuppositions that must always be adequate for the matter to be interpreted. In *Being and Time* it is never a question of getting free from presuppositions but, on the contrary, of securing an adequate presuppositional frame, of seeing to it that the complex of presuppositions one brings to bear on the object is wide enough and sharp enough to give an adequate rendering of the object. It is never a question of assuming too much, but of assuming too little. The "hermeneutic situation," or complex of hermeneutic conditions under which understanding is possible, must accordingly meet three conditions. In the first place, the projective fore-structure must cover the whole range of entities to be understood, which is what Heidegger calls "fore-having," (*Vorhabe*). That is to say, any projective or horizonal framework that we employ must be ample enough to provide for all the phenomena that can appear within its range. Second, the projective understanding must be guided by a certain conception of the kind of being, or categorical type, of the phenomena included within its range; this Heidegger calls "fore-sight" (*Vorsicht*). Finally, the preunderstanding requires an articulate table of categories that unfold, delineate, and provide for an adequate analysis of, the mode of being caught sight of in fore-sight; this is called "fore-grasping" (*Vorgriff*) (cf. SZ, 150–51/191–92, 231–33/274–75).

Together these conditions supply the hermeneutic forestructures, the anticipatory conditions of possibility, the required presuppositions that constitute the preunderstanding under which explicit understanding is possible. *Being and Time* clearly defends a holistic, horizonal interpretation of understanding according to which the character of an individual act of understanding is set by the constellation of presuppositions that condition it and make it possible. These presuppositions hang together in a system so that there are no isolated, atomic acts of understanding, on the one hand, nor noncontextualized objects on the other. It is just this projective, horizonal theory of understanding that we need to characterize more carefully in order to grasp Heidegger's theory of science in *Being and Time*.

In the first place, the hermeneutic fore-structuring of human understanding is an ontological condition rooted ultimately, according to Heidegger, in the temporality of human existence as a being pointed toward the future. That is why talk about "escaping" this fore-structuring—which is what "freedom from presuppositions" would mean—makes no sense. On the one hand, such "escape" would be ontologically impossible; one would have to escape from one's condition as a temporal being. And, on the other hand, such talk of escape misunderstands the very nature of understanding that proceeds, not by means of a presuppositionless blank stare, but by projecting the horizonal preunderstanding that befits a given category of entities (SZ, 153/194–95).

Accordingly, the hermeneutic make-up of the sciences is but a particular instance of the universal hermeneutic structure of all understanding. The Diltheyan opposition of explanatory-causal natural sciences and clarifying-hermeneutic human sciences makes no sense in *Being and Time*.

Both explaining and clarifying, both natural and human science, are possible only on the basis of a prior hermeneutic projection that constitutes the field of objects carved out by that discipline. Heidegger shows little interest in, and put little emphasis on, the distinction between the natural and the human sciences in *Being and Time*. Indeed, he takes all the sciences to have the same ontological weight. All science, as science, is a projective determination of beings in terms of some categorical framework or another—be it natural, social or human. What does interest Heidegger, however, is a different and, for him, far more important distinction—between science itself and prescientific life. It is that distinction that his existential genealogy seeks to establish and clarify; it is on that distinction that he thinks everything depends.

Let us examine this genealogy more carefully. *Being and Time* emphasizes the primacy of our concrete involvement in the sphere of historical and cultural practices and the secondary or derivative status of abstract, theoretical investigations. The concrete historical world is first of all the world of "instruments" whose primary character is their "being ready to hand" (*Zuhandensein*), that is, their availability for use. *Zuhandensein* is to be differentiated from the complementary concept, *Vorhandensein*, which refers to the world in its "objective presence," its sheer reality apart from our use, which is the way it is considered in the natural sciences. However, we must resist the temptation to give the theoretical-scientific relationship to the world a purely negative genealogy, as if it arises as the simple privation of our primary practical engagement with the world (= "circumspective concern"):

> it would be easy to suggest that merely looking at entities is something which emerges when concern *holds back* from any kind of manipulation. What is decisive in the "emergence" of the theoretical attitude would then lie in the *disappearance of praxis*. (SZ, 357/409)

That would flatly contradict the universally projective nature of understanding, suggesting that, although our concrete involvement with the world is hermeneutic and presupposition-bound, the natural sciences treat the world in an objective, presuppositionless way.

Consider the case of the hammer. The hammer that is used *as* a hammer is grasped only in the using; one cannot explain in theoretical terms its "feel," its balance, its aptness as an instrument, its place in the instrumental system. One knows how to use a hammer only by trying his hand at hammering with it. But one can step back from using the hammer and describe its "properties" in a series of "assertions," indicating, for example, that "the hammer is heavy." In that case, the hammer shows itself in a new light, but not merely because we no longer use it:

Not because we are keeping our distance from manipulation, nor because we are just looking *away* (*absehen*) from the equipmental character of this entity, but rather because we are looking *at* (*ansehen*) the ready-at-hand thing which we encounter, and looking at it "in a new way," as something present-at-hand. The *understanding of Being* by which our concernful dealings with entities within-the-world have been guided *has changed over*. (SZ, 361/412)

As an object of the theoretical attitude, the hammer ceases to be a hammer-in-use and is *recast* as a thing-with-properties, obedient to the laws of gravity, measurable in mathematical space, and so on. It is projected positively and anew, literally re-cast, re-projected, now no longer on the horizon of its readiness-at-hand, but rather of presence-to-hand. (One can imagine a graded series of such assertions, from the after-hours talk of two carpenters comparing the objective properties of various sorts of hammers, all the way to the considerations of a physicist who would treat it purely in terms of its mass and velocity.)

Hence, it is a mistake to think that in the scientific attitude things lose their interpreted character, their character *as* something (which Heidegger calls their "as-structure"). It is a mistake to think that the ready-to-hand is treated *as* a tool while the present-at-hand has to do with things *in themselves*, free of any "as." On the contrary, we have simply switched from one "as" to another (SZ, 157–58/200–01). We have simply shifted projective frameworks, hermeneutic forestructures. We have ceased to regard the hammer as a tool and now regard it as an entity with mass, shape, gravity, and the like. In other words, "objective presence" is a positive projection of the Being of entities, a hermeneutic-interpretive act. (It should also be clear from this passage that there is nothing mystical or mystifying about the word "Being" as Heidegger uses it, particularly in *Being and Time*; it simply refers to the conceptual framework, the horizon, in terms of which a thing is grasped [= "projected"]).

Furthermore, one should not treat the genesis of the scientific attitude from our concrete involvement with the world as a passage from practice to theory. For that presupposes a hard and fast distinction between practice and theory—as if theory did not have a praxis of its own, and practice did not have a "sighting" of its own—a distinction that Heidegger thinks cannot be defended. On the contrary, scientific work is a complex praxis that requires skilled investigators, and this is no incidental feature of science, but of the utmost importance to it:

> Reading off the measurements which result from an experiment often requires a complicated "technical" set-up for the experimental

> design. Observation with a microscope is dependent upon the production of "preparations." Archaeological excavation, which precedes any interpretation of the "findings," demands manipulations of the grossest kind. But even in the "most abstract" way of working out problems and establishing what has been obtained, one manipulates equipment for writing, for example. However "uninteresting" and "obvious" such components of scientific research may be, they are by no means a matter of indifference ontologically. The explicit suggestion that scientific behavior as a way of Being-in-the-world, is not just a "purely intellectual activity," may seem petty and superfluous. If only it were not plain from this triviality that it is by no means patent where the ontological boundary between "theoretical" and "atheoretical" really runs! (SZ, 358/409)

It is also important to see that Heidegger does not think that science is exhausted by, or restricted to, the change-over in conceptual frameworks from readiness-to-hand to objective presence. For that change-over explains the possibility of the natural sciences, but not of the social and humanistic sciences. Science is constituted, not only when cultural objects (tools) are treated as physical objects, but also when tools are grasped *in their tool-ness* and made objects of scientific inquiry: "The ready-to-hand can become the 'Object' of a science without having to lose its character as equipment" (SZ, 361/413). Hence economics is a science of the goods men make, and the laws of supply and demand by which they are regulated, which preserves the character of the tool as tool. Economics is a science of the ready-to-hand *as* ready-to-hand.

The genealogy of the scientific attitude in *Being and Time* therefore involves two things: (1) the change-over from the attitude of immediate involvement ("concernful dealing") to a certain distancing objectivity; and (2) the positive projection of a kind of Being that befits the beings under investigation, which will supply a horizonal or hermeneutic framework within which they can be investigated. In the natural sciences, that means the projection of things in terms of their objective presence, but in the social and human sciences it obviously does not. Economics, politics, literary, and historical science would cast things in their objective presence only at the cost of their existence as meaningful inquiry. The natural and human sciences arise from a different sort of projective horizon, with different conceptions of time, space, law, and meaning. These differences amount, however, to "ontic" differences in *Being and Time*, interdisciplinary differences about their presuppositions (their fore-having, fore-sight and fore-grasping). They do not affect the ontological make-up of science as science, but the intra-scientific differentiation of the sciences among themselves.[7]

Heidegger's concern in *Being and Time* is to see to it that the natural, social and human sciences—which together are characterized as "positive" or "ontic" sciences, for they have to do with entities, with posited, existing things—do not get mixed in, indiscriminately, with the ontological science which he is conducting in this work. His ultimate interest in *Being and Time* is in Being, not beings, that is, in the ontological framework that renders possible every human practice, both scientific and prescientific, or, as he puts it, every ontic comportment with beings (*Verhalten zu Seiendem*, SZ, 4/23). The aim of *Being and Time* is to reach a determination of Being as the horizon, not of any particular region, but of all horizons, and this he will argue is "time." It was thus of the utmost importance that the inquiry into "Dasein," which serves as the point of departure in this ontological investigation, not be taken in ontic terms, as if it were an anthropological, psychological, or sociological science. And although he wished to keep this purely ontological interest in human being in *Being and Time* distinct from the positive-ontic sciences of human being it was not his intention to question the validity of the latter. He wanted neither to demean the natural sciences in the light of the human sciences, nor to demean the particular, positive sciences in the light of fundamental ontology. On the contrary, he meant to offer an ontology of human being as a hermeneutic being.

The genesis of science is to be located, not in its object, but by the projective attitude with which it carves out and constitutes a field of objects and defines a formal approach to them. When an historian trains a student to look for certain patterns, to value certain documents, to be on the alert for certain clues, he is engaged in establishing the projective standpoint that constitutes and defines a scientific inquiry into history. And the same thing is true of the literary critic teaching students to thematize poetic imagery or the construction of English novels in the nineteenth century, or of a physicist working with assistants in the laboratory.

Science is objectification, and objectification, which is the constitution of any sort of scientific object, whether natural, social or human, is effected by the change-over (*Umschlagen*), a new way of projecting beings in their Being. The piece of equipment is released from its place within the equipmental totality and treated as a detachable thing, isolated from its equipmental context. But if it is decontextualized as equipment it is recontextualized as an object within a new theoretical framework and made the object of a new and more sophisticated praxis. It makes its appearance as a scientific object only by assuming its place within a new horizonal setting.[8]

It is also clear from this hermeneutic or projective theory of science, that Heidegger is committed to denying "bare" or uninterpreted facts of the matter. Indeed, for Heidegger, the selection of facts in any science is a function of its capacity to discover a way to project things; a scientist can pick out facts only in virtue of a prior frame that the scientist "has in advance." Take the case of physics, which is for Heidegger a "paradigmatic" or exemplary case of scientific projection. Mathematical physics proceeds by projecting a strictly quantitative nature, a nature cast or projected in terms of mass, location, velocity, and so on, which thereby enables it to discover facts:

> The "grounding" of "factual science" was possible only because the researchers understood that in principle there are no "bare facts." In the mathematical projection of nature, moreover, what is decisive is not primarily the mathematics as such; what is decisive is that thus projection *discloses something that is* a priori. Thus, the paradigmatic character of mathematical natural science does not lie in its exactitude or in the fact that it is binding for "Everyman.'" It consists rather in the fact that the entities that it takes as its theme are discovered in it in the only way in which entities can be discovered—by the prior projection of their state of Being (SZ, 363/414)

This is one of the most significant points of contact between Heidegger and the recent rereading of the history and philosophy of science. There are no facts except within the pregiven horizon that enables them to appear in the first place.

However, one should not conclude from Heidegger's denial of bare facts that he regards all projective horizons as arbitrary or pragmatic fictions. On the contrary, projection has for him a "disclosive" power. Understanding discloses the world as the sort of world it is; it renders it manifest in a certain way. And disclosure, manifestness, is what "truth" means in *Being and Time* (SZ, sec. 44c). Projections are not arbitrary; we must involve just the right kind of framework to free beings up for the kind of Being that befits them. The sciences do not traffic in "free-floating constructions" (SZ, 28/50) but in fact seize on something in the things themselves. This Heidegger holds *both* that it is impossible to gain access to bare and uninterpreted facts of the matter—which is to reject any notion of objectivism or absolutism, as if we could jump out of our skins and make some absolute contact with things—*and* that our hermeneutic constructions, when they are well-formed, do capture something about the world—which is to provide for the objectivity of knowledge. Thus, in one stroke Heidegger provides for the possibility of science while delimiting the claims of objectivism. He has no reason to think that science does not seize something about the world, even while he thinks it nonsense to suggest that this constitutes a break in the hermeneutic circle.

SCIENTIFIC CRISES

Up to this point we have treated Heidegger's views on what Kuhn would call normal science. However, there is also a clear picture of "revolutionary" science in *Being and Time*, in section 3, where Heidegger treats of the "crises" and "radical revisions" of "fundamental concepts" that periodically shake the sciences. The sciences, as we have seen, are guided in advance by certain basic concepts (*Grundbegriffe*), the interpretive fore-structures that constitute or project the field of objects that belong to that science:

> Basic concepts determine the way in which we get an understanding beforehand of the area of subject-matter underlying all the objects a science takes as its theme, and all positive investigation is guided by this understanding. (SZ, 10/30)

"Basic concepts" are a good deal like Kuhnian paradigms—guiding fore-structures that guide a whole scientific practice. The difference is that speaking of "concepts" tends to obscure Kuhn's emphasis on the practical character of the paradigm, an emphasis, however, which Heidegger clearly would share (SZ, 358/409). Now just as the transition from everyday concern to science is effected by a fundamental change-over (*Umschlagen*), which is a shift from a prescientific to a scientific projection, so *within* each science certain fundamental shifts of projective understanding are possible which result in revolutionary changes within that science:

> The real "movement" of the sciences takes place when their basic concepts undergo a more or less radical revision which is transparent to itself. The level which a science has reached is determined by how far it is *capable* of a crisis in its basic concepts. (SZ, 9/29)

At the time he was writing *Being and Time*, Heidegger thought that any number of such fundamental revolutionary movements were underway. He speaks of a crisis in mathematics, in the dispute between formalists and intuitionists; in physics, as a result of relativity theory; in biology, because of the dispute between vitalism and mechanism; and in theology, because of Luther's fundamental insights about the nature of faith. So the work of a Galileo, Newton, or Einstein, of an Aquinas or Luther, of a Smith or a Keynes, or any of the founding or revolutionary geniuses of the respective disciplines, consists not so much in making new factual discoveries, or generating new information, as in effecting certain fundamental conceptual breakthroughs, in revising radically the fundamental terms in which the practitioners of a discipline think about a field of objects, in thoroughly "re-casting" them.

Inasmuch as Heidegger holds to a "horizonal" or holistic theory of understanding, he regards progress in a science as possible on two levels. In the first place, the scientist can continue to fill in the existing horizon, building up in a continuous way the known body of information (confirming predictions, refining calculations, etc.). This is what Husserl called the "fulfillment" (*Erfüllung*) of a predelineated horizonal scheme (*Logical Investigations*, investigation VI, section One, chapter 1). And it is also, of course, the everyday business of what Kuhn called normal science. But it is possible—and sometimes necessary—that the horizon itself undergoes revision, and that can occur only by a discontinuous revision or shift of horizons. Certain fundamental thinkers in a discipline, working at its boundaries, force a reorganization of the whole field of disciplinary activity. In the language of phenomenology, their work is carried out on the level of "regional ontology," that is, of the ontologically guiding and horizonal concepts within which all the work in their field is conducted.[9] The great creative geniuses work on and at the horizons within which their more pedestrian colleagues labor unquestioningly. The phenomenology of horizons explains the phenomenon of scientific revolutions.

The history of science is punctuated by these horizonal shifts that reorganize the data contained within the horizon. How, then, are we to think of scientific "progress"? Heidegger wrote in 1938, more than two decades before Kuhn:

> When we use the word "science" today, it means something essentially different from the *doctrina* and *scientia* of the Middle Ages, and also from the Greek *episteme*. Greek science was never exact, precisely because, in keeping with its essence, it could not be exact and did not need to be exact. Hence it makes no sense whatever to suppose that modern science is more exact than that of antiquity. Neither can we say that the Galilean doctrine of freely falling bodies is true and that Aristotle's teaching, that light bodies strive upwards, is false; for the Greek understanding of the essence of body and place and of the relation between the two rests upon different interpretations of beings and hence conditions a correspondingly different kind of seeing and questioning of natural events. No one would presume to maintain that Shakespeare's poetry is more advanced than that of Aeschylus. It is still more impossible to say that the modern understanding of whatever is, is more correct than that of the Greeks. Therefore, if we want to grasp the essence of modern science, we must first free ourselves from the habit of comparing the new science with the old solely in terms of degree, from the point of view of progress.[10]

The individual scientist works within a projective horizon that sets forth its own standards of what is reasonable and scientific. Scientific theories belonging to different projective structures, operating within different constellations of basic concepts, may be compared and contrasted as to their presuppositions, but one cannot be labeled more rational or progressive than the other.

On a horizonal theory of the understanding, therefore, scientific development is not uniform and continuous, but at certain times must be shocked by more or less radical revisions of the existing horizons. But what in particular occasions such crises? What are the reasons that lead scientists to abandon the old horizon and adopt a new one? What place can the data gained under the old horizon assume within the new horizonal framework? Heidegger does not answer, or even ask, these questions in *Being and Time*. That is why I think that Kuhn's conception of scientific revolutions is not only congenial to the standpoint of *Being and Time*, but in fact elucidates, works out and corrects what is only a seminal suggestion in Heidegger. Indeed Kuhn's approach is so preeminently in keeping with the hermeneutic conception of science that we are defending here that he himself has been led to describe his view as "hermeneutic."[11]

If Heidegger failed to investigate the character of the decision that the scientist makes at the point of crisis, when the old horizon wavers in instability and a basic shift is about to be made, Kuhn's account of this moment created a storm of criticism, including Lakatos' famous observation that Kuhn reduced scientific decision making to mob-psychology.[12] Kuhn might have avoided this criticism had he at his disposal the hermeneutic conception of *phronesis* that Gadamer developed by listening to the young Heidegger's lectures on the *Nicomachean Ethics* and whose counterpart in *Being and Time* is the theory of *Verstehen*.[13] The Aristotelian model shows that intelligence does not always have explicit rules to fall back on, that at certain critical points it is left to its own devices to grasp what the situation demands. This does not mean that for Aristotle we are sometimes driven to act irrationally. On the contrary, the understanding is never more faithful to its nature than it is in these moments. For it is precisely the work of understanding to make the first cut into the complexity of the concrete world, to find the nerve of intelligibility that runs throughout it. It is precisely at these moments, when it lacks rules to fall back on, that intelligence must be what it most essentially is: insightful, capable of grasping what is demanded in a concrete setting. It is to Gadamer's credit to have elaborated the notion of *phronesis* in an admirable discussion in *Truth and Method*,[14] and in so doing to have set forth an essential implication, not only of Aristotle, but of *Being and Time*, for the theory of the sciences.

We should not fail to notice, moreover, that Heidegger's examples of crises in fundamental concepts in the sciences cut across the realm of natural, social and human sciences, including in one sweep physics, economics and theology. Such revolutions affect *Wissenschaft* itself,[15] in the wide sense that this term has in German, and which for Heidegger means any disciplined investigation, in which a particular region or field of objects has been staked off and thematized (objectified) in terms of a certain projection or conceptual organization. That happens both in theology and physics, whose respective histories are punctuated by basic upheavals in their forestructures. Accordingly, one requires *phronesis*—hermeneutic insight and understanding, a feel for what one is about, which Polanyi calls "personal knowledge"—across the board: in the natural, social and human sciences.

It is clearly a myth to think that there is hostility or denigration of the sciences in *Being and Time*. On the contrary, Heidegger brings to fruition in this book the work he began as a student of Rickert who, as a leader of the Baden school of Neokantianism, had written extensively on the theory of the sciences, a point clearly reflected in Heidegger's *Antrittsrede* at Freiburg in 1915.[16] Heidegger is interested, not in undermining the sciences, but in providing them with a hermeneutic accounting. In particular, he wants to show that scientific activity, of whatever disciplinary type—natural, social or human—is nourished by a prescientific, historical life which is its matrix and point of departure. He wants to explain how science is "derived" from historical life (= "Being-in-the-world"), how it is ontologically generated from our concrete entanglement with the world (viz., by a horizonal changeover from our primary and inescable "concern" with the world to a relatively disengaged projection of the world as a field of objects). He wants to show that, although legitimate in its own sphere, any such scientific projection is limited. Scientific projections are theoretical constructions aimed at elucidating a world from which we cannot finally or wholly extricate ourselves and to which we belong more primordially than science itself can say. It is always that prescientific belonging to the world that has primacy for Heidegger. It is that world that Heidegger ultimately wants to elucidate. In so doing, he does not intend to hold science in contempt, but only in check. He delimits its claims by subordinating science to the world in which we live and which has a prior claim on us. If science is made possible by a hermeneutic projection of a sphere of objects, it is also limited by the hermeneutic horizon of the scientist. A *pure* scientific standpoint is thus an illusion; science is always a projective undertaking of Being-in-the-world.

THE "ESSENCE OF TECHNOLOGY" IN THE LATER HEIDEGGER

Heidegger's later writings on science and technology are marked by a dramatic change of tone.[17] The sober hermeneutic analyses, suffused with the technical vocabulary of phenomenology, give way to the voice of protest. The determination of the first, hermeneutic essence of science is displaced by a deconstructive critique of the *Wesen* of technology, of what is coming to pass, what is coming about, in a technological world. There is a tone of urgency and of shrill protest against a danger that he takes to be imminent. He is no longer concerned with a logic of science—not even an existential or hermeneutic logic, a genea-logic. Science belongs together with technology as inseparable forms of the will-to-power as knowledge, of the will-to-know, the will to dominate and manipulate, of what Foucault calls "power-knowledge." The metaphysics of the will-to-power stamps our age, marks our epoch, dominating all the phenomena of our time—political and social, scientific and artistic. We are in danger of being swept up in an enormous totalitarian and totalizing movement that aims to bring every individual, every institution, every human practice under its sway.

His protest, as we have said is directed, not against science and technology themselves, but against their "essence," which means, what is coming to pass in a world that gives science and technology paradigmatic status. He is not concerned with science and technology, but with the way understanding being, human being and truth their enormous success have induced in us, an understanding that he thinks is marked by a preoccupation with control, manipulation and power. The metaphysics that articulates the guiding conception of modern technological civilization is to be found in Nietzsche's conception of the will-to-power. A dangerous momentum has been set loose in the modern world, one that conceives nature, and human being itself, as the raw material of a manipulative technology, that conceives all problems—political, social, personal—as technological problems for which an appropriate technology of behavior is required. The world has become the raw material for the various technologies of power—political technologies that manipulate and control public opinion and policy; social technologies that manipulate and control personal mores and standards of conduct; educational technologies that insure the normalization and regulation of educational practices. It is not only nature that must submit to our control, but education, sexuality, the political process, the arts, in short the whole sphere of human practices. This is what is coming to pass (*Wesen*) in science and technology, no in the sense that it is explicitly taught by science, but in the sense that this is the frame of mind of a culture dominated by the success and prestige of science and technology.

We are caught up in a momentum that is running out of control, and Heidegger's later work is a philosophy of liberation, emancipation, protest against that momentum. It has, I think, a profound kinship with the work of Derrida and Foucault in France, with Adorno's negative dialectics—despite the latter's well-known diatribes against Heidegger—in German social theory, and in the United States perhaps with Feyerabend. It is an eloquent protest against normalization, regulation, manipulation, against the rule of the police. It is a philosophy of freedom and letting-be, which wants to let the world, to let human being, to let the gods, be.

Unfortunately, in the course of lodging a protest which could not be more salutary against what is coming to pass in and with science, against its *Wesen* in this special sense, Heidegger loses sight of his own more careful, more perspicuous characterization of science itself—it's "essence" in the first, ordinary or hermeneutic sense. Thus, in his later writings Heidegger speaks of science as if it were an unbending method that knows only how to apply fixed rules to unchanging circumstances, as if it were, by its nature, by its first essence, part and parcel of the disciplinary society. Unhappily, he seems to forget what he himself said in *Being and Time* about understanding as *phronesis*, about the historicity of the scientific investigator, about horizonal shifts and breakthroughs, crises and ground-breaking discoveries. And that is a regrettable development. For his argument is with the totalitarian forces that exploit science and that sweep it up in a vast political and economic armature, a complex of forces and power that knows no bounds. That argument would have been strengthened, not weakened, had he continued to insist, as he did in *Being and Time*, that science itself, in its real practice and practical reality, in its first, hermeneutic essence, is not the rigorously rule-governed, inflexible apparatus it is made out to be. That is why we need both the early Heidegger and the later, or among contemporary post-Heideggerian figures, both Derrida and Gadamer. We need both a positive hermeneutic rendering of science such as we find in Gadamer, who represents something of a right-wing, conservative, early-Heideggerianism, and a vigorous protest against standardization, normalization, and ideological imperialism, such as we find in Derrida, who represents a left-wing Heideggerianism, and in Foucault and recent French philosophy.

Despite this slip in strategy in the later Heidegger's reflections on science—and perhaps, in a philosophy of protest, moderate, balanced accounts are not always what we need!—one can see, nonetheless, the unity of Heidegger's project from *Being and Time* on. He has been concerned throughout with that world which precedes science, which funds it, to which science always returns, and from which the scientist him or herself is never granted leave. In *Being and Time* this prescientific sphere was characterized in terms of the historical, cultural "life-world" of phenomenology, and in the later writings in terms of a more poetically conceived life of "mortals," "under the skies," "before the gods," "upon the earth," a picture inspired ultimately by the poetry of Hölderlin. It is as if the threat had grown so great that we required a more soaring vision, a more powerful mythology, than was provided by the sober analytics of Husserlian phenomenology. But he meant all along to recover the prescientific world that antedates science and that has its hold on us long before science arrives on the scene, and his point has always been to *let* that world *be*.

NOTES

1. Reprinted from *Rationality, Relativism and the Human Sciences*, eds. J. Margolis, M. Krausz and R. M. Burian (Dordrecht: Martinus Nijhoff, 1986), 43–60.

2. *Wesen* was used as a verb in Middle High German; whence the modern *west*. The English translators of Heidegger tend to render it as "come to presence."

3. This point runs throughout the work of Theodore Kisiel, who has done the best work on Heidegger's philosophy of science. See his "Heidegger and the New Images of Science," *Research in Phenomenology* 7 (1977), 162–81; "New Philosophies of Science in the USA: A Selective Survey," *Zeitschrift für allgemeine Wissenschaftstheorie* V (1974), 138–91; also see note 5 below. On Heidegger and Polanyi, see Robert Innis, "Heidegger's Model of Subjectivity: A Polanyi Critique," in *Heidegger: The Man and the Thinker*, ed. Thomas Sheehan (Chicago: Precedent Publishing Co., 1981), 117–30.

4. As in the current discussion of what Popper calls the "logic of discovery," Heidegger thinks that the mistake of the philosophy of science up to now has been to concentrate on the finished results of science, on science as a body of established propositions, rather than in the process by which such propositions arise. But the danger entailed by the genealogical approach—which wants to avoid being a pure logic—is psychologism and relativism, as Husserl pointed out in the first volume of the *Logical Investigations*. As an application of phenomenology, Heidegger's hermeneutic genealogy intended to steer a middle course between the two. As a hermeneutic, it is no pure logic, but historical and genetic; as a phenomenological ontology, it is concerned, not with human psychology, but with the ontological structure of understanding [of "Dasein," not of "man."] The disagreement between Popper and Kuhn in *Criticism and the Growth of Knowledge* [ed. Imre Lakatos and Alan Musgrave (Cambridge: University Press, 1970)] is illustrative. Kuhn takes Popper's concerns to be too strongly dictated by an historical logic (21–22) and Popper takes Kuhn's views to be relativistic and psychologistic (55–58). Kuhn would have done better in responding to the charge of psychologism to argue that his work is hermeneutic. To argue, however, that it is "social" and not "individual" psychology does not answer Popper's charge but simply confirms and refines it. Theodore Kisiel makes this important point in his "Scientific Discovery: Logical, Psychological or Hermeneutical?" *Explorations in Phenomenology*, ed. David Carr and Edward Casey (The Hague: Martinus Nijhoff, 1973), 263–84; See also his "The Rationality of

Scientific Discovery" in *Rationality Today/La Rationalité Aujourdhui*, ed. Theodore Geraets (Ottawa: University Press, 1977), 401–11.

5. If the positivists supported a "unity of the sciences" program inasmuch as they wanted to reduce all the science to the method of the natural sciences, which was the very thing Dilthey opposed, Heidegger treats all science as a hermeneutic unity (i.e., an exercise in projective understanding). That is not to reduce all science to the method of one privileged science but rather to make all sciences generally conform to the ontological-hermeneutic structure of the understanding.

6. Gestalt psychology is a common point of reference for Kuhn, Polanyi, Merleau-Ponty, Husserl, and Heidegger.

7. It is important for us to observe that Heidegger never questions the legitimacy of the social and human sciences, or that they play a valid role in the work of *Wissenschaft* as a whole.

8. It is not clear to me that Hubert Dreyfus takes this into account in his various treatments of Heidegger's philosophy of science in *Being and Time*. He seems to me overly fond of emphasizing the decontextualization of the ready-to-hand without insisting on the concomitant recontextualization or hermeneutic projection without which understanding is impossible for Heidegger.

9. Heidegger seems to think that such fundamental conceptual breakthroughs would be effected by regional ontologists; this makes sense so long it is recognized that the revolutionary figures in the disciplines themselves are their own regional ontologists. For Kuhn, they are made by people working at the most advanced and specialized level of puzzle solving in that discipline.

10. Heidegger, *The Question Concerning Technology and Other Essays*, trans. W. Lovitt (New York: Harper & Row, 1977), 117–18.

11. Kuhn seems to have in mind, however, only the Diltheyan sense of hermeneutics as historical empathy. He does not at all intend the Gadamerian view that scientific understanding depends on *phronesis*. See *The Essential Tension* (Chicago: University Press, 1977), xiii, xv.

12. Imre Lakatos, "Falsification and the Methodology of Scientific Research Programmes," in Lakatos and Musgrave, eds. (1970), 178. This volume contains a series of mostly critical responses to Kuhn; cf. the contributions by Popper, Toulmin, and Watkins. Feyerabend, on the other hand, "defends" Kuhn by saying that science is indeed at least (and in fact even more) irrational than Kuhn holds, a defense that Kuhn describes as "vaguely obscene" (264) in his instructive "Reflections on my Critics" at the end of the volume.

13. Hans-Georg Gadamer, *Philosophical Hermeneutics*, trans. David Linge (Berkeley: University of California Press, 1976), 201–02. I have learned a great deal about a rapprochement between Kuhn and Gadamer from Richard Bernstein's insightful application of hermeneutics to the problems of philosophy of science in *Beyond Objectivism and Relativism* (Philadelphia: University of Pennsylvania Press, 1984).

14. Hans-Georg Gadamer, *Truth and Method*, trans. G. Barden and J. Cumming (New York: Seabury, 1975), 274–305.

15. The English word that seems to me to cover the range of all the sciences in the manner of the German *Wissenschaft* is "discipline." This word, of course, has a Foucauldian ring that, although foreign to *Being and Time*, would become a welcome nuance in the later Heidegger.

16. Heidegger's *Antrittsrede* at Freiburg in 1915, entitled "The Concept of Time in the science of History," differentiated the historian's and the physicist's conceptions of time. See *Frühe Schriften* (Frankfurt: Klostermann, 1972), 413 ff. and my review, "Logic, Language and Time," *Research in Phenomenology* 3 (1973), 147–56.

17. See the first two essays in The Question Concerning Technology and Other Essays.

DEVELOPMENTS AND IMPLICATIONS

Trish Glazebrook

In the contemporary global context, there are at least two lines of inquiry that arise directly out of Heidegger's critique of science. The first concerns ecophenomenology, for which Heideggerian analysis demonstrates the inadequacy of the term *environmental ethics*. Moreover, his reading of Aristotle on teleology provides a conceptual basis for scientific and technological practices that might be ecologically sound, in contrast to the logic of domination by which, ecofeminists argue, they are currently informed. And his notion of dwelling suggests that human ways of knowing can respect, care for, and safeguard nature, in contrast to the destruction enacted through contemporary science and technology. The second is the question of the social obligations of the sciences. Heidegger argues that science warrants reflection that is nontechnical and nonscientific in order to evaluate its thinking and paradigmatic epistemological function. Both issues are taken up here, first by analysis of Heidegger's view and development of its direct consequences, and second, through application of the developed view to pressing issues of global capital that are central both to environmental sustainability and social justice.

HEIDEGGER'S ENVIRONMENTAL PHENOMENOLOGY

As several of the contributions to this volume have already suggested, Heidegger's critique of science has significant implications for environmental philosophy. Ecophenomenologists draw substantial theoretical support from his work, despite Michael Zimmerman's worry that Heideggerian ecology may be prone to ecofascism.[1] In contrast to Zimmerman's repudiation, John Llewellyn argues that Heidegger's notion of ontological responsibility is a basis for ecological conscience,[2] and I have shown that reading Heidegger in the context of ecofeminism precludes ecofascism.[3] In fact, his critique of science, especially the analysis of teleology in Aristotle's *Physics* in contrast to Newtonian mechanism, and the vision of an alternative dwelling in nature, constitute an incipient and constructive environmental phenomenology that promotes both sustainability and environmental justice.

Environmental Ethics

In the post-positivist atmosphere that still lingers in many philosophy departments, ethics arouses suspicion with respect to its philosophical validity. For philosophy proper aims at universal truths, while the truths of ethics are not always amenable to such abstraction. Thus, ethics is labeled "soft philosophy." So-called "applied" ethics are even more suspect as they are further confined to particularized contexts (e.g., business ethics, computer ethics, medical ethics). Yet each of these subdisciplines cannot function without raising fundamental questions concerning ontology and epistemology. Likewise, "environmental ethics" poses serious challenges and makes significant contributions to ontological questions of what "nature" is and what it means to be human, and to epistemological questions concerning the nature, value, and limitations of human ways of knowing.

Thus, Richard Sylvan argues that an environmental ethic requires radical break from all previous ethics that cannot simply be revised to accommodate environmental issues: The history of ethics has anthropocentrism at its very core, and environmental ethics cannot accept anthropocentrism.[4] Similarly, Heidegger attempts in "Letter on Humanism" to think what it means to be human "without elevating man to the center of beings" (W 353/255). In the modern predicament of the oblivion of being, "the greatest care must be fostered on the ethical bond at a time when technological man, delivered over to mass society, can be kept readily on call only by gathering and ordering all his plans and activities in a way that corresponds to technology" (W 353/255). Ethics are paramount for responding to the "uprooting of all beings" (W 353/255) in modernity and the corresponding appropriation of human being by the essence of technology. Thus, he continues his analysis by arguing that *"Êthos* means abode [*Aufenthalt*], dwelling place" (W 354/256).

Ethics is, then, in one sense a philosophical subdiscipline, but in a more originary sense, it is ontology: the thinking of being. For in the human orientation to being, ethical decisions have always already been made. Environmental philosophy is thus much more than "environmental ethics" as traditionally conceived (i.e., a marginalized academic subdiscipline in applied ethics). Rather, it is an urgently needed philosophical contribution to global attempts to avert impending and ongoing environmental catastrophes (in the forms of pollution, climate change, species extinction, and exceeded carrying capacities) by means of sustained reflection on the human place in the natural order. It is "ethics" only in the Heideggerian sense of *ethos*: a reflection on human dwelling.

Teleology and the History of Physics

Heidegger's contrast of Aristotle's physics against Newton's is a basis for understanding historical transformations in the concept of nature that culminate in a logic of domination. As discussed previously, natural entities are for Aristotle self-moving. Motion is the actualization of potentialities (201a11) and natural entities reach actuality without external impulse. This is not to say that they do not draw from surrounding resources, but rather, that they require no external efficient cause to achieve their *telos*. For example, the acorn, if it becomes anything, becomes an oak tree. Thus nature is for Aristotle teleological (198b10) and he gives four arguments to this effect at *Physics* 2.8. Accordingly, the final cause is crucial to Aristotle for understanding the movement of nature, which is most significant as growth (198b4). In contrast, definitive of artifacts is their conception in the mind of the artist prior to production. Artifacts are ontologically dependent on an efficient cause, the artist, who creates using material taken from nature. Thus, natural generation is necessarily prior to production for Aristotle. Hence, Heidegger notes that artifacts are derivative, and natural generation is absolutely different from and irreducible to human production (W 288/259).

Subsequent to Aristotle, however, *physis* is understood by analogy to *technê* (W 288–89/259; 292/262), that is, as a special kind of artifact. In medieval ontology, nature is divinely crafted; in the secular physics of modernity, it is self-making (W 255/234; 292/262). Nature remains teleological in the medieval conception, but as Lynn White Jr. has pointed out, natural teleology is subordinated to human ends: "God planned all this explicitly for man's benefit and rule: no item in the physical creation had any purpose save to serve man's purposes."[5] In Newton's physics, natural teleology no longer figures. His laws reduce motion to locomotion (FD 68/287) and give efficient causes (i.e., forces, explanatory weight).[6] Thus, the modern scientific model of the universe is mechanistic: Nature is like a big machine. Accordingly, natural entities are reduced to scientific objects and stripped of their *telos*. Having no end of their own, they are laid bare for appropriation to human ends.

The modern scientific ontology of nature as object therefore conduces its technological (and economic) appropriation, as if natural entities are nothing more than inert matter standing ready for the imposition of human purpose. Heidegger's analysis accordingly provides an explanation of how the history of science plays into manipulative assault on nature by revealing all that is encountered as a standing-reserve to be stockpiled for human consumption. Heidegger calls "the revealing [*Entbergen*] that rules in modern technology . . . a challenging [*Herausfordern*]" (VA 18/14) that "sets upon [*stellt*] nature" (VA 18/15) such that "even the Rhine itself appears as something at our command" (VA 19/16). Karen Warren characterizes this contemporary, and now globally enforced, way of thinking about nature as a logic of domination, and Val Plumwood likewise characterizes science and technology in terms of mastery.[7]

Heidegger's concern is that technology threatens to drive out "every other possible way of revealing" (VA 31/27). Certainly, policymakers, corporate and multinational leaders, and institutions like the World Bank, the International Monetary Fund, and the International Fund for African Development, for example, repeatedly demonstrate in their so-called "structural adjustment programs" and development initiatives (which ecofeminists rename "maldevelopment") that they are unable to conceive of nature as anything but resources to be exploited regardless of the social and environmental costs. Attempts to resist this ideology lead to the persecution, torture, and murder of environmental justice activists. The film *Life and Debt* is an explicit demonstration of how the Jamaican economy and resource-base has been gutted by enforced globalization that treats both ecosystems and people as nothing more than resources standing by for exploitation.[8] Deane Curtin shows further how indigenous women's traditional environmental epistemologies are marginalized and displaced by science-based approaches that trivialize their expertise as "mere 'wives' tales.'"[9] Alternative epistemologies that may have functioned sustainably for centuries are driven out by profit-driven strategies in international capitalism that bring the very techno-scientific ideology of which Heidegger was so critical ruthlessly to bear on governments and peoples in lesser developed countries.

Heidegger argues for the possibility of other ways of revealing than the setting upon that is essential to modern science and technology. His argument is not, however, for return to some premodern, idyllic existence. He reads Aristotle not nostalgically, but in order that possibilities covered over in the historical development of Western conceptions of nature might be retrieved toward a new beginning. The Aristotelian insight he uncovers is that nature is teleological, not in some theological sense of divine plan, but at the much more modest level of the organism. In *Die Frage nach dem Ding*, he discusses Aristotle's conception of violence (*bia*). Natural motion is in accord with a body's nature (e.g., fire moves upward, whereas a rock falls down to the earth). Violent motion is forced movement against a body's natural tendency, and "has its cause in the force that affects it" (FD 66/285). Thus, Newton's doctrine of motion, he argues, eliminates the distinction between natural and violent motion (FD 68/288). He says he cannot "set forth here the full implications" of the shift to Newtonian physics, yet the displacement of Aristotelian teleology and consequent loss of his concept of violent motion have obvious implications for contemporary environmental destruction. Scientific approaches and technologies that acknowledge and respect, rather than make invisible, natural teleology might prove more sustainable than contemporary practices. That is to say, rather than move by a logic of domination that assaults nature, science and technology can be practiced *consistently with* natural *teloi*.

For example, corporate, large-scale, industrialized, and science-based (so-called "Green Revolution") agricultural practices entail flooding fields with fertilizers and growing monoculture cash crops. Analysts argue that such practices may have increased production in the short term, but at the price of creating a "bubble economy" that is on the verge of popping: The human species has already exceeded the carrying capacity of the planet, and is borrowing against the natural capital of the future.[10] For in the long term, these practices lead to soil exhaustion, deforestation, erosion, and desertification. Alternatively, traditional agricultural practices of leaving a field fallow and planting to promote biodiversity respect natural teleology and promote ongoing regeneration of soil toward indefinite sustainability. Women of the Deccan Plateau in India, for example, have remediated fields abandoned as useless by corporate, cash-crop farmers by means of simple practices that stop erosion, promote biodiversity in cropping, and replenish seed banks.[11] In the process, they have empowered themselves. They are no longer forced to farm for someone else as wage-laborers, lacking benefits and subject to unwanted sexual exploitation, but are an autonomous political group that provides job security and support (e.g., through maternity or sickness) to women in the community. Their seed banks make them independent from the likes of Monsanto, for example, whose Terminator seed is engineered to grow but produce no seeds of its own in order that DNA from the genetically modified plant not become entangled with indigenous varieties at unknown risks. The reality is that Terminator seeds will eventually give Monsanto control of the seed market as farmers, who can no longer collect seeds during harvest to replant next season, must buy from them again each growing season. The women of the Deccan Development Society respect natural teleology by working with natural processes rather enacting a logic of domination. In doing so, they promote both sustainability and environmental justice. Their agriculture is poetic in Heidegger's sense: a bringing-forth rather than a challenging-forth. Not *Ge-stell* but *Entbergung*, their practices reduce neither human being nor nature to *Bestand*.

Flowforms are a second example. A Flowform is a series of sculpted bowls through which a cascade of water is passed.[12] John Wilkes, a sculptor drawing on the work of hydrologist Theodor Schwenk and mathematician

George Adams, "wondered if it might be possible to design a sequence of forms through which water could fulfill its potential to manifest an orderly metamorphic process . . . [that] might bring to physical expression the delicate potential for ordered movement that appears to be inherent in the nature of water."[13] Flowform-treated water turns out not only to be "penetrated by rhythmical movements in support of biological processes" but also is highly oxygenated.[14] Thus, Flowforms are beautiful, but also functional technologies that can be used to treat stored or desalinated water. They remediate water for swimming pools in Sweden, and produce water used in beer brewing and bread making in Germany. They have further applications in irrigation, pharmaceuticals, and aquaculture systems.[15] They do not project Ge-stell onto water, but achieve human ends by working cooperatively with water's natural tendencies.

Heidegger's reading of Aristotle *Physics* accordingly offers conceptual content to the deep ecologist's notion of "inherent" or "intrinsic value."[16] The phrase is intended to capture the idea that nature has value in itself beyond its instrumental value, but pressing deep ecologists for what it means usually results in recourse to, as Robert Elliot puts it, "beauty, diversity, richness, integrity, interconnectedness, variety, complexity, harmony, grandeur, intricacy, and autonomy," and "magnificence, splendor, awesomeness."[17] A deeper account saying in what intrinsic value consists can be given by appealing to teleology: The good of a natural entity is its self-governing function as part of an interdependent ecosystem. Paul Taylor draws upon teleology similarly in his "ethics of respect for nature": What is good for an organism promotes its flourishing.[18] Heidegger's contribution is a reading of the history of physics that shows how the Aristotelian concept was lost, and pinpoints specifically the ideology of science and technology as the location where the teleological conception can be reintroduced to promote sustainable practice. Thus, Heidegger offers a new beginning for human dwelling that is ethical in his fullest sense.

Nature as Dwelling

Heidegger offers an alternative epistemology, in contrast to the representational thinking of objectivity, based on dwelling in nature rather than on consumption of resources. In *Being and Time*, Dasein's *Unheimlichkeit* is its anxiety at not being at home in the world. (SZ189/233). In *The Basic Problems of Phenomenology*, Heidegger is already analyzing being-in-the-world in terms of what it means to be at home, "*zu Hause*" (GA 24, 244–46/172–73). In "Letter on Humanism," he uses "*Heimat*," home, "with the intention of thinking the homelessness of contemporary [human being]" (W 338/241). In the *Nietzsche* volumes, he diagnoses "the organized global conquest of the earth" (GA 6.2, 358/248) as symptomatic of, but also causally active in human being's homelessness. He has thus realized across these texts that homelessness is not a necessary part of the human condition, but rather belongs specifically to modernity in the human alienation from nature. For nature is the ground on which all human living takes place. He subsequently explores an alternative possibility for human belonging: dwelling.

Dwelling takes place in what Heidegger calls the fourfold: earth and sky, mortals and gods. Rosemary Radford Ruether argues similarly "Mother and nature religion traditionally have seen heaven and earth, gods and humans, as dialectical components within the primal matrix of being."[19] For both thinkers, human being can be at home in nature. Reading Hölderlin, Heidegger calls *Heimat* "the power of the earth" (GA 39, 88). He says elsewhere that the earth is "the building bearer, nourishing with its fruits, tending water and rock, plant and animal" (VA 176/178) and "the serving bearer, blossoming and fruiting, spreading out in rock and water, rising up into plant and animal" (VA 149/149). Dwelling is "cultivating and caring [*Pflegen und Hegen*]" (VA 191/217) and he describes it in terms of peace, preservation, sparing and safeguarding; it is "the manner in which mortals are on earth" (VA 148/148). Human beings:

> dwell in that they save the earth. . . . To save the earth is more than to exploit it or even wear it out. Saving the earth does not master the earth and does not subjugate it, which is merely one step from spoliation. (VA 150/150)

As Will McNeill puts it, dwelling "means protecting the fourfold, saving the earth and heavens in letting them be."[20]

This notion of letting be permeates Heidegger's work. Concerning the human other, he argues that in guest-friendship, "lies the resolve . . . to let the foreigner be the one he is" (GA 53, 175–76; my translation) and McNeill reads love as "the desire that the beloved remain the one that it is."[21] Letting be is not, however, limited to human relations. Heidegger argues for letting beings be throughout *Gelassenheit*, and in "On The Essence of Truth," he displaces freedom from transcendental subjectivity to beings exactly in terms of letting beings be (W 187–91/124–28). Human thinking can assault its object, whether human or otherwise, or give it the freedom to speak for itself. That the sciences have the potential to treat nature as something to be listened to rather than manipulated and exploited by instrumental reason is explicit in the work of contemporary geologists. Christine Turner, a consultant formerly with the U.S. Geological Survey, argues that rocks talk, and that listening is a crucial part of scientific process.[22] Victor Baker, past president of the Geological Society of America, argues for what he calls "Earth-directed" science (i.e., science that is in conversation with the Earth).[23] Listening as a strategy for science uses experience to accommodate theory to context, rather than imposing abstract, a priori and universalized conceptions onto what is encountered in the natural world. To give natural entities the freedom to speak for themselves is to let them be what they are without reductively appropriating them as object and resource.

In order to express this vision of dwelling as an alternative to global conquest of the earth, Heidegger borrows from Hölderlin the phrase that "poetically man dwells . . ." (VA 181–98/213–29). He argues that the "poetic is the basic capacity for human dwelling" (VA 197/228). In "The Question Concerning Technology," he binds together Aristotle's four causes by means of the poetic. The causes are ways things are brought into being, and *poiêsis* is any creative act.[24] Thus "*physis* also . . . is a bringing-forth, *poiêsis*"; in fact, argues Heidegger, "*physis* is indeed *poiêsis* in the highest sense" (VA 19/10). For if poetry in the original Greek sense means any creative act, then nature is the master-poet. *Homo faber* depends on nature's poetic power insofar as technologically manipulated materials have natural origin. The capacity to produce artifacts is itself in the nature of human being. Heidegger's argument that human being dwells poetically is the claim that human being can be at home in nature thoughtfully, creatively and symbiotically, rather than exploitatively and destructively.

Heidegger's conception of dwelling in nature resonates with indigenous environmental movements. The Chipko women of Northern India exclaim, "The forest is our home." Environmental justice analysts argue that control of land should be taken from multinationals, who are rapidly destroying the global resource base, and returned to local, indigenous communities whose care for the land promotes long-term sustainability. Robert Weissman argues precisely that because members of these communities dwell on the land, their practices are a sustainable care taking rather than a liquidation for profit.[25] To say they dwell in the full Heideggerian sense is not to patronize them or idealize their practices—traditional lifestyles can be fraught with social inequities, particularly with respect to gender. But it is to acknowledge that indigenous communities have a right to their land, and that their practices have epistemic validity. Thus, local communities are fully entitled to a voice in environmental and development policy. Western/Northern scientists and technologists can learn from them rather than devaluing their knowledge and practices.

Heideggerian ecophenomenology accordingly turns on a notion of dwelling that recognizes both nature and people as more than standing-reserve. Nature is a home that houses human worlds. The mutually complicit scientific and technological *a priori* conceptions of nature as object and resource respectively fall short of such dwelling. Ethical, that is, phenomenological, practices of science and technology would rise to contemporary challenges in dwelling by taking responsibility for their consequences to both humans beings and ecosystems.

THE SOCIAL OBLIGATIONS OF SCIENCE

Francis Bacon promised at the beginning of the modern epoch that the sciences can improve the human condition.[26] Heidegger's call is to hold them to that promise. In his analysis, Dasein is essentially the inquirer (SZ 7/27), and historical epochs are determined by the particular paradigms of knowledge that inform inquiry. In 1929, he argues that the sciences stand at the heart of the university (W 103/94), and in 1938, that science is the issue from which modernity can be understood (GA 5, 76/117). Science is the modern realization of the human urge for knowledge and sets an epistemological standard that thoroughly permeates the human experience. Thus, the sciences do not simply provide descriptions of "the way the world is," but establish knowledge in situated paradigms that inform human experience. Science is accordingly much more than a body of objective theory. It is a social force. The sciences are therefore open to social and political analysis, and their responsibilities extend beyond technical excellence. In Heidegger's sense of originary ethics, scientists are responsible for evaluating the social consequences of their epistemology, ontology, and practice.

For there are questions about science that cannot be answered scientifically (e.g., what is worth knowing?). In "Die Bedrohung der Wissenschaft," an address to the Faculty of Medicine at Freiburg given in 1937, Heidegger argues:

> [with] respect to the question of the character of reflection on science is above all to be noted a basic fact, which we cannot think through often enough. Namely: *No science can know by itself about its own fulfillment as knowledge*. We cannot reflect on physics as a science with the help of the method [*Vorgehen*] of physics. The essence of mathematics lets itself neither be determined mathematically nor even raise questions about mathematical methods [*Methode*]. Geology does not let itself be investigated geologically, as little as philology [does] philologically.[27]

In the 1950s, he argues that the sciences uncover truths concerning their object, but no science can question "the essence and essential origin of the manner of knowing which it cultivates" (WD, 57/33). In "Science and Reflection," he calls the sciences one-sided for this reason, and names this aspect of science "*das Unumgängliche*," that which cannot be gotten around (VA, 62/177). Scientists cannot question their *a priori* ontology and epistemology, except by stepping outside their technical practice. This analysis echoes his earlier claim that movement takes place in the sciences "when their basic concepts undergo a more or less radical revision" (SZ, 9/29). There, he meant much the same as Kuhn intended in his analysis of revolutionary science: Radical revision happens when the conceptual framework of a science undergoes a paradigm shift. But by 1937 and on into the 1950s, Heidegger's conception of radical revision developed beyond shift in particular theoretical concepts. Rather, he extends analysis to the epistemological question of how the sciences are a fulfillment of knowledge, and thus to how they fulfill the public good.

For he argues in 1938, long before theorists in science, technology, and society programs make this claim, and likewise well before Lakatos' argues that science is practiced in research programs, that science has become an industry. In "The Age of the World Picture," he notes that contemporary research is determined by its *Betriebcharakter*" (GA 5, 77; 90/124; 138–39) (i.e., it has the character of ongoing activity and industriousness). The researcher is always on the move between conferences and congresses, and publishers determine along with researchers which books are to be written. This is not because publishers necessarily have "the best ear for the needs of the public" (GA 5, 90/139) but because they "bring the world into the picture for the public and confirm it publicly" (GA 5, 91/139). Researchers and publishers collaborate in a public confirmation of the "objectification of whatever is" (GA 5, 86/126) and the "certainty of representation" (GA 5, 80/127). The institutionalization of knowledge thus generates a solidarity across specialized disciplines that secures "the precedence of methodology over whatever is (nature and history), which at any given time becomes objective in research" (GA 5, 78/125). The humanities become more like physics than traditional humanistic disciplines when the beings toward which they are oriented are projected as objects, and when objectivity becomes also their epistemological standard. Yet "ongoing activity," Heidegger argues, "becomes mere busyness whenever, in the pursuit of its methodology, it no longer keeps itself open on the basis of an ever-new accomplishing of its projection-plan, but only leaves that plan behind itself as a given" (GA 5, 90/138). The knowledge industry risks taking its project of objectivity for granted. Only when this modern essence of science is taken seriously do researchers and the sciences "offer themselves for the common good" (GA 5, 79/126). Likewise, ecofeminists argue that putting objectivity into question promotes the human good: Sustainable practices require an attitude of caring beyond the impartiality traditional to objectivity.

Thus, Heidegger calls in "Science and Reflection" for reflection [*Besinnung*] upon the sciences that Lovitt says "involves itself with sense and meaning" (QCT 155, note 1) and Heidegger himself describes as a "calm, self-possessed surrender [*Gelassenheit*] to that which is worthy of questioning" (VA, 64/180). Although he holds that such reflection cannot take place within technical practice, he does not believe paternalistically that the task of such reflection belongs just to the philosopher. Rather, he argues that "every researcher and teacher of the sciences, every [one] pursuing a way through science, can move, as a thinking being, on various levels of reflection" (VA, 66/181–82). He is arguing for a revolutionary moment in the history of science that requires, beyond technical expertise, reflection on its purpose and social consequences. As a crucial component of human history, the sciences have, consistent with the infamous *Rektoratsrede*, an obligation to the people. Indeed, scientists are in the best position of all to evaluate both their goals, and the value and limitations of their ontology and epistemology.

In support of Heidegger's analysis, the Duhem-Quine thesis can be interpreted such that the under-determination of theory by evidence implies that theories are selected on the basis of other than technical factors. That societal evaluation of the sciences may thus be appropriate has horrified traditional philosophers of science who stick by Michael Polanyi's position from 1945 that social constraint on the sciences impedes their growth by compromising their spirit of free enquiry.[28] In contrast, Maria Lugones and Elizabeth Spelman argue that, although objectivity has as its standard truth and falsity, theories can and should also be judged according to whether they are "useless, arrogant, disrespectful, ignorant, ethnocentric, imperialistic,"[29] and I would add, environmentally destructive or irresponsible. The deeper truth of science lies not in the correctness of its facts, but in the way that it opens an *êthos* for human being to make its home.

Thus, Heidegger asks in 1937, which is preferable, researching hand grenades, or fertilizers?[30] In its contemporary institutionalization, research is made possible by either government or corporate funding, so such questions are answered by scientists, policymakers, government leaders, and corporate heads. Yet if science is not value-free, but an historically and culturally located orientation toward both being and knowledge, then the validity and usefulness of particular funding initiatives warrant broader discussion than that provided by these sources. Government-funded science is paid for by citizens, and there is an expectation of concern with public interest in government policy, so the case for citizen voice in directing research is not hard to make. But what of privately funded research? In 1984, the Hastings Report suggested that corporate funding is a preventative against science being co-opted by government interests. Yet corporations are motivated by profit, regardless of arguments that might be made for the so-called "Triple Bottom Line," that attends to profit, social consequences, and environmental impacts, and have repeatedly demonstrated that profit trumps the public good. To cite just one example among many, in the pharmaceutical industry, drug trial results historically have been kept under wraps when not consistent with "product development initiatives." Controversy recently has erupted because data concerning the possibility of increased suicide, particularly among teenagers, in consequence of taking selective serotonin reuptake inhibitors (SSRIs), a kind of antidepressant was not made available. Corporate science has substantial consequences for human lives, so the private funding argument does not justify the exclusion of citizen voice from research policy.

Furthermore, citizens are surely entitled to a say concerning which methodologies are practiced on their behalf. Animal experimentation, for example, has received widespread public criticism, particularly in the United Kingdom. Those who protest animal testing are dismissed as radical insurgents, and Britain has introduced specific legislation to allow the arrest and prosecution of identified activists simply on the basis of proximity to research facilities. Yet in the pharmaceutical industry, cross-species transferability issues remain: Nine out of ten experimental drugs are not brought to market because of reactions that failed to surface in animal tests.[31] Cosmetics testing regularly involves trivial changes that allow marketers to claim products are new and improved at the expense of animal suffering and death. The infamous LD50 test was removed from the Organization for Economic Cooperation and Development's international guidelines for product safety in consequence of technical criticisms voiced by the scientific establishment concerning repeatability of results, as well as scientists' humanitarian objections.[32] Public resistance to genetically modified organisms (so-called "Frankenfoods") has in large part appealed to the precautionary principle on the basis that not enough is known about environmental and health consequences, which could be severe and irreversible. Recent scientific reports have likewise called for caution and urged that more research is needed. Given agreement on the part of the scientific establishment with long-term protest on the part of ordinary citizens, why should the latter voice be dismissed or, worse yet, demonized?

The old prejudice that the ordinary citizen is not sufficiently informed to contribute constructively to such discussion simply preserves the cult of the expert inappropriately. Ducks Unlimited provides an example of the nonprofessional contribution to technical science, and in astronomy likewise substantial significant discoveries have been made by amateurs. Nonspecialists can contribute technically to science, but also are capable of providing thoughtful input at the level of policy. For example, the recent court case in Pennsylvania disallowing the teaching of intelligent design in biology classes was initiated by members of the general public who felt their religious commitment was being compromised by a collapse of church and state. Likewise, the intervention of the gay rights lobby into policy concerning human drug trials is an example of useful public response that reoriented the medical industry away from elite science and toward patient care through participatory policy. Many contemporary challenges require collaboration between a variety of stakeholders because nontechnical considerations play a significant role in their resolution. Philosophers in particular can contribute useful critical analysis of the limits of objectivity, but more broad, democratic debate involving scientists, philosophers, policymakers, and community members is needed. Such collaborative scrutiny of the sciences is incipient in Heidegger's call for reflection on science. It puts into question the ways in which the sciences fulfill the human urge for knowledge, and encourages a renewed investment in their social obligation that could lead directly to better science, that is, science that more effectively responds to current human needs.

NOTES

1. Michael Zimmerman, "Rethinking the Heidegger-Deep Ecology Relationship," *Environmental Ethics* 15, no. 3 (1993), 195–224: 205.

2. John Llewellyn, *The Middle Voice of Ecological Conscience* (New York: St. Martin's Press, 1991), 355.

3. Trish Glazebrook, "Heidegger and Ecofeminism," *Re-Reading the Canon: Feminist Interpretations of Heidegger*, eds. Nancy Holland and Patricia Huntington (University Park: The Pennsylvania State University Press, 2001), 221–51.

4. Richard Sylvan, "Is There a Need for a New, an Environmental, Ethic?" *Environmental Philosophy: From Animal Rights to Radical Ecology*, ed. Michael Zimmerman et al., 3rd ed. (Upper Saddle River, NJ: Prentice Hall, 2001), 17–25.

5. Lynn White, Jr. "The Historical Roots of Our Ecological Crisis," *Environmental Ethics: What Really Matters, What Really Works*, eds. David Schmidtz and Elizabeth Willott (New York: Oxford University Press, 2002), 7–14: 11.

6. W 248–249/228–229; H. S. Thayer, ed., *Newton's Philosophy of Nature: Selections from his Writings* (New York: Hafner Press, 1953), 25–26. Cf. VA, 11–14/6–10; 26–27/23; and 46/160.

7. Karen Warren, "The Power and Promise of Ecological Feminism," *Environmental Ethics* 12, no. 2 (Summer, 1990) 125–46; Val Plumwood, *Feminism and the Mastery of Nature* (London: Routledge, 1993), see esp. Chapter 5.

8. *Life and Debt*, dir. Stephanie Black. A Tuff Gong Pictures Production, 2001.

9. Deane Curtin, "Recognizing Women's Environmental Expertise," *Environmental Philosophy: From Animal Rights to Radical Ecology*, eds. Michael Zimmerman et al., 3rd ed. (Upper Saddle River, NJ: PrenticeHall, 2001), 305–21: 313.

10. Linda S. Hjorth, et al. "Population Carrying Capacities: Four Case Studies—Pakistan, India, Ethiopia, Mexico," *Technology and Society*, 2nd ed. (Upper Saddle River, NJ: PrenticeHall, 2003), 296–311.

11. V. Rukmini Rao, "Women farmers of India's Deccan Plateau: Ecofeminists Challenge World Elites," *Environmental Ethics: What Really Matters, What Really Works*, eds. David Schmidtz and Elizabeth Willott (New York: Oxford University Press, 2002), 255–62.

12. Cf. Trish Glazebrook, "Art or Nature? Aristotle, Restoration Ecology, and Flowforms," *Ethics and Environment* 8, no. 1 (2003), 22–36.

13. Mark Riegner and John Wilkes, "Flowforms and the Language of Nature," *Goethe's Way of Science: A Phenomenology of Nature*, David Seamon and Arthur Zajonc, eds. (Albany: State University of New York Press, 1998), 239.

14. Riegner and Wilkes (1998), 243.

15. Ibid., 245–47.

16. On "inherent value" and "intrinsic value," see Baird Callicott, "Intrinsic Value, Quantum Theory, and Environmental Ethics," *Environmental Ethics* 7 (1985), 257–75; Arne Naess, "The Deep Ecological Movement: Some Philosophical Aspects," *Philosophical Inquiry* 8, no. 1–2 (1986), 10–31; and Warwick Fox, "A Postscript on Deep Ecology and Intrinsic Value," *The Trumpeter* 2 (1985), 20–23.

17. *Nature* 395, no. 6701 (October, 1998), 428–29.

18. Robert Elliot, *Faking Nature: The Ethics of Environmental Restoration* (New York: Routledge, 1997), 59 and 63.

19. Paul W. Taylor, "The Ethics of Respect for Nature," *Environmental Ethics: What Really Matters, What Really Works*, eds. David Schmidtz and Elizabeth Willott (New York: Oxford University Press, 2002), 83–95.

20. Rosemary Radford Ruether, *New Woman/New Earth: Sexist Ideologies and Human Liberation* (New York: The Seabury Press, 1975), 194.

21. Will McNeill, "Heimat: Heidegger on the Threshold," *Heidegger Toward the Turn: Essays on the Work of the 1930s*, ed. James Risser (Albany: State University of New York Press, 1999), 326.

22. McNeill (1999), 344.

23. Christine Turner, "Messages in Stone: Field Geology in the American West," *Earth Matters: The Earth Sciences, Philosophy, and the Claims of Community*, ed. Robert Frodeman (Upper Saddle River, NJ: Prentice Hall, 2000), 51–62.

24. Victor Baker, "Conversing with the Earth: The Geological Approach the Uniderstanding," Frodeman (2000), 2–10.

25. Cf. Plato, *Symposium*, tr. Alexander Nehamas and Paul Woodruff (Indianapolis: Hackett Publishing, 1989), 205b–d.

26. Robert Weissman, "Corporate Plundering of Third-World Resources," *Toxic Struggles: The Theory and Practice of Environmental Justice*, ed. Richard Hofrichter (Gabriola Island, BC: New Society Publishers, 1993), 186–96.

27. Francis Bacon, *The Great Instauration and New Atlantis*, ed. J. Weinberger (Arlington Heights, Illinois: Harlan Davidson, Inc., 1980).

28. Martin Heidegger, "Die Bedrohung der Wissenschaft," *Zur philosophischen Aktualität Heideggers, Band 1*, ed. Dietrich Papenfuss und Otto Pöggler (Frankfurt: Vittorio Klostermann, 1991), 5–27: 12. (My translation)

29. See Robert Klee, *Introduction to the Philosophy of Science: Cutting Nature at its Seams* (New York: Oxford University Press, 1997), 64–67 and 73–77 for a useful survey of debate concerning this interpretation of the Duhem-Quine thesis. Pierre Duhem's initial presentation of the thesis is in *The Aim and Structure of Physical Theory* (Princeton, NJ: Princeton University Press, 1982). Larry Laudan ("Demystifying Underdetermination," *Minnesota Studies in the Philosophy of Science, Vol XIV*, ed. C. Savage (Minneapolis: University of Minnesota Press, 1990), uses this interpretation to argue that the thesis is philosophically invalid and dangerous in the hands of social critics of science who use it in just this way. Michael Polanyi, "Rights and Duties of Science," *Society for Freedom in Science Occasional Pamphlet No. 2* (June 1945), 1–18.

30. Maria Lugones and Elizabeth Spelman, "Have We Got a Theory for You! Feminist Theory, Cultural Imperialism and the Demand for 'The Woman's Voice,'" *Women and Values*, ed. Marilyn Pearsall, 3rd ed. (Belmont, CA: Wadsworth Publishing, 1999), 18–29: 24.

31. Heidegger (1991), 27.

32. "Tested on Humans," *New Scientist* 189, no. 2535 (January 21, 2006), 6.

33. LD50 means "lethal dose 50%." The test establishes the dosage at which 50% of animal test subjects die. In many cases, the product tested is simply not that lethal, and animals die through drowning or suffocation because of the volume required to achieve mortality.

LIST OF CONTRIBUTORS

Babette E. Babich studied in the United States, Belgium, and Germany, and received her PhD from Boston College. She has published approximately forty articles and several books, most recently, *Words in Blood, Like Flowers: Philosophy and Poetry, Music and Eros in Hölderlin, Nietzsche, Heidegger* (State University of New York Press, 2006), and her work has been translated into several languages. She currently is professor of philosophy at Fordham University in New York, and executive editor of *New Nietzsche Studies*.

Christina Behme received an MSc in marine biology from the University of Rostock, Germany, and an MA in philosophy from Dalhousie University in Halifax, Nova Scotia. She currently is a PhD candidate at Dalhousie University.

David R. Cerbone received his PhD from the University of California at Berkeley, and is associate professor in the Department of Philosophy at West Virginia University. He specializes in Heidegger, Wittgenstein, and the history of analytic philosophy, and has published various articles on these authors and topics, as well as *Understanding Phenomenology* (Acumen, 2006).

John D. Caputo is Thomas J. Watson Professor of Religion and Humanities at Syracuse University and David R. Cook Professor Emeritus of Philosophy at Villanova University. He has published twenty-two books and approximately two-hundred articles throughout his career. Books about his work include *A Passion for the Impossible: John D. Caputo in Focus*, *Religion With/out Religion: The Prayers and Tears of John D. Caputo*, and *The Very Idea of Radical Hermeneutics*.

Robert P. Crease is professor of philosophy at Stony Brook University and the historian at Brookhaven National Laboratory. His books include *The Prism and the Pendulum: The Ten Most Beautiful Experiments in Science*, *Making Physics: A Biography of Brookhaven National Laboratory, 1946–1972*, and *The Play of Nature, Experimentation as Performance*. He writes "Critical Point," a monthly column about science and society issues in *Physics World*, and is also known for his writings on jazz and dance.

Trish Glazebrook received her doctorate from the University of Toronto. She has published *Heidegger's Philosophy of Science* (Fordham University Press, 2000), and is currently completing *Eco-Logic: Erotics of Nature*. She has published papers on Heidegger, ecofeminism, philosophy of technology, and international issues in environmental phenomenology. She is professor of philosophy at Dalhousie University with cross appointments in International Development Studies, Women and Gender Studies, and the School for Resource and Environmental Studies, and chair of Philosophy and Religion Studies at the University of North Texas.

Ute Guzzoni is retired professor of philosophy at the University of Freiburg, Germany. Her recent publications include *Sieben Stücke zu Adorno* (2003), *Hegels Denken als Vollendung der Metaphysik* (2005), *Wasser. Das Meer und die Brunnen, die Flüsse und der Regen* (2005), and *Weiße Tautropfen*, three-hundred Haikus on rain, cloud, and sea, edited by Ute Guzzoni and Michiko Yoneda (2006).

Lawrence J. Hatab received his PhD from Fordham University and is chair of the Department of Philosophy and Religious Studies at Old Domninion University in Norfolk, Virginia. He has published approximately forty articles, and his books include *Myth and Philosophy: A Contest of Truths* (Open Court: 1990), *Ethics and Finitude: Heideggerian Contributions to Moral Philosophy* (Rowman & Littlefield, 2000), *Nietzsche's Life Sentence: Coming to Terms with Eternal Recurrence* (Routledge, 2005).

Patrick A. Heelan is the William A. Gaston Professsor of Philosophy at Georgetown University in Washington, DC, and a member of the Jesuit Order. He holds a doctorate in philosophy from the University of Louvain, and a doctorate in theoretical physics. He studied with Erwin Schroedinger at the Dublin Institute for Advanced Studies, and Eugene Wigner at Princeton University. He has published approximately a hundred papers on the philosophy of the quantum theory, contextual logic, the hermeneutics of theory and experiment, and the geometric structure of pictorial and other visual spaces, and is best known for his book *Space-perception and the Philosophy of Science* (University of California Press, 1983; reprinted in 1987).

Theodore Kisiel is Distinguished Research Professor at Northern Illinois University in DeKalb, Illinois. He received his PhD from Duquesne University, and has for decades been a central figure in Heidegger scholarship as a writer, translator, and editor, particularly on questions pertaining to the sciences. Recent publications include *The Genesis of Heidegger's Being and Time* (University of California Press, 1993), and *Reading Heidegger from the*

Start: Essays in his Earliest Thought (State University of New York Press, 1994), co-edited with John van Buren.

William J. Richardson, S. J. received his doctorate from the University of Louvain, and is professor of philosophy at Boston College. He has published more than seventy papers on Heidegger and psychoanalysis, and his book, *Heidegger: Through Phenomenology to Thought* (Martinus Nijhoff, 1963), remains a foundational treatment.

Ewald Richter taught for many years at the University of Hamburg. His publications include *Heideggers Frage nach dem Gewährenden und die exakten Wissenschaften* (1992), *Ursprüngliche und physikalische Zeit* (1996), and *Wohin führt uns die moderne Hirnforschung? Ein Beitrag aus phänomenologischer und erkenntniskritischer Sicht* (2005).

Gail Stenstad received her PhD from Vanderbilt University and is professor of philosophy at East Tennessee State University. She has published many articles on Heidegger, anarchism, feminism and other issues, as well as *Transformations: Thinking After Heidegger* (University of Wisconsin Press, 2006).

James R. Watson is emeritus professor at Loyola University, New Orleans. He is author of *Thinking with Pictures* (Art Review Press, 1990) and *Between Auschwitz and Tradition* (Editions Rodopi, 1994), editor of *Portraits of American Continental Philosophers* (Indiana University Press, 1999), which includes his photographs, and co-editor with Alan Rosenberg of *Contemporary Portrayals of Auschwitz and Genocide: Philosophical Challenges* (Humanity Books, 2000). He is president of the Society for the Philosophical Study of Genocide and the Holocaust.

INDEX OF HEIDEGGER TERMS

". . . Poetically Man Dwells . . . ," in PLT, 213–29. »Logos (Heraklit, Fragment 50),« 199–221; ("Logos (Heraclitus, Fragment B 50),"), trs. David Farrell Krell and Frank A. Capuzzi in *Early Greek Thinking* (San Francisco: HarperCollins, 1984), 59–78:
17–18, 23–24, 28, 30, 33, 40–41, 51–53, 62, 74–75, 84–85, 88, 90, 115, 122, 128, 165–169, 172–176, 184, 193–196, 201–206, 211–214, 220, 284, 287–293

Aus der Erfahrung des Denkens (Frankfurt: Vittorio Klostermann, 2002):
200

Beiträge zur Philosophie. (Vom Ereignis) (Frankfurt: Vittorio Klostermann, 2003) *Contributions to Philosophy (From Enowning)*, trs. Parvis Emad and Kenneth Maly (Bloomington: Indiana University Press, 1999):
18, 57–58, 76, 83–85, 195, 206–221

Bremer und Freiburger Vorträge (Frankfurt: Vittorio Klostermann, 1994) »Die Kehre,« 68–77; "The Turning," in QCT, 36–49:
180, 184, 196–197, 202, 254

Der Satz vom Grund (Frankfurt: Vittorio Klostermann, 1997) *The Principle of Reason*, tr. Reginald Lily (Bloomington: Indiana University Press, 1991):
196

Die Frage nach dem Ding, 3. Auflage (Tübingen: Max Niemeyer Verlag, 1987) Section B.I.5.a)–f3) (S. 50–83) is translated as "Modern Science, Metaphysics, and Mathematics" in BW, 271–305, which reprints with minor changes and deletions the translation at *What Is a Thing?* trs. W.B. Barton, Jr. and Vera Deutsch (Chicago: Henry Regnery Co., 1967). Translation citations are to BW:
17, 21, 23, 28, 128, 283–284

Die Frage nach dem Ding. Zu Kants Lehre von den transzendentalen Grundsätzen (Frankfurt: Vittorio Klostermann, 1984) Section B.I.5.a)–f3) is translated as "Modern Science, Metaphysics, and Mathematics" in BW, 271–305, which reprints with minor changes and deletions the translation at *What Is a Thing?* trs. W.B. Barton, Jr. and Vera Deutsch (Chicago: Henry Regnery Co., 1967). Translation citations are to BW:
48–49

Die Grundbegriffe der Metaphysik. Welt—Endlichkeit—Einsamkeit (Frankfurt: Vittorio Klostermann, 2004) *The Fundamental Concepts of Metaphysics: World, Finitude, Solitude*, trs. William McNeill and Nicholas Walker (Bloomington: Indiana University Press, 2001):
93–102, 159–162

Die Grundprobleme der Phänomenologie
(Frankfurt: Vittorio Klostermann, 1975)
The Basic Problems of Phenomenology,
tr. Albert Hofstadter (Bloomington:
Indiana University Press, 1982):
    15, 22, 77–80, 109, 226, 242, 256,
    286

Einführung in die Metaphysik, 5. Auflage
(Tübingen: Max Niemeyer Verlag,
1987) *An Introduction to Metaphysics*,
tr. Ralph Manheim (New Haven: Yale
University Press, 1959):
    159, 185, 250, 254, 256, 258

*Einführung in die phänomenologische
Forschung* (Frankfurt: Vittorio Klostermann, 1994):
    244

Einleitung in die Philosophie (Frankfurt:
Vittorio Klostermann, 2001):
    73, 86, 241–242

Frühe Schriften (Frankfurt: Vittorio
Klostermann, 1978):
    19, 21, 185, 240

Gelassenheit, 10. Auflage (Pfullingen:
Verlag Günther Neske, 1992) *Discourse
on Thinking*, trs. John M. Anderson and
E. Hans Freund (New York: Harper &
Row, 1966):
    54–55, 60, 180–185, 194, 198–203,
    205, 214, 220

*Grundfragen der Philosophie. Ausgewählte
»Probleme« der »Logik«* (Frankfurt: Vittorio Klostermann, 1984) *Basic Questions of Philosophy: Selected "Problems"
of "Logic,"* trs. Richard Rojcewicz and
André Schuwer (Bloomington: Indiana
University Press, 1994):
    208

Grundprobleme der Phänomenologie
(1919/20) (Frankfurt: Vittorio Klostermann, 1993):
    69–70

Hölderlins Hymne »Der Ister« (Frankfurt: Vittorio Klostermann, 1993)
Hölderlin's Hymn "The Ister," trs.
William McNeill and Julia Davis
(Bloomington: Indiana University
Press, 1996):
    287

Holzwege (Frankfurt: Vittorio Klostermann, 2003) »Der Ursprung des
Kunstwerkes,« 7–68; "The Origin of
the Work of Art," in PLT, 17–87. »Die
Zeit des Weltbildes,« 75–113; "The
Age of the World Picture," in QCT,
115–54:
    49–50, 68, 115, 127, 134, 253–254,
    289–290

Holzwege, 7. Auflage (Frankfurt:
Vittorio Klostermann, 1994) »Der
Ursprung des Kunstwerkes,« 1–74;
"The Origin of the Work of Art," in
PLT, 17–87. »Die Zeit des Weltbildes,« 75–113; "The Age of the World
Picture," in QCT, 115–54:
    28–34, 170–175, 178, 189, 195–196

Identität und Differenz, 5. Auflage
(Pfullingen: Verlag Günther Neske,
1976) *Identity and Difference*, tr. Joan
Stambaugh (New York: Harper & Row,
1969):
    81

Kant und das Problem der Metaphysik
(Frankfurt: Vittorio Klostermann, 1991)
Kant and the Problem of Metaphysics, tr.
Richard Taft (Bloomington: Indiana
University Press, 1990):
    179

Logik. Die Frage nach der Wahrheit (Frankfurt: Vittorio Klostermann, 1976):
69, 71

Metaphysische Anfangsgründe der Logik im Ausgang von Leibniz (Frankfurt: Vittorio Klostermann, 1978) *The Metaphysical Foundations of Logic*, tr. Michael Heim (Bloomington: Indiana University Press, 1984):
20, 95

Prolegomena zur Geschichte des Zeitbegriffs (Frankfurt: Vittorio Klostermann, 1979) *History of the Concept of Time: Prolegomena*, tr. Theodore Kisiel (Bloomington: Indiana University Press, 1985):
47, 72, 74, 257

Reden und andere Zeugnisse eines Lebensweges 1910–1976 (Frankfurt: Vittorio Klostermann, 2000):
249, 255

Schelling. Vom Wesen der Menschlichen Freiheit (1908) (Frankfurt: Vittorio Klostermann, 1988):
259

Schelling: Zur erneuten Auslegung seiner Untersuchungen über das Wesen der menschlichen Freiheit (Frankfurt: Vittorio Klostermann, 1991):
77

Sein und Zeit, 16. Auflage (Tübingen: Max Niemeyer Verlag, 1986) *Being and Time*, trs. John Macquarrie and Edward Robinson (New York: Harper & Row, 1962):
20–23, 38, 48, 68–71, 79–81, 88–89, 94–96, 115, 199, 122, 127–128, 152, 159–163, 185, 243–246, 263–271, 289–290

The Question Concerning Technology and Other Essays, tr. William Lovitt (New York: Harper & Row, 1977):
49, 291

Über *Logik als Frage nach der Sprache* (Frankfurt: Vittorio Klostermann, 1998):
164

Unterwegs zur Sprache (Frankfurt: Vittorio Klostermann, 1985) *On the Way to Language*, tr. Peter D. Hertz (New York: Harper & Row, 1971), which does not include »Die Sprache,«; "Language," tr. Albert Hofstadter in PLT, 189–210:
149–150

Vorträge und Aufsätze (Frankfurt: Vittorio Klostermann, 2000) »Die Frage nach der Technik,« 7–36; "The Question Concerning Technology," in QCT, 3–35. »Wissenschaft und Besinnung,« 39–65; "Science and Reflection," in QCT, 155–82. »Bauen Wohnen Denken,« 147–64; "Building, Dwelling, Thinking," in PLT, 145–61. »Das Ding,« 167–187; "The Thing," in PLT, 165–86. ». . . dichterisch wohnet der Mensch . . . ,« 191–208; ". . . Poetically Man Dwells . . . ," in PLT, 213–29. »Logos (Heraklit, Fragment 50),« 213–34; ("Logos (Heraclitus, Fragment B 50),"), trs. David Farrell Krell and Frank A. Capuzzi in *Early Greek Thinking* (San Francisco: HarperCollins, 1984), 59–78:
135–140, 146–151

Vorträge und Aufsätze, 8. Auflage (Pfullingen: Verlag Günther Neske, 1997) »Die Frage nach der Technik,« 9–40; "The Question Concerning Technology," in QCT, 3–35. »Wissenschaft

und Besinnung,« 41–66; "Science and Reflection," in QCT, 155–82. »Bauen Wohnen Denken,« 139–56; "Building, Dwelling, Thinking," in PLT, 145–61. »Das Ding,« 157–79; "The Thing," in PLT, 165–86. ». . . dichterisch wohnet der Mensch . . . ,« 181–98; *Was Heisst Denken?*, 5. Auflage (Tübingen: Max Niemeyer Verlag, 1997) *What Is Called Thinking?*, tr. J. Glenn Gray (New York: Harper & Row, 1968):
18, 26, 150, 165–166, 289

Wegmarken (Frankfurt: Vittorio Klostermann, 1967) »Was ist Metaphysik?« 103–22; "What Is Metaphysics?" in BW, 93–110. »Vom Wesen des Grundes,« 123–75; "On the Essence of Ground," in P, 97–135. Truth," in BW, 115–38. »Vom Wesen und Begriff der *physis*. Aristoteles' Physik B, 1,« 239–301; "On the Essence and Concept of *physis* in Aristotle's *Physics* B, I" in P, 183–230. »Brief über den Humanismus,« 313–64; "Letter on Humanism," in BW, 217–65. »Zur Seinsfrage,« 385–426; "On the Question of Being" in P, 291–322:
24–26, 33, 39, 42, 108–109, 115, 168

Zur Bestimmung der Philosophie (Frankfurt: Vittorio Koostermann, 1999):
77, 241, 247

Zur Sache des Denkens, 3. Auflage (Tübingen: Max Niemeyer Verlag, 1988) *On Time and Being*, tr. Joan Stambaugh (New York: Harper & Row, 1972):
67, 82, 175

INDEX

abandonment, 168, 196, 207, 210, 213–218, 241
abgehoben, 67
Abgrund, 161
abode, 282
abortion, 124
absehen, 267
absence, 51, 83, 106, 109, 133, 207, 210, 251
absolutism, 270
abyss, 24
academic, 168, 172–173, 191, 251, 282
Adorno, 5, 168, 179, 198, 202, 204, 276, 298
Aeschylus, 272
aesthetic, 149–150
African, 284
agrarian, 182
agriculture, 121, 254, 285
aletheic, 115–117, 127, 163
algorithm, 87, 105, 111
alienation, 39, 48–49, 287
alterity, 226
Amsterdam, 185
analytical, 100, 104
anarchism, 299
Anaximander, 208
Ancient, 24, 61, 169, 171, 175, 240–241
   Ancient Science, 21, 28
animal, 4, 2–26, 84, 93, 96–102, 107–111, 161, 182–183, 287, 292–293, 295
anthropocentrism, 179, 282
anthropology, 179
antiquity, 188, 237, 272

antirealism, 13, 19–21
   (*see* Realism)
apprehension, 72, 226, 232
apriori, 128
aquaculture, 286
Aquinas, 271
Arbeit, 254
archaeology, 14–15, 268
Archimedes, 174, 190
architecture, 122
Arendt, 168
Aristotle, 7, 15, 21–22, 25–26, 28, 30–32, 35, 80, 99, 104–105, 110–111, 126–127, 165, 167, 185, 208–209, 211, 272–273, 281, 284–286, 288, 294
   Aristotelian Science, 30–31, 167
   Nicomachean Ethics, 26, 273
   On Motion, 21, 212, 183–285
arithmetic, 86–87
art, 14–15, 26–27, 40–42, 49, 54, 62, 67, 85, 95, 107, 121 122, 134–135, 137, 171, 200, 221, 226, 275, 294, 299
artifacts, 21–22, 236, 283, 288
artificial, 2, 105, 111, 124, 215, 255
artist, 22, 172, 283
asceticism, 61, 174
astrology, 167
astronomy, 122, 292
astrophysics, 254
atmosphere, 107, 282
atom, 51, 103, 120–121, 137, 191, 254
atomic physics, 33, 52–53, 176, 254
atomistic, 99, 103
Auschwitz, 299

authenticity, 26, 105, 115–116
authority, 31, 71, 84–85, 132, 146
automation, 60
autonomous, 143, 285
awareness, 43, 101, 123, 179, 203, 208, 210, 218
awe, 216, 218, 235
axiomatic, 21

Babich, 5, 8, 25, 115, 159, 185–186, 188–191, 236, 297
Bachelard, 167–168, 174, 185–186, 188
Bacon, 30–31, 169, 211, 289, 294
Baden, 274
Badiou, 188
Baldwin, 111
Baudrillard, 173–174
beautiful, 35, 74, 229, 286, 297
becoming, 50, 135, 148, 245
beginnings, 168, 213, 217
beingness, 50, 208–210, 215, 217
Berkeley, 19, 64, 127, 191, 221, 236–237
Besinnung, 5, 9, 164, 194, 242, 250, 259
Bestand, 18, 24, 53, 56, 59, 115, 119, 122, 255, 285
Bestellen, 59
biodiversity, 285
biology, 4, 7, 14–15, 17, 22, 25, 65, 93, 96–97, 99–103, 105–111, 121–123, 127, 160–162, 186–187, 191, 233–234, 243, 271, 186, 293, 297
bodily, 18, 23, 50, 59, 61, 74, 85, 124–125, 127, 132, 137–139, 153, 174, 232–235, 272, 277, 284, 289
borders, 197, 202
boundary, 268, 272
Bourdieu, 184
brain, 85, 111, 127, 132, 248
building, 119
business
  scientific business, 141, 170, 250, 272

calculability, 170, 176, 196, 202–203
  calculation, 23, 29, 31, 41, 49, 88–89, 118, 134, 159, 161, 170–171, 173, 176–178, 183, 211, 213, 218, 221, 255
Callicott, 294
calling, 107, 136, 149, 214
capitalism, 253, 284
capitalist, 3, 60, 188
Caputo, 6, 8, 245, 257, 261, 297
Cartesian, 14, 33–34, 37, 49–50, 55–56, 160–161, 211, 228, 231, 234
Cassirer, 113, 179
categories, 20, 35, 94, 96, 103, 199, 265
Catholic, 50
causality, 16, 51, 64, 94, 176, 195, 211
  cause, 58, 64, 177, 211, 283–288
caution, 88, 292
Cerbone, 4, 8, 131, 297
CERN, 48
certainty, 3, 23–24, 34, 49, 59, 62, 85, 167, 174, 195, 209–210, 215, 290
certitude, 34–35, 37
chance, 244
chaos, 16–17, 25, 189
chemistry, 160, 162, 171, 185–186
chemotherapy, 186
China, 254
christian, 61, 197
Churchill, 65
circularity, 108, 173
classic
  classical physics, 3, 33, 4760, 86, 176–178
  classical science, 116
climate change, 282
cloning, 124
cogito, 23, 34
cognition, 70, 72, 103
coherence, 17, 24, 52, 88, 176
cohesiveness, 137, 140, 153
colonization, 182
communism, 250, 253
communist, 250
comportment, 100–101, 180, 184, 203, 215, 269

computers, 32, 105, 111
concealment, 36, 101, 110, 246
conformity, 36–37, 42
consciousness, 19–20, 47, 79, 83, 127, 132–133, 165
constructivism, 87, 172
consumption, 249, 284, 286
contemplation, 21, 135–136, 152, 176
Copernicus, 51, 122
corporate, 256, 284, 291
    corporate research, 182
    corporate science, 285
    corporation, 212, 256, 292
correctness, 3, 79–82, 87, 163, 195, 291
correspondence, 50, 53, 61, 74, 76, 79, 81–82, 98
cosmology, 118, 122
culture, 4, 7, 25, 37, 42, 93, 102, 105–108, 111, 115–116, 118, 125–126, 165, 167, 172–173, 188–189, 191, 235, 261, 275
cybernetics, 2, 8
Czech, 161

Darwin, 102, 110, 111, 188
Dasein, 2, 4–6, 9, 19–20, 38, 40, 47–48, 54, 68–69, 71–72, 75, 77–78, 83, 93–98, 101–103, 106, 108, 110, 115, 163, 242–243, 249, 269, 277, 286, 289
Davidson, 294
Dawkins, 105, 110
death, 37, 53–54, 101, 108, 132, 203, 221, 254, 292
decadence, 171
deforestation, 285
Deleuze, 167
Denken, 9, 18, 90, 188, 190, 197–198, 204, 298
Dennett, 103, 105–106, 110–111
Derrida, 24, 26, 262, 276
Descartes, 16, 20, 23, 33–35, 37–38, 43, 50–51, 57, 62, 72, 76, 111, 126, 174–175, 189–190, 211
destiny, 18, 75, 78, 196, 249

development, 7–19, 90, 169, 176, 181–186, 212, 251, 2 81–298
Dilthey, 160, 185–187, 239, 241–243, 257, 264, 278
dimensionality, 100, 105, 201, 203
disease, 118, 124, 231
DNA, 103, 285
dozen, 154
Dreyfus, 2, 111, 152, 226, 230, 236, 278
dwelling, 54, 77, 84, 96, 119, 127, 133, 135, 146–148, 151, 194, 197, 200, 203, 220, 281–283, 286–289

ecofeminism, 220, 281–284, 293–294, 298
ecological, 26, 100, 103, 160, 186, 220–221, 281, 293–294
ecophenomenology, 7, 13, 281, 288
education, 14, 252, 275
Einstein, 51, 58, 63–64, 237, 243, 271
electromagnetism, 56, 123, 228–230, 236
eliminativism, 133, 137
embryology, 186
Emerson, 127
enframing, 17–18, 24, 172, 178–180, 213–214, 218–219, 221–222
engage, 68, 154, 207, 216, 218, 235, 248
Engels, 188
Enlightenment, 174, 182, 211, 243
Entbergen, 78, 84–85
    Entbergung, 9, 23, 85, 285
Environmental, 13, 19, 26, 59, 72, 80, 98, 100–101, 103–105, 108–109, 118, 121, 123, 182–183, 212, 227, 230, 294
    environmental science/studies, 9
    environmental philosophy, 13, 19, 183, 281–292
epistemology, 3, 9, 15–16, 18, 23, 51, 53, 59, 61–62, 163, 166, 263–264, 286, 289, 291
Ereignis, 9, 62, 84, 216–217
essentialism, 103, 105

ethics, 5, 7, 26, 95, 110–111, 186, 205–206, 220, 273, 281–283, 286, 289, 293–294, 298
Ethiopia, 294
evolution, 40, 93, 99, 101–103, 105, 110–111, 127–128, 228, 233
existentialism, 26, 236
experiment, 42–43, 55, 60–64, 73, 89, 170
  experimentation, 29–32, 49–51, 54, 60–62, 117, 187, 228–229, 267, 292

facticity, 232, 237, 243, 245
farmers, 249, 285, 294
Feenberg, 221
feminism, 16, 25, 293–299
Feyerabend, 153, 167, 169, 187–188, 239–240, 261, 276, 278
Feynman, 228
food, 124, 182, 212, 254
forestructures, 265, 267, 274
forgetting, 62, 97, 210, 214–215
formulism, 43
Foucault, 167, 177, 184, 191, 275–276
freedom, 97, 115, 203, 248, 265, 276, 287–288, 295
Frodeman, 25, 294

Gadamer, 25, 126–127, 188, 262, 273, 276, 278
Galileo, 15, 17, 21–22, 25–26, 51–52, 122, 128, 176
Geisteswissenshaft(en), 2, 14, 239, 264
Gendlin, 49
genetics, 4, 102–106, 124, 182, 243, 245, 277
geology, 14, 25, 287, 289, 294
geometry, 50, 62, 122–123, 161–162
Gestell, 9, 53–54
Glazebrook, 10, 13, 25–26, 65, 67, 90, 126, 152, 185, 189–191, 240, 245, 255–259, 281, 293–294, 298
globalization, 54, 167, 247, 256, 284
Goethe, 184, 294
gravity, 17, 267

Greeks, 15, 23, 27–28, 84, 175, 190, 200, 207–210, 214–215, 248, 272

Hackett, 294
Hacking, 240
Haeckel, 160
Heelan, 7, 10, 44, 57, 64, 113, 115, 126–129, 180, 185, 189–191, 229, 236, 245, 298
Hegel, 1, 174, 263
Heisenberg, 33, 42–44, 47, 51, 56–57, 60, 63–64, 189, 233
hermeneutics, 6–8, 20–21, 75, 94–96, 107, 117–118, 125–126, 128–129, 185, 188, 221, 236–237, 239, 241, 243, 245, 248, 257, 262, 264, 273–278, 297–298
  hermeneutic phenomenology, 48, 70, 245
Hitler, 250, 253, 259
humanism, 8, 42, 44, 168, 174–175, 179, 182–183, 185, 282, 286
Husserl, 1, 36, 38, 79–80, 113–116, 125–128, 160, 162, 241, 243, 245–247, 272, 277–278

idealism, 13, 19–21, 230, 246, 249
Ihde, 115, 128, 190–191, 221
imperialism, 26, 276, 295
indeterminacy, 33, 43, 51, 56, 120, 142–143, 154
industry, 172, 181–182, 196, 205, 212, 218, 251, 254, 290, 292–293
insects, 108
internet, 172, 195
interpretation, 24, 69, 75, 79–80, 94, 97, 123, 125, 210, 245–246, 262, 265
intersubjectively, 195
intuition, 79, 244–246, 258

Jaspers, 225, 243
Jung, 2

Kant, 28, 48, 62, 72, 89, 126, 128, 163, 179, 187, 191, 200, 204, 246, 255
Kaufmann, 152

INDEX 309

Kepler, 51

Lacan, 168, 188
language, 4–5, 7, 9, 15, 40–41, 43, 70–71, 75, 101, 107–111, 136–138, 148–152, 164, 167–168, 171, 176, 200, 214–218, 220–222, 232, 240–241, 244, 249, 272, 279
Lefebvre, 174
Leibniz, 255
lifeworld, 4, 7, 9, 15, 20, 73, 96, 103, 115–127, 226–227
linguistics, 40, 127, 191
logic, 2, 1, 3, 6, 20, 25, 51, 54, 56, 90, 95, 160, 163, 174, 189, 196, 221, 241, 243–244, 257, 262–263, 275, 277, 279, 281, 283–285, 298
logos, 193, 196, 241, 248
Lyotard, 167, 188

machination, 6, 75, 82, 85, 90, 183, 207, 209–211, 214–215, 218–219, 221–222
maldevelopment, 284
  see development
Malpas, 26, 188
marxism, 188
maternity, 125, 285
mathematical, 3–5, 14–17, 22–25, 28–30, 32, 41, 49–51, 58, 61–62, 72, 86–89, 100, 104, 113, 115, 120, 122, 126–128, 138–139, 160, 162, 170, 189–190, 196, 222, 226, 236, 240, 242–243, 247, 255, 257, 267, 270, 289
McCluhan, 173
McNeill, 287, 294
measurement, 3, 13, 28–30, 40, 49, 51–52, 56, 58–60, 63, 74, 86, 88, 117, 120–122, 127, 170, 189, 230, 236
medicine, 6, 118, 121, 123, 125, 129, 161, 186, 251, 289
memory, 136, 153, 201
Merleau-Ponty, 114, 125, 170, 225, 235–236, 278

modernity, 3, 5, 14–16, 18, 26, 115, 175, 187, 221, 255, 282, 287, 289
Monsanto, 285
motion, 21, 43, 51–52, 88, 160, 212, 283–285
Mussolini, 259
mythology, 277

Nagel, 132–133, 139, 152, 190
naturalism, 4, 96, 103, 105, 131, 133, 137, 140, 146–148, 152–154, 189
Naturwissenschaft, 14, 64, 26, 204
Neo-Kantian, 113, 274
neuroscience, 127
Newton, 15–16, 21–22, 25, 29–30, 50–52, 59, 86, 163, 176, 271, 283–284, 293
  Newtonian (Physics), 3, 7, 16–17, 28, 43, 49, 55–57, 59, 86, 88–89, 163, 282, 285
Nicomachean Ethics, 26, 273
Nietzsche, 25, 37–38, 62, 131, 152, 159, 161–162, 165, 167, 169, 171–172, 174–175, 182–185, 187–190, 243, 259, 275, 286, 297–298
Nichts, 38–39
nihilism, 37–39, 54, 105, 174, 250
normativity, 205–206
nostalgia, 148
(The) nothing, 38–39, 139, 186, 197–1999, 216, 218
nothingness, 24, 37
nuclear, 26, 53, 55, 58, 181

objectivity, 3, 15–16, 18, 23–25, 52, 74, 88, 93–95, 105, 132, 169–170, 175, 189, 236, 268, 270, 286, 290–291, 293
oikos, 100
oil, 191
ontopoiesis, 127
ontotheology, 105
otherness, 4, 101, 106, 108–109

Parmenides, 175
Pascale, 190

pharmaceutical, 121, 292
phenomenology, 2–3, 15, 21, 113–116, 126, 132, 160–161, 241, 243, 245–248, 272, 275–277, 281
   Heidegger's, 2–3, 93–95, 102–103, 106, 109, 245–248, 288
   see ecophenomenology, hermeneutic phenomenology
philosophy
   Analytic Philosophy, 106, 114, 188, 240, 297
   Continental Philosophy, 8, 225–227, 229, 235–236, 241, 299
phronesis, 126, 273–274, 276, 278
physics
   see Newtonian physics, quantum physics, classical physics
physis, 3, 15, 22, 25, 28, 209, 211, 214–215, 221, 283, 288
Plank, 71
Plato, 50, 208–209, 215, 294
poetry, 141, 145–146, 148–151, 155, 168, 188, 200, 272, 277, 288, 297
politics, 14, 25, 114, 121, 124, 186, 248, 258, 268
pollution, 183, 282
population, 103, 166, 294
positivism, 61, 179, 244, 250
postmodern, 24
poverty, 5, 43, 97, 101, 179, 247
prescientific, 54, 123–125, 239, 263, 266, 269, 271, 274, 276–277
presencing, 40, 75, 82, 109, 148, 176
propaganda, 252
protoscience, 241
psychiatry, 2, 40, 42, 74
   psychotherapy, 8
psychology, 14, 40, 110, 133, 139, 178, 247, 257, 264, 277–278
pythagorean, 17, 139, 151, 183

quantum physics, 16–17, 28, 58–60, 118, 178

racism, 26

radiation, 43, 56–57, 121
rationalism, 211
rationality, 106, 244, 257, 277–278
realism, 3, 9–10, 13, 19–20, 25–26, 51, 61, 93–95, 153, 226, 230–231, 237
   see anti-realism

reason, 23, 50, 86–87, 122, 242, 287
reductionism, 4, 15, 94, 104–106, 161, 221
relativity, 16, 49, 128, 153, 243, 271
religion, 14, 95, 107, 121, 174, 183, 287, 297
   religiousity, 180
repeatability, 292
representation, 24, 52, 55, 62, 80, 89, 115, 120, 135, 184–185, 209, 231, 240, 290
   representational thinking, 5, 8–9, 16, 23, 197
reproduction, 103–104, 203
revelation, 3, 36, 42, 115, 120, 136
Richardson, 7–10, 17, 27, 44, 64, 126, 152, 247, 299
robotics, 127
Romanticism, 184, 221
rupture, 53

Scholasticism, 7, 44, 64, 126
Schroedinger, 298
scientism, 18, 93, 179, 250
Seamon, 294
Seinsvergessenheit, 49, 210, 214
selfhood, 84, 100–101
sensuality, 201
sexual, 98, 102, 275, 285
Shakespeare, 131, 152, 272
Sheehan, 188, 277
Simondon, 174
socialism, 8, 174, 188, 250, 253, 258
sociobiology, 105, 110–111
sociology, 173, 178, 191
Socrates, 184
solidarity, 290
spatiality, 120, 207

sports, 252
standing reserve, 3, 18, 23, 53, 56, 115, 176–179, 212–214
stakeholders, 293
standardization, 119, 276
Stellen, 88, 195
  Bestellen, 59
  Sicherstellen, 74
  Sicherzustellen, 85
  Nachstellen, 85
structuralism, 106
subatomic, 33
subjectivism, 37, 95, 175
suicide, 6, 253, 292
sustainability, 7, 16, 26, 182, 282, 285, 288

technē, 22, 24–25, 127, 191, 209, 211, 214–215, 283
teleology, 105, 111, 281–283, 285–286
television, 163
temporality, 6, 78, 109, 207, 229, 231–232, 245–246, 265
theology, 14–15, 242–244, 271, 274
Thompson, 26
Thomson, 152
totality, 55, 165, 187, 199, 229, 236, 269
traces, 137, 143, 245, 248
transcendence, 38, 48, 107, 241–242

uncertainty, 56–58, 60, 63, 120, 124, 249
unconcealment, 77–78, 82, 245–246

uniformity, 175, 195, 202–203
unveiling, 232
Utilitarianism, 250–252

Vattimo, 188
violence, 187, 190, 195, 284

Wahrheit, 77
war, 65, 114, 181, 250, 253–254, 256
water, 84, 124, 166, 183, 213, 285–287
wave, 55–56, 58, 120, 236, 248
weapons, 122
Weber, 104, 171, 250
Weimar, 243
Weltanschauung, 257, 259
Wesen, 26, 41, 208, 211, 261–262, 275–277
wisdom, 145, 165, 200
Wissenschaft, 6, 14–15, 25, 165, 185, 188–190, 244, 258–259, 274, 277–278, 294
witnessing, 108, 228
Wittgenstein, 110, 155, 297
women, 1, 128, 221, 284–285, 288, 293–295, 298
worldhood, 48, 54, 72
worldview, 6, 33, 50, 54, 59, 163, 169, 173, 185, 250, 252–253

Yale, 64, 258

Zarathustra, 188
zero, 56
zoology, 22, 96